Atomic Scale Structure of Interfaces

MATERIALS RESEARCH SOCIETY SYMPOSIUM PROCEEDINGS   VOLUME 159

Atomic Scale Structure of Interfaces

Symposium held November 27-29, 1989, Boston, Massachusetts, U.S.A.

EDITORS:

R.D. Bringans
Xerox Palo Alto Research Center, Palo Alto, California, U.S.A.

R.M. Feenstra
IBM T.J. Watson Research Center, Yorktown Heights, New York, U.S.A.

J.M. Gibson
AT&T Bell Laboratories, Murray Hill, New Jersey, U.S.A.

MRS MATERIALS RESEARCH SOCIETY
Pittsburgh, Pennsylvania

CAMBRIDGE UNIVERSITY PRESS
Cambridge, New York, Melbourne, Madrid, Cape Town,
Singapore, São Paulo, Delhi, Mexico City

Cambridge University Press
32 Avenue of the Americas, New York NY 10013-2473, USA

Published in the United States of America by Cambridge University Press, New York

www.cambridge.org
Information on this title: www.cambridge.org/9781107410343

Materials Research Society
506 Keystone Drive, Warrendale, PA 15086
http://www.mrs.org

First published 1990
First paperback edition 2012

Single article reprints from this publication are available through
University Microfilms Inc., 300 North Zeeb Road, Ann Arbor, MI 48106

CODEN: MRSPDH

ISBN 978-1-107-41034-3 Paperback

This work was supported by the Office of Naval Research under Grant Number
N00014-90-J-1389. The United States Government has a royalty-free license throughout
the world in all copyrightable material contained herein.

Contents

*Invited Paper

PART II: HETEROINTERFACE STRUCTURE

*Invited Paper

PART IV: OXIDE INTERFACES

PART V: CLEAN SURFACES AND CHEMISORBED LAYERS

*Invited Paper

*Invited Paper

Preface

This volume contains papers presented at the symposium on "Atomic Scale Structure of Interfaces" at the 1989 Fall Meeting of the Materials Research Society. The symposium was supported financially by the Office of Naval Research.

The atomic structure of interfaces has received less study than either surface or bulk structure, partly because it is difficult for most experimental and theoretical techniques to provide information that is localized to the interface. In the symposium, we attempted to bring together researchers with different approaches to this problem. As a result, there were many papers which examined buried interfaces, grain boundaries, encapsulated surfaces and interface roughness. A number of papers also deal with the determination of interface structure by measuring bulk lattice positions far from the interface. At the opposite extreme, several sections of the book cover work done on surfaces with thin overlayers so that the structure of interfaces can be observed at the time they are forming.

Particular examples of interfaces received particular attention. Part III of this book deals with silicide interface reactions and part IV covers work on the interface formed by oxides on silicon. The interface between GaAs and Si and the related studies of the interaction of group III and group V atoms with silicon were also common themes.

The study of interfaces really does need an interdisciplinary approach and participants at the symposium brought many different perspectives. We would like to thank all of them for contributing to an exciting meeting.

We are grateful to the staff of the MRS for their considerable help before, during and after the meeting. We would also like to thank Diana Angelini and Elizabeth Plowman for their assistance with the manuscripts.

R.D. Bringans
R.M. Feenstra
J.M. Gibson

PART I

Initial Stages of Heteroepitaxy

Theoretical Studies of GaAs on Si

John E. Northrup

Xerox Palo Alto Research Center

3333 Coyote Hill Road, Palo Alto CA 94304

Abstract

The energies of various two-dimensional GaAs on Si films have been calculated using the first principles pseudopotential method and density functional formalism. For GaAs on Si(111), the structures formed by adding a bilayer of GaAs to Si(111)1x1:As are shown to have positive formation energies, even after exchange reactions which eliminate the interface dipole are allowed. For GaAs on Si(100), the dependence of the formation energy of the films on the chemical potentials of the atomic constituents has been calculated. In the limit where $\mu_{As} = \mu_{As(bulk)}$, and assuming the films have equilibrated with a bulk GaAs reservoir ($\mu_{Ga} + \mu_{As} = \mu_{GaAs\ (bulk)}$), the lowest energy film is found to be the Si(100)2x1:As surface. In the opposite limit, $\mu_{Ga} = \mu_{Ga(bulk)}$, the lowest energy film is the Si(100)2x1:(GaAs) surface. A new metastable structure obtained by adding 1/2 monolayer of Ga to Si(100)2x1:As has been studied.

Introduction

There is currently a large effort underway to understand and control growth of GaAs on Si. Most of the effort involves experiments[1-10] designed to characterize the morphology and atomic structure of the films with techniques such as transmission electron microscopy, Auger electron spectroscopy, photoemission, scanning tunneling microscopy, and x-ray standing wave measurements. One of the challenges for theory[11-17] is to determine the growth mode and the atomic structure of the film-vacuum and substrate-film interfaces from first principles. This is an especially difficult problem for GaAs on Si, where we are dealing with a three-constituent system, and with an interface between polar and nonpolar materials. Moreover, the relative importance of kinetics vs. energetics in the growth is unclear. While a first principles theory of the kinetic aspects of growth is not possible at the present time, it is possible to determine the relative stability of possible structures. The question of whether the system can actually achieve the structure which minimizes the free energy globally is more problematical. The purpose of this paper is to present the results of total energy calculations[14-16] performed using the first-principles pseudopotential method and local density approximation for a set of two-dimensional films of GaAs on Si(111) and Si(100).

For GaAs on Si(111), calculations have been perfomed for overlayers formed by adding one bilayer of GaAs to the Si(111)1x1:As surface. The calculations indicate that the interface dipole occuring in the Si(111)AsGaAs film leads to a very high formation energy. While exchange reactions leading to an atomically mixed interface reduce the energy by a significant amount, the formation energy of the film remains positive with respect to the As-terminated Si(111) surface and large GaAs islands. This result is in accord with the observation that GaAs islands co-exist with an As-terminated Si(111) surface[10]. The calculations indicate that interfacial Si-Ga exchange reactions are exothermic and that the driving force for these reactions is the reduction of the interface electrostatic dipole.

The growth mode of GaAs on bare Si (100) is Stranski-Krastonov (SK), and TEM work indicates that the island formation begins at a very early stage[1,2]. The problem has been to determine the GaAs layer thickness where the transition from two-dimensional to three-dimensional growth occurs. To address this problem, total energy calculations have been performed for various possible two-dimensional films. These calculations indicate that the films constructed by adding one or two bilayers of GaAs to the Si(100)2x1:As surface have positive formation energies with repect to the Si(100)2x1:As surface plus a large GaAs island acting as a reservoir. These results indicate that if the deposited material could reach equilibrium, then island formation would occur before a complete bilayer of GaAs covered the As-terminated surface.

It should be emphasized that the structures obtained following molecular beam epitaxy (MBE) are not necessarily the ones which minimize the free energy globally. Recent Auger measurements by Fenner et al[7-8] indicate that for substrate deposition temperatures of 400° C the equivalent of ~2 bilayers of GaAs is present between islands on the surface. The local atomic structure of this material is unknown. It is possible that the material between the islands is trapped in structures which corresponds to a local, but not global, minimum of the total energy.

Methodology

A determination of the relative stability of two-dimensional films consisting of a variable number of Si, Ga, and As atoms requires calculation of the thermodynamic potential[18] $\Omega = U - TS - \Sigma \mu_i n_i$. Here U is the total energy, μ_i is the chemical potential of the ith constituent, and n_i is the total number of the ith constituent. In this work we neglect the entropy term, TS because it is expected to contribute very little to the difference in energies for various films. Using this formalism we can determine the chemical composition and structure of the film which minimizes Ω for a given set of chemical potentials. The surface energy is given by $E_s = \Omega/A$ where A is the area of the surface.

In general, the chemical composition of the minimum energy film will change depending on the chemical potentials. Therefore it is important to know the appropriate values of the μ_i, or at least place limits on their allowable range[19]. Since the films are in equilibrium with bulk Si we have $\mu_{Si} = \mu_{Si\ (bulk)}$. Since we are considering the relative stability of two-dimensional phases which are assumed to be in equilbrium with a large GaAs island, we also have the constraint $\mu_{Ga} + \mu_{As} = \mu_{GaAs\ (bulk)}$. This equality holds only for large islands. For small islands the effect of the island surface energy can be significant. The difference $\Delta\mu = \mu_{Ga} - \mu_{As}$ is limited to an energy range of $2\Delta H$ where ΔH is the heat of formation of GaAs from bulk Ga and bulk As. This fact is a consequence of the two inequalities $\mu_{Ga} \leq \mu_{Ga\ (bulk)}$ and $\mu_{As} \leq \mu_{As\ (bulk)}$. The chemical potentials for As and Ga can not rise above the bulk values. These two inequalities, together with the definition $\Delta H = \mu_{Ga(bulk)} + \mu_{As(bulk)} - \mu_{GaAs(bulk)}$, and the constraint $\mu_{Ga} + \mu_{As} = \mu_{GaAs\ (bulk)}$ place the limits on the range over which $\Delta\mu = \mu_{Ga} - \mu_{As}$ can vary. Specifically,

$$\mu_{Ga\ (bulk)} - \mu_{As\ (bulk)} - \Delta H < \Delta\mu < \mu_{Ga\ (bulk)} - \mu_{As\ (bulk)} + \Delta H.$$

In the present work we calculate Ω for various phases as a function of $\Delta\mu$.

During deposition, the system is obviously not in equilibrium: there is a net flow of matter from the gas phase to the surface. The equilibrium thermodynamic description outlined above is only approporiate after equilibration has taken place between the various structures formed in the deposition process. The rate at which equilibration occurs may be extremely slow compared to the time between the end of deposition and experimental characterization of the film.

The total energies U were calculated with the local-density-functional[20], scalar-relativistic-pseudopotential method[21-22]. The Kohn-Sham equations were solved in a plane-wave basis[23] containing plane waves with kinetic energies up to 10 Ry. The energy of each of the surface structures was calculated with a centrosymmetric supercell containing 24 atoms. For each surface, the forces on the atoms were calculated via the Hellmann-Feynman theorem and used to obtain the set of positions which minimizes the energy for that particular bonding topology. The total energies of the bulk phases of Si, GaAs, As, and Ga were also calculated in order to establish the appropriate chemical potentials.

Results for GaAs on Si(100)

As a first step let us consider the four structures illustrated schematically in Fig. 1. These 2x1 symmetry structures are formed by terminating the Si(100) surface by As-As, Ga-As, Ga-Ga, or Si-Si dimers. In each case the atomic positions were determined by force calculations. The surface energies of these four structures are plotted in Fig. 2, as a function of $\Delta\mu$ over its maximum allowed range, $2\Delta H$. The bulk calculations give ΔH = 0.92 eV whereas the experimental value of ΔH is 0.74 eV[24]. The dashed

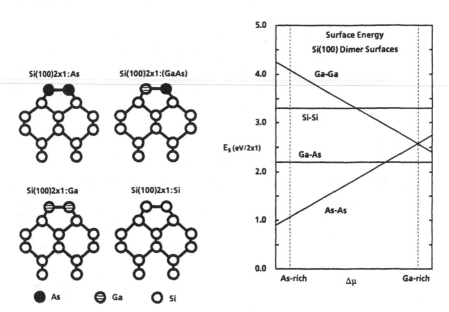

Fig. 1. Schematic views of 4 simple dimer terminated Si(100) surfaces.

Fig. 2. Surface energy E_s vs. Δμ for the structures shown in Fig. 1. The surface energy is given in eV/(2x1 cell).

lines in Fig. 2 indicate the limits for Δμ corresponding to the experimental value of ΔH. Clearly, the relative stability of these four structures depends on the chemical potentials. If the As chemical potential is near its maximum, $μ_{As(bulk)}$, and the Ga chemical potential is correspondingly low, then the Si(100)2x1:As surface has the lowest surface energy. The Si(100)2x1:As surface has been studied extensively[25]. However, if the Ga chemical potential is near its maximum, $μ_{Ga(bulk)}$, and the As chemical potential is correspondingly low, then the Si(100)2x1:GaAs structure has a lower surface energy than Si(100)2x1:As. An interesting result is that the Ga-Ga dimer surface is unstable on a Si(100) surface on which GaAs islands are present. Even in the Ga-rich limit, the Ga-As dimer is energetically more favorable than the Ga-Ga dimer. Fig. 2 indicates that the Ga-Ga dimer surface is stable in the absence of As, as implied by experiment[26].

In the presence of As and/or Ga, the clean Si(100)2x1 surface is energetically very unfavorable. The surface energy of the clean surface is 3.3 eV/(2x1 cell) or equivalently 1.65 eV/(dangling bond). This energy is about 0.2 eV/(dangling bond) higher than that for the relaxed Si(111)1x1 surface. One reason for the higher surface energy on the (100) dimer surface may be the strains which are induced in the underlying layers of the (100) substrate by the dimer formation.

Now consider structures which can be obtained by adding Ga and As to Si(100)2x1:As. Fig. 3 illustrates two of the many possibilities. Structures formed by

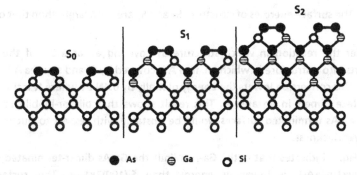

● As   ⊖ Ga   ○ Si

Fig. 3. S_0 is the As terminated Si(100) surface. S_1 and S_2 are formed by the addition of 1 or 2 bilayers of GaAs to S_0.

adding 1, 2, 3, or 4 GaAs pairs per 2x1 unit cell have been considered[16]. In all cases the formation energy is positive, in the As-rich limit, with respect to the As-terminated Si(100)2x1 surface and a large GaAs island acting as a reservoir for GaAs. The structures shown in Fig. 3, S_1 and S_2 correspond to adding 1 or 2 bilayers of GaAs to the As-terminated Si surface (S_0). In both S_1 and S_2, a Si-Ga exchange reaction has occurred to reduce the interface dipole and lower the total energy[16]. The surface energies of the structures S_1 and S_2 are plotted in Fig. 4 as a function of $\Delta\mu$. Note that since $\Omega = U - TS - n_{Si}\mu_{Si(bulk)} - \frac{1}{2}(n_{Ga} + n_{As})\mu_{GaAs(bulk)} - \frac{1}{2}(n_{Ga} - n_{As})\Delta\mu$, the surface energies of structures with the same areal density of $(n_{Ga} - n_{As})$ have the same dependence on $\Delta\mu$. The results shown in Fig. 4 indicate that it is energetically more favorable to add the GaAs to the islands than to cover the As terminated Si surface with 1 or 2 bilayers of GaAs.

We can also consider the stability of the S_1 film in terms of atom conserving reactions which do not involve interchange of atoms with a bulk GaAs reservoir. We

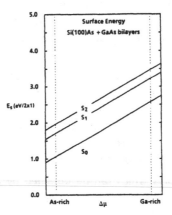

Fig. 4. The surface energies of structures S_1 and S_2 are both larger than that of S_0.

consider the reaction in which structure S_1 covering an area 2A of the surface converts into a structure in which an area A is covered by S_0 and an area A is covered by S_2. This conversion lowers the energy globally by 0.4N eV where N is the number of surface dimers in an area A. This result shows that bilayer-by-bilayer growth of ...AsGaAs terminated surfaces would be unstable with respect to fluctuations in the layer thickness.

Fig. 2 indicates that in the Ga-rich limit the Ga-As dimer-terminated surface, Si(100)2x1:(GaAs), is lower in energy than Si(100)2x1:As. This surface, like Si(100)2x1:As, also appears to be stable with respect to the addition of GaAs. Removing a Ga-As pair from the islands and forming Si(100)(AsGa)(GaAs), illustrated in Fig. 5, increases the surface energy by 0.47 eV/dimer. As shown in Fig 6, the

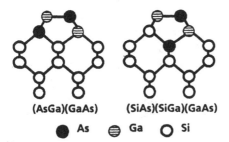

(AsGa)(GaAs)   (SiAs)(SiGa)(GaAs)

● As   ⊜ Ga   ○ Si

Fig. 5 Two simple structures containing equal numbers of Ga and As atoms. Both of these structures have higher surface energies than the GaAs dimer terminated surface.

surface energies of the GaAs dimer terminated surface (2.21 eV), and the (AsGa)(GaAs) surface (2.68 eV) are independent of Δμ. The energy of the Si(100)-

(SiAs)(SiGa)(GaAs) structure, also illustrated in Fig. 5, is 0.50 eV/dimer higher than Si(100)2x1:(GaAs).

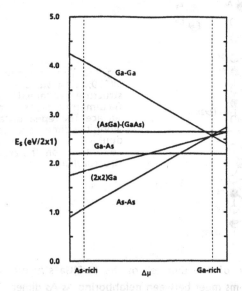

Fig. 6 Surface energies as a function of $\Delta\mu$ for 5 structures which are close in energy in the Ga-rich limit. The Ga-As dimer terminated surface has the lowest energy in this limit.

An interesting result recently obtained by Biegelsen et al [9] is the observation with a scanning tunneling microscope of a 2x3 (or equivalently 3x2) structure obtained following deposition of GaAs onto a vicinal Si(100)2x1:As surface. A possible model for this structure[9] corresponds to a mixture of 2x2 and 2x1 building blocks as shown in Fig. 7. The 2x2 block corresponds to adding one Ga-Ga dimer to each 2x2 unit cell of the As-terminated Si surface. This structure will be denoted the (2x2)Ga structure. The added Ga breaks the As-As dimer bonds, but the As atoms remain threefold coordinated. The Ga atoms form Ga-Ga dimers and are also threeefold coordinated. The 2x3 appears to correspond to an ordered mixture of the 2x2 and 2x1 structures. Calculations for the (2x2)Ga structure indicate that it is 0.1eV/(2x1 cell) lower in energy than the Si(100)2x1:As surface in the Ga-rich limit. This corresponds to an energy difference of 0.2 eV/(Ga dimer). Calculations for the 2x3 structure have not yet been performed, but it is possible that the observed 2x3 is preferred over the 2x2 because of an effective attractive interactions between the 2x2 and 2x1 building blocks. Since the 2x2 and 2x1 blocks are very close in energy, the optimum mixture will depend on the interactions between the blocks.

As shown in Fig. 6, the (2x2)Ga structure is higher in energy than the Si(100)2x1(GaAs) structure in the Ga-rich limit. We should note, however, that a significant amount of bond-breaking, surface diffusion, or As desorption would be required to obtain the Si(100)2x1(GaAs) structure starting from Si(100)2x1:As. On

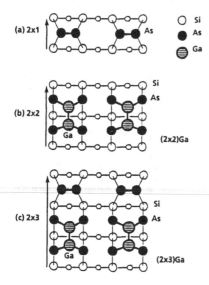

(a) 2x1

(b) 2x2

(2x2)Ga

(c) 2x3

(2x3)Ga

○ Si
● As
⊜ Ga

Fig. 7 (a) Top view of the Si(100)2x1:As surface. (b) A 2x2 structure is obtained by adding Ga dimers to the As terminated surface. The repeat distance along the dimer row direction is doubled. (c) A 2x3 structure can be formed from the 2x2 and 2x1 structural units.

the other hand, formation of one unit cell of the (2x2)Ga structure can occur whenever diffusing Ga atoms meet between neighboring As-As dimers. Thus, this structure and the 2x3 are formed easily by deposition of Ga onto the Si(100)2x1:As surface.

A possible scenario for the growth sequence is as follows: (1) In the first stage Ga atoms diffuse on the Si(100)2x1:As surface until they are trapped into the (2x3)Ga structure. (2) After the (2x3)Ga structure is completed the Ga atoms diffuse on the surface without trapping and equilibrate with Ga droplets. (3) As is incoporated into the Ga droplets resulting in the formation of GaAs islands. This picture is consistent with experiments which indicate the presence of both Ga and As in the "sea" regions between GaAs islands. However, Auger and RBS experiments[7] suggest that there may be more Ga and As in the sea than what would be expected in this scenario. Thus, an interesting question is the rate at which additional Ga and As can be incorporated into the sea regions and the rate at which the (2x3)Ga phase converts into the equilibrium Si(100)2x1(GaAs) phase.

Results for GaAs on Si(111)

Fig. 8(a) illustrates the Si(111)1x1:As surface[27-28], and 8(b) illustrates a structure obtained by adding a bilayer of GaAs and allowing Si-Ga exchange reactions to occur between the second and the fourth layer. These exchange reactions are essential in order to reduce the interface charge transfer dipole[14-15], and to achieve the lowest possible formation energy for the bilayer. In general we can create an overlayer with

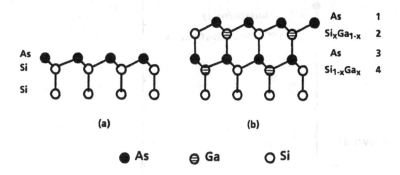

Fig. 8 (a) Si(111)1x1:As surface. (b) ...SiSi(Si$_{1-x}$Ga$_x$)As(Si$_x$Ga$_{1-x}$)As surface.

a fraction x of Si atoms in the second layer and a fraction 1-x of Si atoms in the fourth layer. Correspondingly, there are a fraction x of Ga atoms in the fourth layer, and a fraction 1-x of Ga atoms in the second layer. The total energies of these structures were calculated for values of x equal to 0, 1/4, 1/2, 3/4, and 1. The results are shown in Fig. 9. These calculations required a 2x2 unit cell containing 48 atoms, and the plane wave kinetic energy cutoff for the electronic wave functions was restricted to 7 Rydbergs. The results indicate that the unmixed interface (x=0) has by far the highest total energy. In such an interface we would have a SiAs bilayer buried underneath a GaAs bilayer. In that case one extra electron (per 1x1 surface unit cell) in the SiAs bilayer has to be transferred to the surface GaAs bilayer to fill the As-lone pair band. This charge transfer makes the abrupt interface energetically unfavorable relative to the atomically mixed interfaces. The lowest energy was obtained for the fraction x=3/4. Interestingly, this corresponds to having zero electrostatic dipole relative to the Si(111)As surface. Harrison et al[11] argued that such atomic mixing is necessary to explain the absence of interfacial charge at a Ge(100)-GaAs(100) interface. What has been shown here[14-15] is that the atomically mixed structures are significantly lower in total energy, even for the thinnest possible overlayers, and even before the interface dipole has increased to the point where interface charging occurs via creation of electrons or holes. Thus energetics favors the formation of an atomically mixed interface right from the start.

The formation energy of these bilayer films is positive with respect to the As-terminated Si(111) surface and a large GaAs island acting as a bulk reservoir of GaAs. Thus, island formation is favored rather than formation of a complete bilayer. This result is consistent with photoemission experiments[10] which indicate that the area between the islands is predominantly Si(111)As.

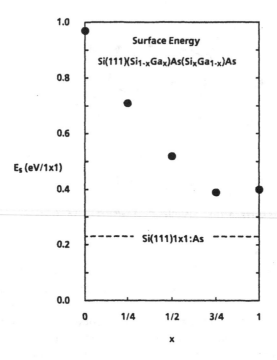

Fig. 9 Surface energies of the structures Si(111)(Si$_{1-x}$Ga$_x$)As(Si$_x$Ga$_{1-x}$)As for x=0, 1/4, 2/4, 3/4, and 1. The dashed line denotes the surface energy for As terminated Si(111). The chemical potential for As is that of bulk As. The formation energy of the lowest energy bilayer film is 0.16 eV/(1x1cell).

REFERENCES

[1]   D. K. Biegelsen, F. A. Ponce, A. J. Smith, and J. C. Tramontana, J. Appl. Phys. **61**, 1856 (1987).

[2]   R. Hull, A. Fischer-Colbrie, Appl. Phys. Lett. **50**, 851 (1987).

[3]   D. K. Biegelsen, F. A. Ponce, B. S. Krusor, J. C. Tramontana, R. D. Yingling, R. D. Bringans, and D. B. Fenner, Mater. Res. Soc. Symp. Proc. **116** (1988).

[4]   J. R. Patel, P. E. Freeland, M. S. Hybertsen, D. C. Jacobson, and J. A. Golovchenko, Phys. Rev. Lett. **59**,2180 (1987).

[5]   M. Zinke-Allmang, L. C. Feldman, and S. Nakahara, Appl. Phys. Lett. **52**, 144, (1987).

[6]   R. D. Bringans, M. A. Olmstead, F. A. Ponce, D. K. Biegelsen, B. S. Krusor, and R. D. Yingling, J. Appl. Phys. **64**, 3472 (1988).

[7]   D. B. Fenner, D. K. Biegelsen, B. S. Krusor, F. A. Ponce, J. C. Tramontana, M. B. Brooks, and T. W. Sigmon, (to be published).

[8]   D. B. Fenner *et al*, (these proceedings).

[9]   D. K. Biegelsen, R. D. Bringans, J. E. Northrup, and L. E. Swartz, (these proceedings).

[10]   R. D. Bringans, M. A. Olmstead, R. I. G. Uhrberg and R. Z. Bachrach, Appl. Phys. Lett, $\underline{51}$, 523, (1987); Phys. Rev. B, $\underline{36}$, 9569, (1987).

[11]   W. A. Harrison, E. A. Kraut, J. R. Waldrop and R. W. Grant, Phys. Rev. B $\underline{18}$, 4402 (1978).

[12]   H. Kroemer, J. Vac. Sci. Technol. $\underline{B5}$, 1150 (1987).

[13]   M. H. Grabow and G. H. Gilmer, Mater. Res. Soc. Symp. Proc. 94, 15 (1987).

[14]   J. E. Northrup, Phys. Rev. B $\underline{37}$, 8513 (1988).

[15]   J. E. Northrup, R. D. Bringans, R. I. G. Uhrberg, M. A. Olmstead, and R. Z. Bachrach, Phys. Rev. Lett. $\underline{61}$ 2957 (1988).

[16]   J. E. Northrup, Phys. Rev. Lett. $\underline{62}$, 2487 (1989).

[17]   E. Kaxiras, O. L. Alerhand, J. D. Joannopoulos, and G. W. Turner, Phys. Rev. Lett. $\underline{62}$, 2484 (1989).

[18]   L. D. Landau and E. M. Lifshitz, *Statistical Physics* (Pergamon, London, 1958)

[19]   Guo-Xin Qian, R. M. Martin, and D. J. Chadi, Phys. Rev. B $\underline{38}$, 7649 (1988).

[20]   W. Kohn and L. Sham, Phys. Rev. $\underline{140}$, A1135 (1965).

[21]   G. B. Bachelet and M. Schluter, Phys. Rev. B $\underline{25}$, 2103 (1982).

[22]   D. R. Hamann, M. Schluter, and C. Chiang, Phys. Rev. Lett. $\underline{43}$, 1494 (1979).

[23]   J. Ihm, A. Zunger, and M. L. Cohen, J. Phys. C $\underline{12}$, 4409 (1979).

[24]   *Handbook of Chemistry and Physics*, 65th ed., edited by R. C. Weast (CRC, Boca Raton, 1984).

[25]   R. I. G. Uhrberg, R. D. Bringans, R. Z. Bachrach, and J. E. Northrup, Phys. Rev. Lett. $\underline{56}$, 520 (1986).

[26]   B. Bourguignon, K. L. Carleton, and S. R. Leone, Surface Science 204, 455 (1988).

[27]   M. A. Olmstead, R. D. Bringans, R. I. G. Uhrberg, and R. Z. Bachrach. Phys. Rev. B. $\underline{34}$, 6041 (1986).

[28]   R. I. G. Uhrberg, R. D. Bringans, M. A. Olmstead, R. Z. Bachrach, and J. E. Northrup, Phys. Rev. B $\underline{35}$, 3945 (1987).

VERY THIN 2D GaAs FILMS ON Si DURING THE EARLY STAGES
OF GROWTH BY MBE

D.B. FENNER[*+], D.K. BIEGELSEN,[*] B.S. KRUSOR,[*] F.A. PONCE,[*]
and J.C. TRAMONTANA[*]
[*]*Xerox Palo Alto Research Center, Palo Alto, CA 94304.*
[+]*Physics Department, Santa Clara University, Santa Clara, CA 95053.*

ABSTRACT

GaAs samples deposited on Si by molecular beam epitaxy (MBE) with a graded thickness of $0-3$ nm initially show the presence of a metastable two dimensional (2D) layer containing Ga and As. In the thicker regions of the wedge samples, islands (3D topography) form in the presence of the 2D sea, i.e., Stranski–Krastanov growth. Compositional profiles of these wedges were made with *in situ* Auger electron spectroscopy (AES) which has allowed the identification of at least four regimes of growth. Lattice images from cross–sectional transmission electron microscopy (XTEM) are consistent with the AES profiles. Substrate temperature during deposition of the films has a strong effect on film topography, as does the beam–flux ratio on film stoichiometry.

INTRODUCTION

The quality of heteroepitaxial GaAs films on Si substrates is significantly compromised by the tendency of the deposited GaAs to form first into clumps or islands and then fail to fully heal the island–induced defects when the film is thickened further.[1]   This troublesome formation of three–dimensional (3D) structures (islands) occurs in many thin–film systems and can be phenomenologically ascribed to a non–wetting film condition occurring when the film's surface free energy is larger than the sum of the substrate and interface surface free energies.[2]   Film deposition by MBE provides sufficient control and resolution to allow atomic–scale studies of the island growth itself, but a complete experimental [3–5] or theoretical [6] understanding has not been reached yet. Thin Ge films on Si are known to grow by a Stranski–Krastanov mode, i.e., 2D layer–by–layer followed by 3D islands.[7]   Here, the AES peak height of the substrate Si line was found to attenuate *linearly* during the 2D growth, with successive break points as each layer reached completion.   Exponential decline of the Si peak height returned once islanding and surface roughening set in.

EXPERIMENTAL

We have grown a variety of GaAs films by MBE on Si(100) wafers which had surfaces tilted 2.5° toward the (011). In order to facilitate the preparation and characterization of films with many thicknesses we have used a sample manipulator that can be translated at a uniform speed during the deposition in such a way that the substrate is increasingly shielded from the molecular beams by a fixed shutter located near the substrate surface. As can be seen in the schematic of Fig. 1, this deposition technique results in a film with a wedge shape. Our wedge samples were typically 3.5 − 10 nm in maximum thickness, and extended over ≈ 7 mm width of the 50 mm diameter wafers. Immediately adjacent to the growth chamber was a UHV surface − analysis chamber with an Auger electron spectrometer making it possible to scan over the width of the GaAs wedges just after their deposition. The ≈ 10 μm AES beam spot size allowed high spatial, and hence film thickness and composition, resolution.

After AES profiling, some samples were capped with amorphous Si and moved to other systems for cross − sectional transmission electron microscopy (XTEM) and Rutherford backscattering spectroscopy (RBS). The RBS was very useful for absolute measurements of the quantity of deposited GaAs, and provided some wedge shape information, within the limits imposed by its ≈ 1 mm beam spot size. This RBS wedge − profiling scheme, and the thickness calculated from the profile for a very thin wedge are shown in Fig. 2.

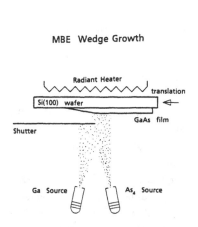

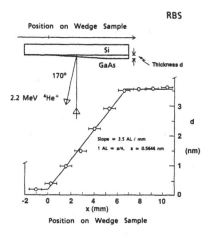

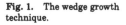

Fig. 1. The wedge growth technique.

Fig. 2. The scheme and results of the RBS profile of a thin GaAs wedge on Si(100).

RESULTS

Wedges deposited under a variety of conditions were AES profiled, and one very thin wedge was profiled with both RBS, Fig. 2, and AES, Fig. 3. Both the AES Si LVV and KLL lines were monitored as well as the Ga and As LMM lines. The Si LVV electrons have a very short escape length (≈ 0.5 nm) and thus are highly surface sensitive, but are easily obscured by nearby Ga and As MMN lines in the spectra of all but the thinnest samples. Thus, we also measured the Si KLL together with the Ga and As LMM lines, as more reliable monitors for thicker films. As seen in Fig. 3, and discussed at more length elsewhere[5, 8], both of the Si–line peak heights decrease linearly in the thinnest portion of this wedge sample, and the Ga and As lines are both seen to rise. We have observed[8] that this is consistent with a simple model of 2D layer growth of a film containing Ga and As uniformly covering the Si at the earliest stages of growth. Based on the RBS calibration, the successive break points, after $x = 0$, then correspond with the completion of ≈ 3 and $\approx 5-6$ atomic layers (AL), where 1 AL = 1/4 the GaAs lattice constant. The various model growth regimes we have proposed for this thin wedge[8] are summarized in Fig. 4, and noted in Fig. 3. Our instrumental resolution of the first 1 AL of deposit near $x = 0$ may not

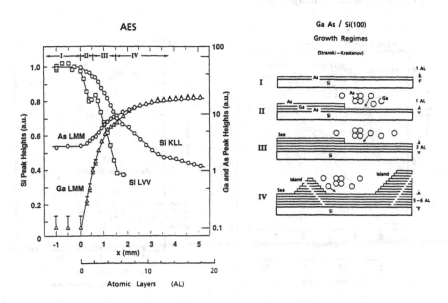

Fig. 3. AES profiles of the same wedge reported in Fig. 2. The Si LVV and KLL peak heights are on a linear scale, and the Ga and As LMM are on a semilog scale.

Fig. 4. Model growth regimes (I – IV) for GaAs on Si(100) consistent with our measurements.

be reliable due to uncertainties in locating the wedge toe and to penumbral effects.

Both the ratio of the Ga $-LMM$ to As $-LMM$ peak heights, and the relative Ga to As RBS peak areas were measured for this wedge sample.[8] These results are shown in Fig. 5 along with a heavy arrow indicating the AES ratio we have measured for the surface of a freshly cleaned GaAs wafer. Clearly the Ga content of the films changes continuously, even at the start of the wedge, with a distinct, but small, plateau in both the AES and RBS curves at the point where we suggest that island growth begins. For ideal layer — by — layer growth of GaAs in the presence of a terminating layer of As, one would expect the [Ga] / [As] ratio to follow a simple n / (n + 1) dependence. This is also shown in Fig. 5, and agrees qualitatively with the data there, especially for the thinnest part of the wedge.

Another GaAs film was deposited under very similar conditions but without the shutter and sample translation. This uniform sample was evaluated by XTEM and RBS. A typical island is shown in the lattice image of Fig. 6. The dark band at the interface was determined to not be a microscope artifact. The band's thickness, combined with the volume of GaAs in the islands, agrees well with the total GaAs measured by RBS, thus we propose that this dark band is the GaAs sea.[8] The islands are extensively faulted, as also noted in the bottom of Fig. 4, but the sea has a high quality (pseudo —) morphology.

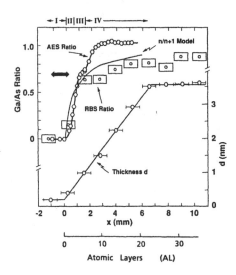

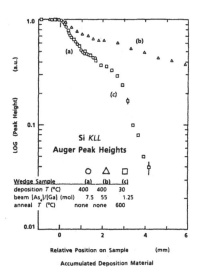

Fig. 5. Measured ratios of Ga to As LMM peak heights, and RBS peak areas for the wedge sample of Figs. 2 and 3. Also, the ratio expected for a simple n/(n + 1) model is indicated.

Fig. 7. AES profiles (Si KLL) for three wedges deposited under the conditions noted. The peak heights are on a semilog scale.

In Fig. 7 are shown the AES Si *KLL* profiles of three wedges, deposited and annealed under the conditions indicated in the figure legend. Fig. 7 (a) is the *KLL* profile of the wedge reported in Fig. 3. Comparing Fig. 7 (a) with (b) it is clear that when more As is available during the deposition, the film is more Si − *KLL* electron transparent, i.e., mainly island growth with little or no sea. The wedge of Fig. 7 (c) was deposited at much lower T and flux ratio than were wedges (a) and (b). At the beginning of wedge (c) its profile follows that of (a), i.e., the 2D and then 2D+3D growth, as modeled in Fig. 4. Then, when the islands in (c) reach a size near which we estimate their coalescence, the Si *KLL* again attenuates rapidly as even more deposited material accumulates. Apparently, once the islands coalesce additional GaAs fills in the thin regions between the islands and the film becomes rapidly more electron opaque, with the profile of Fig. 7 (c) dropping nearly logarithmically. XTEM images of wedge (c) qualitatively verified this structural model. In addition, wedge (c) was AES profiled just after deposition and after anneals to increasingly higher T, but the profiles changed only slightly after each anneal.

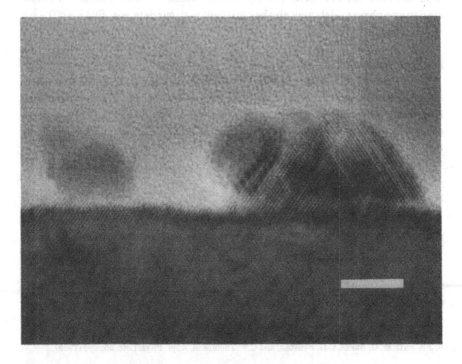

Fig. 6. XTEM micrograph of a thin GaAs film on Si(100). Lattice images of an island, the proposed sea (dark band at the interface), and the Si substrate are seen. A 5−nm scale bar is shown.

Another wedge (not shown) was deposited at 30 °C and a flux ratio of 10, and was found to have approximately the profile shown in Fig. 7 (c) after both deposition and a 500 °C anneal. Then after a 600 °C anneal its profile showed only a slight Si *KLL* attenuation despite substantial Ga and As *LMM*. Again, higher As in the deposition flux facilitated islanding.[9]

CONCLUSIONS

We have shown that 2D films of GaAs are possible on Si(100), at least for the earliest stages of growth in the few − atomic − layer range and at 400 °C. A transition from a 2D to a 3D − growth mode (i.e., Stranski − Krastanov) occurs after about 0.9 nm of 2D film have accumulated. Coexisting 2D and 3D GaAs on Si(100) were found to be (meta −) stable after at least 500 °C annealing. Samples deposited under higher beam flux ratios islanded sooner, while depositions at 30 °C and molar flux ratio near one showed stability against 600 °C annealing and were more opaque to Si *KLL* electrons from the substrate. The 2D and 2D + 3D GaAs structures we have observed are not likely the equilibrium ones.[6] More complex, multiple coexisting − phase models may also be consistent with our data, e.g., Si:As + 2D − sea + 3D − islands.

ACKNOWLEDGMENTS

We greatly appreciate the assistance of Don Yingling with the MBE growths, and Mike Brooks and Tom Sigmon with the RBS measurements. We have benefitted from many conversations with John Northrup, and Ross Bringans. DBF received support from the National Science Foundation (DMR − 8822353) and the Air Force OSR (F49620 − 89 − C − 0017).

FOOTNOTES

1. D.K. Biegelsen, F.A. Ponce, A.J. Smith, and J.C. Tramontana, J. Appl. Phys. **61**, 1856 (1987).
2. M. Copel, M.C. Reuter, E. Kaxiras, and R.M. Tromp, Phys. Rev. Lett. **63**, 632 (1989).
3. S.M. Koch, S.J. Rosner, R. Hull, G.W. Yoffe, and J.S. Harris, J. Cryst. Growth **81**, 205 (1987).
4. R.D. Bringans, M.A. Olmstead, R.I.G. Uhrberg, and R.Z. Bachrach, Phys. Rev. B **36**, 9569 (1987).
5. D.K. Biegelsen, F.A. Ponce, B.S. Krusor, J.C. Tramontana, R.D. Yingling, R.D. Bringans, and D.B. Fenner, in Mat. Res. Soc. Symp. Proc., edited by H.K. Choi, R. Hull, H. Ishiwara, and R.J. Nemanich (Materials Research Society, Pittsburgh, PA, 1988), Vol. 116, p. 33.
6. J.E. Northrup, Phys. Rev. Lett. **62**, 2487 (1989).
7. M. Asai, H. Ueba, and C. Tatsuyama, J. Appl. Phys. **58**, 2577 (1985).
8. D.B. Fenner, D.K. Biegelsen, B.S. Krusor, F.A. Ponce, J.C. Tramontanna, M.B. Brooks, and T.W. Sigmon, (to be published).
9. J.E. Palmer, G. Burns, C.G. Fonstad, and C.V. Thompson, Appl. Phys. Lett. **55**, 990 (1989).

ONSET OF GaAs HOMOEPITAXY AND HETEROEPITAXY

D K. Biegelsen, R. D. Bringans, J. E. Northrup, and L.-E. Swartz
Xerox Palo Alto Research Center, Palo Alto, California 94304

ABSTRACT

We present scanning tunneling microscopy images and atomic models for a variety of GaAs(100) reconstructed surfaces. For homoepitaxial material we show the sequence of phases from c(4x4) through c(8x2) as the As surface concentration is reduced. For the heteroepitaxial GaAs/Si(100) growth we show the first two stages of film development, namely, Si(100):As-(1x2) monolayer coverage followed by adsorption of a submonolayer of Ga dimers. The next stages of film development are discussed.

INTRODUCTION

GaAs is a prototypical, direct-bandgap III-V semiconductor. The GaAs(100) surface forms the foundation of existing and future optoelectronic and high speed switching technologies. As device dimensions parallel and/or perpendicular to the plane of the substrate decrease steadily toward interatomic lengths, the control of growth on an atomic scale becomes crucial. To gain such control and understanding of the fabrication processes, atomically-resolved imaging tools are required. In this paper we describe the application of scanning tunneling microscopy (STM) to visualize various GaAs(100) surfaces. We show the bare GaAs substrate reconstructions which are driven by the difference in chemical potentials of Ga and As. The details of the reconstructions are thought to affect such processes as dopant incorporation, adatom diffusion, atomic layer epitaxy, adatom self-organization, and heteroepitaxy. The homoepitaxial surface thus represents the first step in the creation of an interface on GaAs(100).

In the second part of this paper we will describe some of our work in characterizing the initial stages of the heteroepitaxial growth of GaAs(100) on Si(100). Unlike the case of homoepitaxy, the heteroepitaxial system seems to evolve in character during many layers of the initial deposition. Again STM provides atomic scale real space information about the structures involved.

EXPERIMENT

Samples were prepared in a molecular beam epitaxial deposition system and transported through ultra-high vacuum (UHV) to annealing, Auger or STM stations. The substrates were mounted by tungsten wire clips to molybdenum sample holders. The holders in turn contained integral heat sources, either radiant for GaAs substrates or ohmic (direct current) for the Si substrates. GaAs(100) samples were prepared from nominally on-axis wafers by a sequence of cleaning steps. [1] After entry into vacuum via load-lock, the doped GaAs (10^{18} Si) samples were pre-baked at 500°C for ~24 hours. After transfer to the MBE chamber, surface oxides were sublimed off at 640°C [2] in either an As_4 or As_2 flux. Growth of ~300 nm of nominally undoped GaAs proceeded at 300 nm/hr in an As-stabilized regime. Si samples were cleaned and oxidized by standard solvent and acid dips and the ultra-violet ozone technique. [3] In vacuo cleaning utilized a 24 hr prebake at ≥ 600° followed by oxide sublimation at 1030°C for 5 min or 1200°C for 2 min. After growth the samples were cooled in various As ambients and/or annealed in UHV to obtain the different reconstructed surfaces.

Mat. Res. Soc. Symp. Proc. Vol. 159. ©1990 Materials Research Society

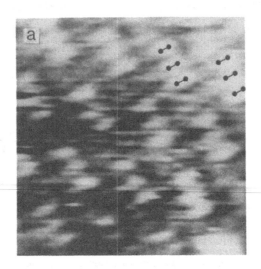

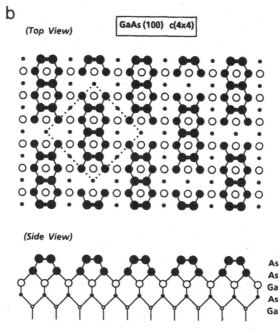

Fig. 1 GaAs(100) c(4x4) reconstruction. (A), STM image of filled states showing three dimers per unit cell and c(4x4) symmetry. (Note: In the gray scale of all STM images shown here white corresponds to tip displacement away from the bulk, i. e. higher); (B), ball and stick model of the c(4x4) surface.

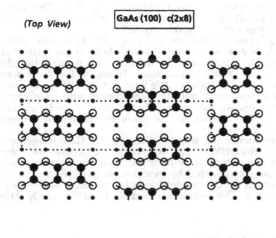

(Top View)

GaAs (100)   c(2x8)

(Side View)

As
Ga
As
Ga

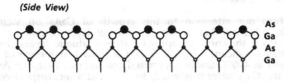

Fig. 2 GaAs(100) c(2x8) / (2x4) reconstruction.   (A), STM image of filled states showing three, two and one dimers per (2x4) unit cell; (B), ball and stick model of the c(2x8) surface with tree dimers per (2x4) sub-cell.

RESULTS

Homoepitaxial GaAs(100)

The ideal GaAs(100) surface is a polar surface and therefore strongly driven to reconstruct in order to neutralize the surface. [4] It is also known that the surface stoichiometry can be varied over a wide range with concomitant phases of structural terminations. [5] Figs. 1(A) through 4(A) are representative images of GaAs surfaces in order of decreasing As concentration. Figs. 1(B) through 4(B) are ball and stick representations of models which are consistent with the STM images and satisfy electron counting (local neutrality) heuristics. [6] All models are referenced to a common complete bulk layer.

Fig. 1 shows the most As-rich surface. It has $c(4 \times 4)$ symmetry. Three dimers are observed per unit cell, with the dimerization direction lying parallel to the [011] direction. The apparently equivalent and symmetric dimers are rotated 90° from the ideal As termination plane and must therefore be an ad-layer. Fig. 1(B) shows the hypothesized model in top and side views. We cannot distinguish between models which contain As atoms in the corner positions of the "rest layer" (layer below the adatom dimers) and Ga atoms at those positions. As shown, the overlayer consists of 1.75 monolayers (ML) of As and 1 ML of Ga. That is, the surface region has layers with compositions (i) 0.75 As, (ii) 1.0 As, and (iii) 1.0 Ga going from surface to bulk. The STM images and electron counting arguments are also consistent with layer (ii) having 0.5 As + 0.5 Ga.

The (2×4) [or $c(2 \times 8)$] structure is shown in Fig. 2(A). This particular image indicates that the (2×4) unit can consist of 3, 2 or 1 As dimer per (2×4) cell. A model for the 3-dimer cell is shown in Fig. 2(B).

Fig. 3(A) is a representative image of the (2×6) surface and Fig. 3(B) a consistent model. We observe much disorder both in the relative alignment of the As dimers and the positioning of unit cells. This parallels the observation in LEED of streaks at the x2 and x3 positions. Finally, in Fig. 4 we present $c(8 \times 2)$ images of the most Ga-rich surface. We have not seen (4x2) subunits with three Ga dimers. These may exist and may well coexist with significant Ga droplet populations. However, we have found Ga rich surfaces to be difficult to image, apparently due to Ga transfer to the tip with subsequent diffusion-induced tunneling noise.

In these experiments we prepare the distinct Ga-rich reconstructions by sublimation of As into UHV. It can be seen in the progression from Figs. 2(B), 3(B) and 4(B) that Ga also must be removed from the reconstructed areas. The excess Ga may be accomodated by evaporation (unlikely on the time scale of these experiments) or by Ga droplet formation or (most likely) by extension of Ga-rich GaAs terraces. This behavior may differ from the steady state of Ga-rich growth.

HETEROEPITAXIAL GaAs/Si(100)

We now turn our attention to the growth of GaAs on vicinal Si(100). Previous work indicated that a two-dimensional layer containing As and Ga covers the Si (to a thickness of approximately 5 atomic layers) before island growth becomes significant. Figs. 5, 7 and 8 show a sequence of the earliest stages of growth at 400°C under Ga-limited conditions. First, the As_4 shutter is opened, then the Ga shutter is opened, for 0, 3 and 6 sec, respectively. Fig. 5 is the Si(100)-(2×1) surface terminated by a monolayer of As becoming Si(100):As-(1×2). (Independent of preparation conditions, we always find in LEED and STM that Si(100):As is (1×2) -- the dimer direction being perpendicular to the step edges and thus rotated 90° from the dimer direction of the clean, vicinal Si(100) reconstruction.) The left and right panels are imaged at the same sample

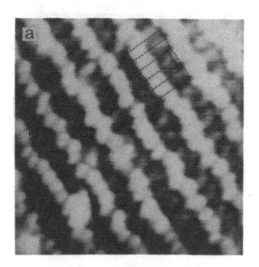

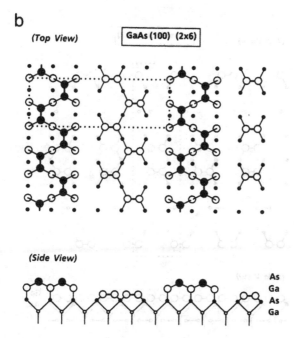

Fig. 3 GaAs(100) (2x6) reconstruction. (A), STM image of filled states; (B), ball and stick model of the (2x6) surface.

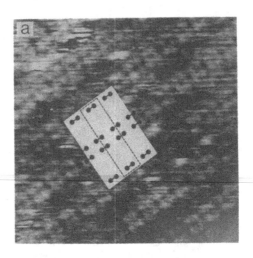

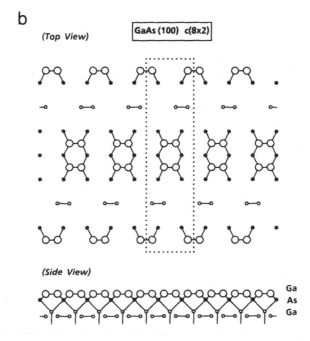

Fig. 4 GaAs(100) c(8x2) reconstruction. (A), STM image of filled states showing two dimers per (4x2) sub-cell; (B), ball and stick model of the c(8x2) surface.

positions but with sample bias varied between -2V and +2V in repeated line scans. Thus, the left panel represents electron tunneling from filled states, whereas the right panel represents tunneling from the tip into empty states of the sample.

In the STM filled state images, the dimers are clearly observable. (The empty states always show less corrugation, independent of tip condition.) Rows of dimers run parallel to the step edges of the (100) terraces. If one lays a grid over the dimers in a given terrace, a phase shift of plus or minus one quarter period is observed for the dimers in neighboring terraces. That both shifts occur can be seen most readily for steps adjoining a terrace in which a dimer stacking fault exists. Fig. 6 shows ball and stick figures representing very low energy models consistent with the images for terraces having even and odd numbers of Si rows. The step in Fig. 6(A), essentially a (111) facet, has the Si with its dangling bond replaced by an As with an energetically favorable lone pair. The quarter period shift can be seen by comparison with the "bulk" lattice. A related configuration, shown in Fig. 6(B), arises for terraces spanned by an odd number of Si rows. In that case we conjecture that, again for electron counting reasons, the Si in the step facet is not replaced by an As atom but dimerizes with the neighboring As. Arsenic termination seems to initiate at the outer step edge of terraces and propagate to the step riser, thereby forcing the necessary step reconstruction.

When the Ga flux is initiated, linear features extend along the center of the As dimer rows (Fig. 7). Nucleation seems to be random within a terrace. Neighboring rows grow coherently so that an ordered overlayer, the symmetry of which is apparantly (3x2), results (Fig. 8). Fig. 9 is our hypothesized model, again consistent with the data. Electron counting implies neutrality for individual dimer units, and therefore any (nx2) arrangement consisting of As-As dimers and Ga-Ga dimers would in principle be possible. The maximum Ga coverage would occur with a (2x2) ordering and correspond to 1/2 monolayer adsorption. The fact that we see (3x2) symmetry at low coverages implies that the Ga-Ga dimer and As-As dimer units are mutually attractive. It is possible that at higher coverages the surface will be driven to a (2x2) arrangement consisting only of Ga-Ga dimers.

Fig. 5 Vicinal Si(100) terminated by one monolayer of As. (100) terraces are bounded predominantly by two-atom high steps with dimer rows running parallel to the step edges. Left panel: sample bias=-2V, tunneling from filled states; right panel: sample bias= +2V, tunneling into empty states.

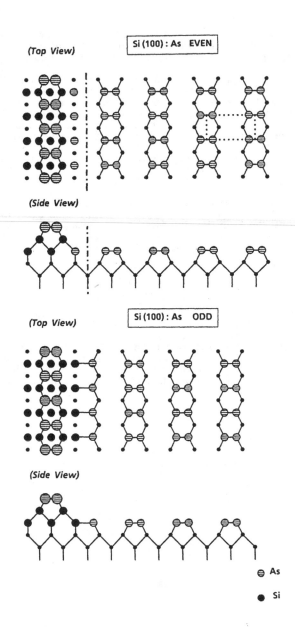

Fig. 6 Ball and stick models of the vicinal Si(100):As surface. (A), an even number of Si rest atoms (topmost small Si balls) are assumed to span the lower terrace. An As atom substituted for a Si in the step facet would result in a low energy step. (B), an odd number of Si rest atoms spanning the lower terrace. For electron counting reasons we guess that a Si-As dimer is formed at the step riser.

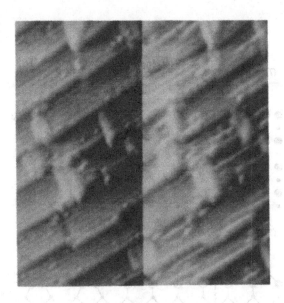

Fig. 7 Vicinal Si(100) with approximately 0.1 monolayer of Ga and As codeposited. The image reveals an As monolayer with a linearly extending overlayer. Left panel: sample bias = -2V, tunneling from filled states; right panel: sample bias = + 2V, tunneling into empty states.

Fig. 8 Vicinal Si(100) with approximately 0.3 monolayer of Ga and As codeposited. Strong correlation is observable between rows. Left panel: sample bias = -2V, tunneling from filled states; right panel: sample bias = + 2V, tunneling into empty states.

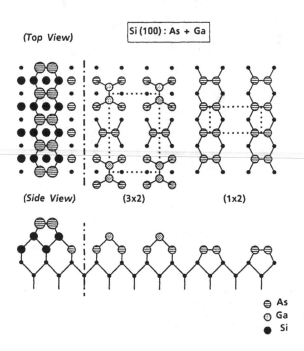

Fig. 9 Ball and stick model of the vicinal Si(100):As + Ga surface showing coexisting Si(100):AS-(1x2) and Ga overlayer units.

When the Ga dimer sites are filled the Ga diffusivity should increase greatly. The next stage of film growth may be a kinetic competition between: (a) Ga cluster formation and As incorporation (3D island nucleation) and (b) Ga and As accretion in the 2D coverage. Recent Auger studies [7] seem to indicate that the latter tends to occur metastably.

We have now observed only the very earliest stages of GaAs/Si heteroepitaxy for only a limited range of deposition/sample parameters. Much work remains to be carried out before a general understanding of the energetics and kinetics can be obtained and utilized.

GaAs(100) homoepitaxy and heteroepitaxy are seen to differ in the earlist stages of development. Homoepitaxy is essentially periodic, modulo a bilayer. Heteroepitaxy of GaAs on Si(100), in contrast, moves through structural phases for many monolayers of deposition before approaching asymptotically to a homoepitaxial, thick film regime. The particular phases generated in both homo- and heteroepitaxy are driven by the relative chemical potentials of As and Ga during deposition and subsequent processing.

SUMMARY

We have presented STM images of the distinct phases of the GaAs(100) homoepitaxial surface as a function of surface stoichiometry. We have also followed the heteroepitaxial growth of GaAs on vicinal Si(100) through the As monolayer and Ga half-layer adsorption during co-deposition. The next stages of growth, namely 2D layer thickening and 3D nucleation are certainly kinetically controlled. We are now using STM to follow the details of growth in these stages.

REFERENCES

1. After solvent dips, samples were etched for 30 sec in 10:1:1 $H_2SO_4:H_2O_2:H_2O$ then irradiated with ultraviolet light for 5 min in air.

2. Auger observation of the oxide desorption was used to calibrate the pyrometric temperature scale; cf. A. J. SpringThorpe and P. Mandeville, J. Vac. Sci. Technol., **B6**, 754 (1988).

3. B. S. Krusor, D. K. Biegelsen, R. D. Yingling, and J. R. Abelson, J. Vac. Sci. Technol. B 7, 129 (1989).

4. D.J. Chadi, J. Vac. Sci. Technol., **A5**, 834 (1987).

5. See for example, A. Y. Cho, J. Appl. Phys. **42**, 2074 (1971); D. J. Frankel, C. Yu, J. P. Harbison, and H. H. Farrell, J. Vac. Sci. Technol. B 5, 1113 (1987); or, J. Y. Tsao, T. M. Brennan, J. F. Klem, and B. E. Hammons, J. Vac. Sci. Technol. A 7, 2138 (1989).

6. W. A. Harrison, J. Vac. Sci. Technol. **16**, 1492 (1979).

7. D. B. Fenner, D. K. Biegelsen, and B. S. Krusor, Proc. Symp. Mat. Re. Soc. (1989); this volume.

μm-SCALE LATERAL GROWTH OF Ga-MONOLAYERS OBSERVED IN-SITU BY ELECTRON MICROSCOPY

J.OSAKA and N.INOUE
NTT LSI laboratories, 3-1 Morinosato-Wakamiya, Atsugi-shi,243-01,Japan

ABSTRACT

An ultra high vacuum scanning electron microscope equipped to an MBE system is utilized to study a transient of a surface atomic structure during MBE growth of GaAs and AlGaAs by the alternate supply method. Lateral growth of a Ga-monolayer over microns is realized utilizing Ga droplets. This is confirmed by discriminating the Ga and As top layer by using the secondary electron intensity difference between the Ga and As top layer. The growth mechanism of the Ga monolayer is discussed based on the results.

INTRODUCTION

In-situ observation of a growing surface provides us with a lot of information to help us understand and control the growth. We have developed a hybrid system for the in-situ observation of the MBE growth of III-V materials using a scanning electron microscope [1-2]. By using this technique, the dynamic growth process has been revealed for the first time [1]. Furthermore, we have found the following characteristics of Ga and Al atoms on a Ga layer advantageous for growth control down to a monolayer:
(1) Even a fractional layer of Ga atoms supplied over a Ga layer can not form a continuous thin film but forms droplets instead. In other words, Ga has a self limiting nature [3].
(2) Ga droplet spacing is as large as 10 microns at 610 C. That is, Ga migration distance over Ga is unexpectedly large. The droplet size and spacing of Ga-Al alloy is almost equal to that of Ga [1-3].
However, there are problems when we want to use these characteristics for the atomic layer controlled growth:
(1) When As atoms are supplied on the Ga layer with Ga droplets, the Ga droplets disappear with the growth. However, mounds are left at the droplet sites. Thus, "smooth growth" is not possible [3].
(2) It may be possible to realize large area lateral growth by using Ga droplets, as a reverse process of Ga droplet formation, but it must be confirmed and control by observation. Therefore, the Ga top layer and As top layer must be discriminated by a electron microscope.
In this paper, we show a solution to these problems. As a result, micron-scale lateral growth of Ga-monolayer is established and confirmed by in-situ observation.

EXPERIMENTAL PROCEDURE

The apparatus was the same as that used in the previous study [1-3]. An ultra high vacuum optic column of a scanning electron microscope is mounted horizontally on the MBE chamber. A focused electron beam with an acceleration voltage of 15 kV and a beam current of about 1 nA scans the surface of a sample at the rate of 10 seconds per frame. Either the intensity of a part of the RHEED pattern on the phosphor screen or that of secondary electrons is used to make topological images of the surface. The glancing angle was about 35 mrad. The resolution is between 500 and 1000 angstroms depending on the magnitude of the sample vibration.
We observed how the surface topography changes when Ga, or Ga and Al, and As atoms were alternately supplied to a GaAs (001) surface at temperatures between 510 C and 610 C. The Ga deposition rate and As

incorporation rate are about one monolayer per 5 seconds and about one monolayer per from 1 to 3 seconds, respectively. The rates were determined from the periods of the RHEED specular beam intensity oscillations [4-5]. When the As effusion cell shutter was closed, the background As_4 pressure near the substrate was 3×10^{-8} Torr. The substrates were etched in a solution of H_2SO_4: H_2O_2:H_2O, and then immersed in HCl to remove surface oxides. They were then mounted with In on Mo blocks. All experiments were performed on surfaces prepared by the epitaxial growth of a buffer layer more than 1000 A thick under As-rich conditions.

DEPENDENCE OF SECONDARY ELECTRON INTENSITY ON SURFACE COVERAGE

In the conventional MBE, the whole surface is always covered with a Ga or As top layer (depending on the III/V ratio). So there should be no contrast except step images. In the alternate supply MBE, some fraction of the surface is covered with a Ga top layer and another is covered with As. To observe the lateral growth, we must discriminate these two top layers. It was found that the secondary electron intensity is different from these two top layers as follows.

The output of a secondary electron detector recorded during the alternate supply MBE growth at 610 C is shown in Figure 1. Initially, the surface showed 2x4 As-stabilized reconstruction under As flux. When the As shutter was closed, the surface changed to 3x1 reconstruction and the secondary electron intensity slightly increased. Then, when the Ga shutter was opened, the intensity increased sharply towards the peak, and then dropped off a little. After the intensity showed the peak, the surface showed 4x2 Ga-stabilized reconstruction. After the total amount of Ga supplied reached about one monolayer, the Ga shutter was closed, but the intensity change was not detected. When the As shutter was opened again, the intensity gradually reduced to the initial value. Then, the surface recovered to 2x4 reconstruction. So, a 4x2 Ga-stabilized layer is easily discriminated from a 2x4 As-stabilized layer by the secondary electron intensity.

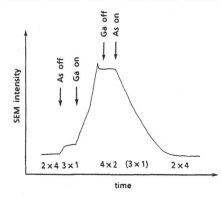

Figure 1. Secondary electron intensity change during alternate supply of As and Ga on GaAs.

An example of SEM micrograph with the growth sequence and surface reconstruction attached to demonstrate how the Ga and As top layer is discriminated is shown in Figure 2. The substrate temperature was 610 C. The electron beam was scanned from upper-left to lower-right for ten seconds. The Ga shutter was opened at the point indicated by "on" and closed at "off". The total amount of supplied Ga corresponds to make one monolayer. After the Ga shutter was opened, the surface started to become brighter, and then, became white at the middle of the micrograph corresponding to the change shown in Figure 1.

From these results, it is concluded that the intensity of secondary electron increases as the density of Ga on the top layer of the GaAs surface increases until the surface changes to a 4x2 Ga-stabilized reconstruction. The mechanism of the secondary electron intensity increase on the Ga surface coverage is not yet fully understood. It is supposed that a work function changes by surface coverage and affects the secondary electron intensity. With our new in-situ method using secondary electron intensity we can observe

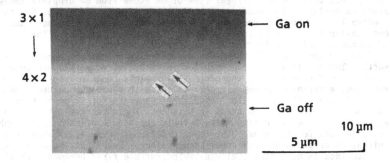

Figure 2. SEM micrograph of surface with As and Ga top layer containing Ga droplets.

not only gross changes, but also local changes in the top layer from moment to moment.

Note that the self-limiting nature of the Ga layer is clearly observed in Figure 2. In the white region of the 4x2 Ga-stabilized reconstruction, there are five black dots which are the Ga droplets. There are smaller black dots indicated by arrows just after the surface becomes white. They are also Ga droplets just nucleated. In the previous paper [3], we showed that the density of the droplets do not depend on the amount of Ga atoms supplied, but on the substrate temperature and that droplets do get bigger as more Ga is supplied. As shown in Figure 2, the Ga coverage of the 4x2 Ga stabilized surface is nearly 1 and that when excess Ga atoms are supplied on a 4x2 Ga-stabilized surface, they cannot form a thin film but form droplets and the surface remains 4x2 reconstruction because the Ga atoms supplied after the droplets nucleation move fast to the droplets. In other words, the Ga-stabilized surface has a self-limiting nature.

OBSERVATION OF A MICRON-SCALE Ga-MONOLAYER LATERAL GROWTH

In the previous paper [3], we showed that the Ga droplets disappear when As is continuously deposited. Before the disappearance of Ga droplets, it was observed that the brighter areas appeared around the droplets. From the result shown in the previous section, it is suggested that Ga lateral growth, the reverse process of Ga droplet formation, may take place. However, the bright areas suddenly disappeared and did not cover the entire surface. This should be due to the fact that the continuously supplied As atoms covered the Ga over-layer. Therefore, we may not realize Ga monolayer growth when As is supplied continuously.

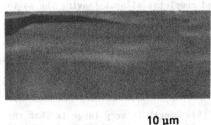

Figure 3. SEM micrograph of stable Ga monolayer (white region) on As top layer.

What happens when As is supplied for one monolayer in pulse? The result is shown in Figure 3. The bright area remains stable, i.e. the co-existence of the Ga and As top layer is observed for the first time. The Ga atoms have enough time to migrate over the Ga top layer without As over layer coverage. It can be seen that the centrifugal flow of Ga atoms from Ga droplets takes place as easily as the condensation of Ga atoms to Ga droplets. So, the problem left is how far does the lateral layer growth takes place, and if we can realize the whole surface coverage. We could utilize here the advantageous self-limiting nature of Ga atoms on a Ga layer.

In the above case the droplets contained less than one monolayer's worth. The result when Ga is supplied in excess of more than one monolayer is shown in Figure 4. So, therefore the droplets contain more than one monolayer's worth. The SEM micrographs shown in the right column show how the Ga droplets shrink and the topography changes after As deposition. Each one took 5 seconds with no break in between. The substrate temperature was about 610 C. The black line lying left in the middle of each micrograph is a pre-made mark for focusing. The top micrograph (Figure 4(a)) shows the initial Ga droplets on a Ga-stabilized surface. Note that the mean distance between them is more than 5 microns. The surface first becomes dark all over by As deposition. But, after the As shutter was closed, bright regions appear around the droplets (Figure 4(b)). In the top of Figure 4(b), it can be seen that these bright regions look like circles at first. Then, they grow and join together as shown in the middle of Figure 4(b). They look like bright bands a few seconds later as the result of fore shortening. These regions grow further (Figures 4(b) and 4(c)) and finally cover the entire surface (Figure 4(d)). The droplets remained, although they shrunk as shown in Figure 4(d).

Ga MONOLAYER GROWTH AND MIGRATION MECHANISM

The illustrations shown in the left column of Figure 4 explain what's happening at each step. The initial Ga droplets on a Ga-stabilized surface are shown in Figure 4(a). The deposited As atoms stick to the Ga surface. Unfortunately, we can not get both a SEM topography and a RHEED pattern from the limited area simultaneously. Therefore, the surface reconstruction of the bright area which grows around the droplets as shown in Figures 4(b) and 4(c) is not determined directly. However, it is clear that those bright areas are 4x2 Ga-stabilized surfaces because their brightness is the same as that of the bright area shown in Figures 4(a) and 4(d), which showed 4x2 reconstruction and that the droplets can exist only on a Ga-stabilized surface. Then, some of the Ga atoms in the droplets migrate onto the As layer in the immediate neighborhood forming bands around the droplets (Figure 4(b')). Now it is possible for other Ga atoms in the droplets to migrate over the preceding Ga atoms to the remaining As surface (Figure 4(c')). These processes correspond to the top end and middle in Figure 4(b), respectively. Finally, the surface is covered entirely by Ga atoms (Figure 4(d')) and the excess Ga atoms still form droplets. Thus, the lateral growth of the Ga monolayer is self limiting and complete, without leaving the As top layer.

In conventional MBE growth, the migration length of Ga atoms is generally on the order of hundreds of an angstrom as estimated from an analysis of RHEED oscillations on the vicinal surface [4]. Recently, it has been reported that Ga atoms can migrate for about 2000 A, and Al atoms can migrate for 300 A during high temperature MBE or MEE [6]. But in the Ga monolayer growth process shown above, the migration length reaches several microns for about ten seconds, and is limited by the distance between the droplets.

The most likely explanation of why this length is very large is that the Ga atoms are migrating on a Ga surface. In the previous study we determined the migration distance of Ga atoms on a Ga layer for the droplet formation process [2]. Both in that process and in the present Ga monolayer growth process, the motive force for the diffusion should be the Ga concentration gradient around the droplets.

However, the situation is more simple in the present case. It can be

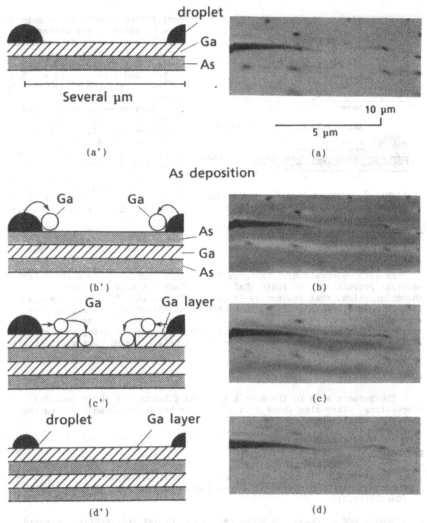

Figures 4:(a)-(d) SEM micrographs of Ga monolayer growth processes formed by pulsed As deposition on a Ga-stabilized surface with Ga droplets. (a')-(d') Illustrations for explaining Ga monolayer growth mechanism.

supposed that there is a saturation concentration of Ga atoms around the droplets, and at the edge of the Ga monolayer, there are, of course, almost no Ga atoms. As a result, there should be Ga concentration distribution like that shown in Figure 5. We estimated the diffusion constant of Ga atoms on a Ga surface using the following equation.

$$<X^2> = Dt$$

where X is a mean distance between the droplets and t is a time needed for covering the entire surface with Ga atoms. A conventional RHEED

38

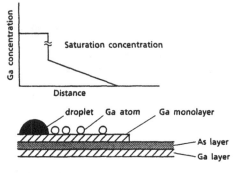

Figure 5. Illustration for explaining the model for Ga monolayer growth.

oscillation technique can not give t, which is the average collision time or lifetime on the surface. In contrast, the present method gives both X and t. We found D to be greater than $10^{-9} cm^2/s$, for the first time, by measuring both of X and t.

We also examined the growth of AlGaAs with an Al content of less than 30%. In that case, the density of the droplets of Ga-Al alloy is almost the same as that of pure Ga droplets. This means that Al atoms can move far as Ga atoms on the Ga-Al alloy surface. This is very important for making a larger, smoother interface with AlGaAs.

SUMMARY

In-situ observation of the growth processes enabled us to measure the material properties. We found that a Ga surface has self-limiting characteristics, that is, excess Ga can not form a thin film, but instead forms droplets. We also found that both Ga and Al atoms are able to move more than a micron on a Ga or Ga-Al surface. Based on this, we have discovered how to grow Ga-monolayers on a micron scale by using Ga droplets. These results may allow us to produce flat AlGaAs surfaces on a micron scale.

ACKNOWLEDGMENTS

The authors wish to thank Dr.K.Wada and K.Yamada for their fruitful discussions. They also thank Dr.A.Ishida for his encouragement during the course of this study.

REFERENCES

1. K. Yamada, N. Inoue, J. Osaka and K. Wada, presented at the 5th International Conference on Molecular Beam Epitaxy, Sapporo, 1988 (unpublished).

2. N. Inoue and J. Osaka, in Technical Digest of 1st Int. Meeting Advanced Processing & Characterization Technologies, Tokyo, 1989, p. 9.

3. J. Osaka, N. Inoue, Y. Mada, K. Yamada and K. Wada, in Proceedings ICCG-9, Sendai, 1989 (to be published in J. Crystal Growth)

4. J. H. Neave, B. A. Joyce, P. J. Dobson and N. Norton, Appl. Phys. A 31, 1 (1983)

5. J. H. Neave, B. A. Joyce and P. J. Dobson, Appl. Phys. A 34, 179 (1984)

6. M. Tanaka, PhD thesis, University of Tokyo, 1988.

A LEED STUDY OF BISMUTH OVERLAYER FORMATION ON InSb(110)

T. Guo, K. J. Wan and W. K. Ford
Advanced Materials Center and the Department of Physics
Montana State University, Bozeman, Montana 59717

ABSTRACT

LEED and Auger electron spectroscopy have been used to studied the interface formation of bismuth on InSb(110). For the first time, well defined superstructures were observed at room temperature on metal/semimetal-III-V semiconductors. A (1x2) phase, at 0.5 ML bismuth coverage, and a (1x3) phase, at 0.4 ML bismuth coverage, appear with different thermal heat cycling. The dynamical origin of these phases is thought to be a repulsive interaction between overlayer chains due to the detailed geometry of the overlayer chain structure and to the overlayer induced strain field within the substrate.

INTRODUCTION

Renewed attention has been given to the interface systems of column-V elements on III-V semiconductor (110) surfaces following the recent discovery of the epitaxial growth of bismuth on the GaAs(110) surface.[1] Although (110) surfaces are convenient to prepare experimentally and to study theoretically, very few elements are known to form ordered overlayers on them. The scarcity of examples of epitaxy reduces the efficacy of the use of the (110) substrate surface to study the microscopic properties of overlayer formation and epitaxial growth, and serves to underscore the importance of detailed bismuth and antimony adatom studies because overlayers of these atoms are epitaxial on several III-V (110) surfaces.[2-4] With the advantages provided by the bismuth and antimony prototype systems studies may proceed that determine more fully the ordering properties, atomic geometries, and chemical bonding in the overlayer.

The atomic and electronic structure of antimony on the GaAs(110) and InP(110) surfaces have been studied extensively in the past.[1,3] Antimony forms an ordered $p(1x1)$ - 1ML overlayer that undoes the substrate reconstruction, leaving the surface in a bulk-terminated geometry with zig-zag chains of antimony atoms running along the [1$\bar{1}$0] direction in the positions normally occupied by gallium and arsenic atoms in the bulk. A clean surface which has not been allowed to relax would have dangling bonds on each of the anion and cation tetrahedral surface sites. This configuration is energetically unstable and a sizable reconstruction occurs for all III-V and II-VI (110) semiconductor surfaces.[5] In the formation of the antimony overlayer geometry, the dangling bonds are bonded to the overlayer adatom chain in such a way so as to stabilize the truncated bulk geometry.[1,3] A recent dynamical LEED intensity analysis of the Bi/GaAs(110) system shows that bismuth forms a similar zig-zag chain structure but that the chains are punctuated by vacancies every six unit cells giving rise to a "(6x1)" surface geometry.[4,6] The vacancies occur to relieve the strain in the interface that arises due to the size mismatch between the bismuth adatoms and the substrate unit cell size. The ordering of the chain vacancies is the subject of another paper in this conference.[7]

In this paper, we extend the investigation of bismuth epitaxy to the InSb(110) surface. This overlayer system is the first reported example of a well defined superlattice that forms at room temperature by metal/semimetal adatoms on the (110) surface of any III-V semiconductor. Our measured LEED intensity profiles show that, unlike the GaAs(110) case, there is no long-range order in the as-deposited bismuth overlayers on InSb(110). However, annealing the film introduces a (1x2) phase that transforms into a (1x3) phase upon further annealing.

EXPERIMENTAL DETAILS

Quantitative low energy electron diffraction (LEED) and Auger electron spectroscopy were used in this experiment to study the interface of bismuth on InSb(110).

Both LEED beam intensity measurements, or I-V curves, and diffraction spot profiling were performed at normal incidence using a newly developed video-based LEED diffractometer.[8] The surfaces were prepared from cleaved InSb(110) n-type doped bars having a carrier concentration of ~1.2 x 10^{16} cm^{-3}. Bismuth films were deposited by sublimation from the bulk at a background pressure of less than 4 x 10^{-10} torr and at a typical rate of less than 1 Å/min. The film deposition was monitored with a quartz crystal oscillator. Rutherford backscattering was used to help calibrate the coverage values. One monolayer (ML) of bismuth adatoms is defined according to the surface density of InSb(110) substrate, i.e. 6.74 x 10^{14} atoms/cm^2 using the two atoms per substrate unit cell convention. Sample cleanliness was verified using Auger electron spectroscopy using the same LEED apparatus in a retarding field analyzer configuration.[8] The samples could be heated to a temperature in excess of 500 C with a button heater mounted on the sample holder. The LEED diffraction profiles and Auger measurements of bismuth coverage as a function of annealing temperature were performed without moving the sample from its position at the center of the LEED optics.

RESULTS AND DISCUSSION

The as-cleaved InSb(110) surface displays a (1x1) diffraction pattern having a slightly smaller reciprocal lattice unit cell than that of GaAs(110), the lattice constants of InSb and GaAs being 6.50 Å and 5.65 Å, respectively. This (1x1) diffraction persists for all bismuth coverages studied for films prepared at room temperature and not annealed. For reference, a photograph of the LEED at 50 eV is given in Fig. 1a for an as-grown 0.8 ML bismuth film and in Fig. 1b for a 2.5 ML film. In the cases of GaAs and InSb the (h0) = (h̄0) diffraction beams from a clean surface have strong intensities, indicating that the clean surface is significantly reconstructed.[8] The h-direction is along the [1̄10] direction of the bulk, along which the surface atoms form zig-zag, anion-cation chains in the (110) surface. If the clean surfaces had a flat, truncated bulk geometry the similarity of the scattering cross-sections of the Ga and As or the In and Sb would lead to near extinction of the (h0) and (h̄0) beams for h an odd integer. An extinction of these beams is observed for Bi/GaAs(110) where the overlayer is now known to remove the reconstruction of the substrate and form flat, zig-zag chains parallel to the surface.[4,6] However, no preferential extinction is observed for the as-grown Bi/InSb(110) films studied up to coverages near 2 ML. The fact that no extinction occurs and that only the (1x1) diffraction occurs over the lower range of bismuth coverages studied indicates that these films do not modify the substrate clean surface reconstruction.

At a higher level of bismuth coverage when three dimensional growth occurs, cf. Fig. 1b, the (1x1) diffraction spots become weaker and convert to streaks along the k direction. The k-direction is along the [001] direction of the bulk. The separation between streaks is along the [1̄10] direction of the substrate whose periodicity of 4.7 Å nearly matches that of bulk bismuth, 4.6Å. Thus, the LEED indicates some registry between the overlayer structures formed at high coverage and the substrate geometry.

A qualitative examination of the LEED diffraction intensities implies that the films are also not ordered. An ordered overlayer of a dissimilar element should introduce changes in the diffraction intensity versus energy data, or I-V curves. In Fig. 2, the (1̄1) diffraction beam intensity is plotted for various bismuth coverages prepared at room temperature without subsequent annealing. The beam intensities were measured after a background intensity generated by incoherent scattering had been subtracted out. The (1̄1) beam is representative of all the diffraction beams observed. No noticeable variation is detected in Fig 2 as the bismuth coverage increases except that the overall intensity of the diffraction is reduced. This trend and the observations from Fig. 1 indicate that bismuth forms a disordered overlayer at room temperature, unlike the GaAs(110) case.[4,6] This overlayer may either be uniform or three dimensional.

Annealing the room temperature prepared films produces a novel surface phase diagram not previously reported for III-V (110) surfaces. The changes introduced in the surface structure are summarized in the photographs of Fig. 1. Fig. 1c shows that half order streaks appeared along the k direction when a 1.5 ML bismuth overlayer was heated to ~ 100 C and then cooled to room temperature. These streaks slowly became sharper with additional annealing at the same temperature. A well defined (1x2) phase

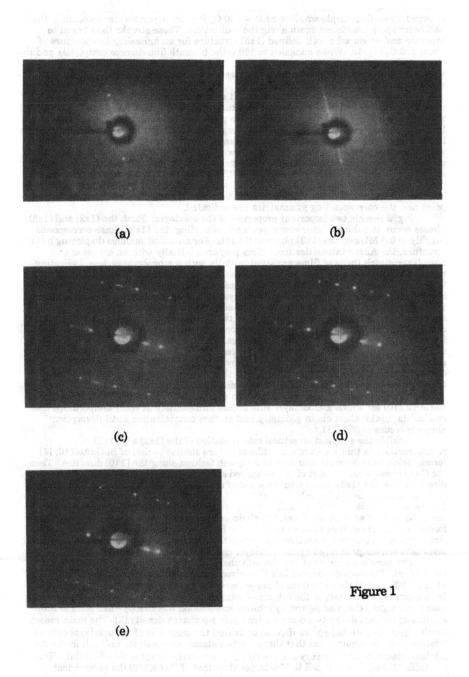

(a)

(b)

(c)

(d)

(e)

Figure 1

emerged when the sample was heated to ~ 180 C, Fig. 1d. Upon further annealing, the half order spots broadened again along the k direction. These streaks then began to separate and produced a well defined (1x3) structure for an annealing temperature of about 270 C, Fig. 1e. When annealed to 350 C the bismuth film desorbs completely and the clean surface diffraction returns, eg. Fig. 1a. The sample was cooled to room temperature in each case before these pictures were collected; the LEED electron gun energy was 70 eV for pictures 1a through 1b.

Auger measurements were carried out as a function of annealing to obtain a coverage estimate for each of the superstructure phases. The results of the Auger measurements for a 0.8 ML bismuth film are presented as a function of annealing temperature in Fig. 3. The changes in coverage indicated was obtained from the changes in the Auger peak-to-peak amplitude of the (unresolved) bismuth 96 eV and 101 eV lines with annealing relative to that of the as-grown film. A linear Auger signal dependence on coverage was deemed adequate for coverages less than 1 ML. The initial coverage of 0.8 ML was determined using the QCO and RBS calibration. The room temperature prepared sample was annealed for ~ 4 minutes at each temperature and then cooled to room temperature before Auger data were taken. LEED patterns were recorded at each point and the corresponding symmetries are indicated.

Fig. 3 reveals two important properties of the overlayer. First, the (1x2) and (1x3) phases occur at submonolayer coverages upon annealing; the (1x2) phase corresponds roughly to 0.5 ML and the (1x3) phase to 0.4 ML. For annealed samples displaying a (1x2) structure, the Auger intensities from films prepared initially with an excess of one monolayer match those of films prepared initially with a monolayer or less, indicating that the overlayer excess desorbs at low annealing temperatures. Therefore, the formation of surface superlattices do not require coverages exceeding one monolayer. Second, a continuous desorption of the overlayer occurs up to a temperature of 350 C and the (1x2) and (1x3) phases can only be obtained in small ranges of annealing temperatures. Although one might envision that bismuth in-diffusion into the substrate could induce the (1x2) and (1x3) reconstructions, one would expect a change in the LEED I-V curves and a residual bismuth Auger signal if bismuth absorption occurred leaving any significant concentration in the surface region. To the contrary, the Auger data showed that bismuth is completely desorbed by annealing to 350 C and the LEED I-V curves measured using surfaces annealed at this temperature were identical to those of the as-cleaved clean surface. Hence, the bonding of bismuth overlayer to InSb(110) is weak and desorption is a continuous function of heating cycle, unlike the case for Bi/GaAs(110) for which a monolayer film orders immediately at room temperature and retains its "(6x1)" short chain geometry and surface concentration until desorption abruptly occurs at 350 C.[1,4,6]

Possibly the simplest structural interpretation of the (1x2) and (1x3) reconstructions is that a sort of chain-like structure similar to that of Bi/GaAs(110) [4] forms, linking the bismuth adatoms in a zig-zag fashion along the [1$\bar{1}$0] direction. Then, the (1x2) phase would consist of chains spaced every other unit cell along the [001] direction, and the (1x3) phase would consist of chains spaced every third unit cell in the [001] direction. The (1x2) would form at 1/2 ML coverage and the (1x3) would form at 1/3 ML coverage. This description is illustrated in Fig. 4. The ordering into these phases might occur for two reasons. First, the chain geometry must be such as to impede the formation of (1x1) islands such as are seen for Bi/GaAs.[6] Second, the (1x2) and (1x3) structures arise due to a repulsive interaction that occurs due the strain in the substrate layer between strips covered by the overlayer chain and those that are not.

The atomic geometry of the bismuth chains on this surface remains to be established. Previous dynamical LEED analyses have determined that for Sb/GaAs(110), Sb/InP(110), and Bi/GaAs(110) the chain geometry is due to a distorted sp^3 bonding.[2-4] In this configuration, two of the column-V adatom valence electrons bond to neighboring atoms in a zigzag chain as p_x and p_y orbitals and another one in a p_z - like bond to the substrate; the remaining two occupy a lone-pair s-p charge density.[2] The main reason for this greatly-distorted sp^3 configuration is that the large size of the overlayer column-V adatom and the requirement that there are two atoms per substrate unit cell forces the adatom-adatom bond to be very close to 90°, the characteristic value for p^2 orbital. Since the InSb(110) surface unit cell is 15% larger than that of GaAs(110) the geometrical

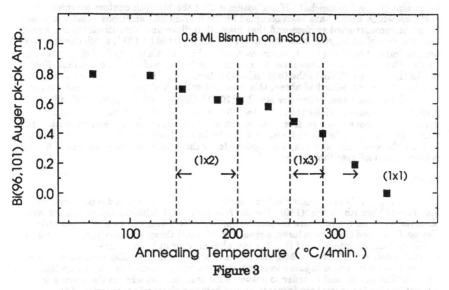

Figure 3

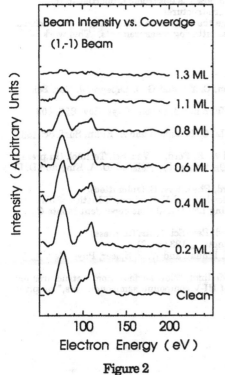

Figure 2

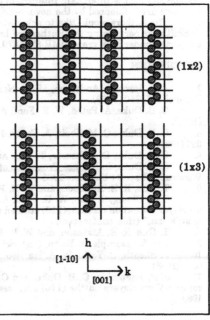

Figure 4

constraint is eased somewhat. If one assumes that the bismuth adatom occupies positions which would have been occupied by the indium and antimony substrate atoms for an unreconstructed surface and that bismuth-indium and bismuth-antimony bonds take their covalent lengths, then the Bi-Bi bond angle would be 103 °, much closer to the tetrahedral sp^3 bond angle. However, we tend to think that a tetrahedral structure is not likely. First, bismuth is known not to form tetrahedrally bonded bulk materials. Second, unlike the case for Bi/GaAs the (h0) diffraction beams for Bi/InSb are intense for each ordered phase. As described above, this is an indication that the surface is not planar.

Other chain geometries are possible.[10,11] One is a p^3 structure originally proposed by Skeath *et al.* for the Sb/GaAs(110) system.[10] The p^3 structure does not necessarily remove clean surface reconstruction. A right angle between bismuth bonds within the chain can be formed, although the Bi-In and Bi-Sb bonds must be stretched. In turn, the stretched bond may be responsible for the weaker bonding between the substrate and adlayer that is observed.

CONCLUSION

Using LEED and Auger electron spectroscopy, we have studied a new ordered interface of bismuth on InSb(110). For the first time, well defined superstructures were observed at room temperature on metal/semimetal-III-V semiconductors. Two ordered phases, (1x2) and (1x3) structures, appear for different thermal annealings. Combined LEED and Auger data showed that the two phases occur at bismuth coverages of approximately 0.5 ML and 0.4 ML, resp. Quantitative dynamical LEED intensity analysis will soon be under taken in order to obtain detailed atomic geometries of this system. More studies are needed in order to answer such questions as what is the driving force responsible for the interchain interaction and how the electronic properties of the interface are influenced by the novel overlayer structures.

The authors would like to acknowledge the assistance of Xu Mingde and Richard J. Smith for performing the Rutherford backscattering measurements. This work is supported under NSF grant DMR-8705879.

REFERENCES

1.   J. J. Joyce, J. Anderson, M. M. Nelson, C. Yu, and G. J. Lapeyre, J. Vac. Sci. Technol. A7, 850 (1989).
2.   C. B. Duke, A. Paton, W. K. Ford, A. Kahn and J. Carelli, Phys. Rev. B26, 803 (1982).
3.   C. B. Duke, C. Mailhoit, A. Paton, K. Li, C. Bonapace and A. Kahn, Surf. Sci.163, 391 (1985).
4.   C. B. Duke, D. L. Lessor, T. Guo, and W. K. Ford, J. Vac. Sci. Technol. (in press).
9.   C. B. Duke, in Surface Properties of Electronic Materials, ed. D. A. King and D. P. Woodruff, Chap. 3 (Elsevier, 1988).
6.   T. Guo, R. E. Atkinson, and W. K. Ford, Phys. Rev. B. (submitted).
7.   S.-L. Chang, T. Guo, W. K. Ford, A. Bowler, and E. S. Hood, "Kinetics of bismuth overlayer formation on GaAs(110) studied using LEED and time dependent Monte Carlo simulation," (this conference).
8.   T. Guo, R. E. Atkinson, and W. K. Ford, Rev. Sci. Instr. (in press).
9.   See, for example, A. Kahn, Surf. Sci. Repts. 3, 193 (1983).
10.   P. Skeath, C. Y. Su, W. A. Harrison, I. Lindau and W. E. Spicer, Phys. Rev. B27, 6246 (1983).
11.   J. P. LaFemina, C. B. Duke, and C. Mailhiot, "New surface atomic structures for column V overlayers on the (110) surfaces of III-V compound semiconductors," preprint, 1989.

BISMUTH OVERLAYER FORMATION ON GaAs(110)

S.-L. Chang, T. Guo, W.K. Ford, A. Bowler, and E.S. Hood
Advanced Materials Center and the Departments of Physics and Chemistry, Montana State University, Bozeman, MT 59715

ABSTRACT

The temperature and coverage dependent ordering of bismuth overlayers on GaAs(110) is examined using low energy electron diffraction (LEED). Sixth order electron diffraction profiles associated with overlayer ordering are observed at coverages of 0.7, 1.0, and 1.5 monolayers (ML) and for temperatures ranging from -110 C to 200 C. The full-width at half-maxima (FWHM) of the sixth-order spots are examined. Profile analysis reveals narrowing widths with increasing annealing temperature, indicating an improvement of the long-range ordering of the overlayers. Differences in inter- and intrachain ordering are observed, analyzed, and discussed.

INTRODUCTION

An understanding of the microscopic formation of metal overlayers on III-V semiconductor surfaces provides valuable insight to a variety of complex metal-semiconductor interface phenomena, e.g., Schottky barrier formation and Fermi level pinning. Not only are the atomic structure, electronic states, and the chemical bonding at the interface important properties that must be characterized, but so is the description of the interface growth kinetics. Antimony on GaAs(110) is one of the interface systems that has been investigated most intensively [1-4]. LEED proved that antimony forms ordered (1x1) overlayers when deposited at room temperature on a GaAs(110) surface up to one monolayer [4]. Antimony atoms bond to the underlying gallium and arsenic atoms by forming linear zigzag chains along the [1$\bar{1}$0] direction [3]. Thermal desorption studies established that the first monolayer is stable against thermal annealing to about 550 C and that the growth mode is Stransky-Kranstanov (SK) [4].

Overlayers formed of bismuth, isoelectronic to antimony, on the III-V (110) surfaces are very little studied. High-resolution photoelectron spectroscopy results of bismuth on GaAs(110) indicated that a modified SK growth mode occurs for coverages up to two monolayers and a three-dimensional island growth occurs at higher coverage [5]. The desorption temperature for the first monolayer was reported to be above 350 C. Scanning tunneling microscopy (STM) results also show two-dimensional layer-by-layer growth for up to one monolayer [6]. Near one monolayer coverage the STM observed chains of bismuth atoms formed along the [1$\bar{1}$0] direction with a typical length 24 Å. These chains are interrupted by a missing bismuth atom due to the larger atomic radii of bismuth compared to that of the antimony.

Two LEED studies of the growth of bismuth overlayers on GaAs(110) prepared at room temperature have been reported which support the SK growth model. The results obtained give evidence of sixth order periodicity in the diffraction and determine the detailed geometry of the adatom chains [7,8]. In this paper, we present a study of the temperature- and coverage-dependent growth kinetics using LEED and Time Dependent Monte Carlo (TDMC) simulation. Our purpose is to study the initial stage of ordered bismuth overlayers formation under different annealing temperature and coverage conditions.

METHOD

The experiments were performed in an ultra-high vacuum chamber with a base pressure of 8×10^{-11} Torr or better after bake. The samples are cleaved from a 4.8 mm x 4.8 mm n-type GaAs post. Sample heating was done with a button heater attached to the sample holder base and cooling was accomplished with three OFHC copper braids attached directly to a liquid nitrogen reservoir. The sample takes about twenty minutes to cool from room temperature to -110 C. Temperatures were measured using a chromel-alumel thermocouple at the base of the sample and values were corrected for the distance between thermocouple and surface.

Both electron diffraction profiles and integrated intensity measurements were made. A detailed description of the video LEED system can be found elsewhere [9]. The freshly cleaved (110) surface displayed a sharp (1x1) LEED pattern with low background intensity. This LEED pattern showed common characteristics of all reconstructed (110) zincblende surfaces, i.e., (hk) = ($\bar{h}$k) symmetry and brighter (10) and ($\bar{1}$0) spots.

The bismuth overlayer was deposited by evaporation of a high purity bismuth metal. A quartz crystal oscillator (QCO) was used to monitor the coverage. In this experiment the bismuth deposition rate was calculated as 0.2 ML/min where one monolayer (ML) is defined as 8.85 x 10^{14} atoms/cm^2, or two atoms per substrate unit cell. The QCO calibration was checked using Rutherford backscattering and is accurate to within 10%. Prior to evaporation the bismuth was degassed for 30 seconds. The cleanliness of this technique in our chamber has been verified in previous studies. [7] The annealing studies were of samples prepared at -110 C and annealed in a stepwise fashion from -110 C to 200 C in steps of 20 degrees. At each step samples were held at temperature for 5 minutes and then cooled to room temperature or below for data acquisition.

Bismuth adatom diffusion and overlayer ordering are examined using a numerical algorithm which incorporates a kinetic treatment in conjunction with a TDMC formalism. The method is based upon a probabilistic description of adatom jump events and the diffusion rate is determined by the energetics of adatom interactions within a local neighborhood on the lattice. The rare event problem associated with other theoretical treatments of surface diffusion is overcome by our unique and highly efficient algorithm. Consequently, we are able to observe events including overlayer ordering and island formation which occur on time scales which are longer by orders of magnitude than simple adatom diffusion. Initial investigations indicate that the diffusion mechanisms and formation of ordered structures of bismuth atoms on GaAs(110) depend strongly on adatom interactions. Surface maps can be produced identifying adatom positions within the overlayer as ordered structures evolve. From these adatoms maps, simple kinematic simulations of LEED measurements can be generated for comparison to experimental studies.

RESULTS AND DISCUSSION

Figures 1a and 1b show the LEED patterns at 136 eV for a 0.7 ML bismuth covered surface at -110 C and after 200 C annealing, respectively. Only a (1x1) pattern is seen for the as deposited film, Fig. 1a. However, after the sample is heated to 200 C, Fig. 1b, a "(6x1)" diffraction pattern appears, as reported previously [7]. These sixth-order spots appear only in the [1$\bar{1}$0] direction, suggesting a superlattice with a unit cell vector which is six times longer in that direction with respect to the substrate unit cell vector. This length represents a periodicity along the adatom chains of 24 Å, or approximately eleven atoms and a vacancy. The shape of the sixth-order spots is elongated in the interchain, or [001] direction, suggesting increased disorder perpendicular to the intrachain, or [1$\bar{1}$0] direction. The sixth order spots are positioned symmetrically about the [1$\bar{1}$0] mirror plane; the measured values of the integrated intensities of the fractional order spots also reflect this symmetry.

A detailed examination of the variation of the sixth-order spots as a function of temperature and coverage is shown in Fig. 2. In Figs. 2a and 2b the spot profiles of the (11) and ($\frac{5}{6}$1) beams along [1$\bar{1}$0] direction at 0.7 and 1.0 ML bismuth coverages at various annealing temperatures are plotted. Note in Fig. 2a that the sixth-order spots do not form for the 0.7 ML film until the sample is annealed to -50 C. However, they appear as a diffuse shoulder adjacent to the (11) beam for 1.0 ML even at -110 C, Fig. 2b. These results suggest that at low coverages and low temperatures, bismuth atoms adsorb on the GaAs(110) surface randomly without preferred sites. Under these conditions adatom interactions are insufficient to generate ordering. However, as the coverage increases or the thermal energy increases, the adatoms tend to align themselves along the [1$\bar{1}$0] direction forming the (6x1) superstructure.

Figure 1

(a)   **(b)**

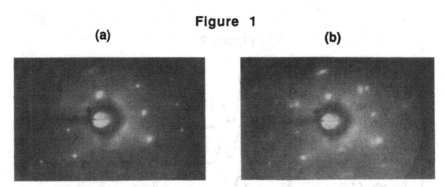

In Fig. 2c the spot profiles of the ($\frac{5}{6}$1) beam along the [001] direction are plotted as a function of annealing temperature for 1.0 ML bismuth coverage. (The LEED intensities have been multiplied by a factor of five with respect to those in Fig. 2b. The corresponding temperatures should be referenced to those in Fig. 2a.) Comparison of Figs. 2b and 2c reveals that the sixth-order spot profiles along this direction are very much broader than those along the [1$\bar{1}$0] direction at low temperatures, indicating greater order in the intrachain direction. If one ignores the complication of atoms adsorbed atop the primary overlayer, our physical picture of the kinetic process is one in which order develops as vacancies 'hop' from site to site within the chain. Intrachain vacancy hopping is energetically more favorable than interchain hopping. Vacancies rearrange along the [1$\bar{1}$0] direction until an average, maximal chain length of eleven atoms is attained.

To quantify the observed trends in the diffraction widths, we have measured the background subtracted full widths at half maximum (FWHM) of the sixth-order spots along the [1$\bar{1}$0] and [001] directions vs. annealing temperature for three different coverages. This data is shown in Figs. 3a and 3b, respectively. The FWHM of the profile taken in each direction is seen to decrease as a function of the annealing temperature. The FWHM of the 0.7 ML overlayer is greater than that of the other two coverages over the entire temperature range studied. Even at 200 C, the profile width is greater than the LEED instrumental transfer width [9]. Thus, the two-dimensional islands formed are on average somewhat smaller than 120 Å, the transfer length of the LEED instrument [9]. This consequence can occur at 0.7 ML because the adatom concentration is not adequate to complete the overlayer. Additionally, the 1.5 ML LEED profiles are always somewhat narrower than the profiles of the 1 ML overlayer. This difference can be attributed to the contribution of second layer atoms to the degree of ordering in the (epitaxial) first overlayer.

As displayed in Fig. 3a, the changes in profile widths along the chain direction appear to occur in three stages. These changes can be observed clearly for bismuth coverage of 1.5 ML. The first stage, occurring for annealing temperatures below -60 C, is accompanied by a rapid decrease in profile width. Atoms initially adsorbed in the second layer migrate to fill vacancies in the first layer. Recall that only the first layer is epitaxial [7,8]. In the second stage, occurring over a temperature range from -50 C to 50 C, no significant change in FWHM is observed. At these temperatures insufficient thermal energy is available to induce vacancy migration and improve overlayer ordering. The final stage, occurring at temperatures above 50 C for the 1.5 ML film, is characterized by rapid decrease in the FWHM of the LEED profiles. At these temperatures the vacancies become mobile and maximal ordering occurs. (The plateau which appears above 120 C for the 1.5 ML in Fig. 3a probably occurs due to our finite instrumental response function.) Recall that the maximum annealing temperature of 200 C is well below the thermal desorption temperature of 350 C for the last monolayer. However, some desorption of bismuth from the second adlayer cannot be discounted. The temperature ranges associated with these three ordering stages are coverage dependent, appearing to decrease with increasing coverage.

Figure 2

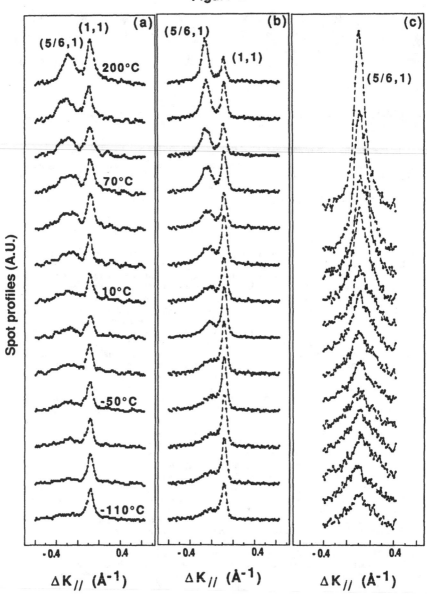

Figure 3

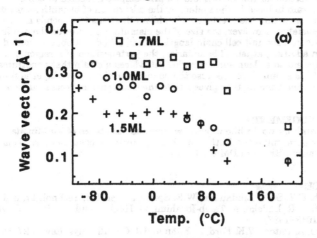

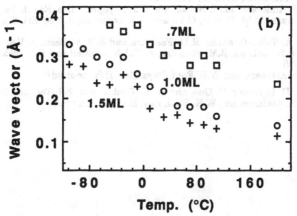

No corresponding stages are observed for ordering among the chains. Referring to Fig. 3b, the profile widths in the interchain direction decrease uniformly with increasing annealing temperature.

The temperature dependence of the FWHM of the LEED profiles contains information regarding the average chain length. In Fig. 3a, the LEED profile widths change by factors between two and three over temperatures ranging from -110 C to 200 C. The magnitude of the change depends upon the coverage examined. A preliminary analysis indicates that these FWHM changes correspond to an increase of 30% to 35% in average chain length. STM studies suggest that eleven atom chains should dominate the fully annealed bismuth overlayers [6]. Thus, annealing induces the growth of adatom chains from approximately seven atoms to eleven atoms in length. A detailed, microscopic TDMC analysis to provide quantitative verification of this model is currently in progress.

The overlayers exhibiting the greatest ordering were those annealed to 200 C. However, a significant difference is observed in the degree of ordering for each of the two substrate directions. Comparison of Figs. 3a and 3b shows that the minimum widths along the interchain direction were approximately twice those associated with the intrachain direction for the 1.0 ML and 1.5 ML samples annealed to 200 C.

The strain induced in the system by the adsorption of bismuth atoms drives the ordering process within the overlayer. The bismuth atoms form bonds as dictated by the substrate geometry. However, the size of the bismuth atom crowds the overlayer. This strain results in a six unit cell chain length in the [1$\bar{1}$0] direction, terminated by a vacancy. In addition, at island boundaries, the juxtaposition of a region containing adatoms adjacent to a clean surface region introduces a strain that penetrates into the substrate. This strain arises because the bonding within the overlayer is fundamentally different from the bonding that gives rise to the clean surface reconstruction [7].

ACKNOWLEDGEMENTS

The authors would like to acknowledge the assistance of Xu Mingde and Richard J. Smith for performing the Rutherford backscattering measurements. This work is supported under NSF grant DMR-8705879.

REFERENCES

1. P. Skeath, C.Y. Su, I. Lindau and W.E. Spicer, J. Vac. Sci. Technol. 17, 874 (1980).
2. F. Schaffler, R. Ludeke, A. Taleb-Ibrahimi, G. Hughes and D. Rieger, J. Vac. Sci. Technol. 5, 1048 (1987).
3. C.B. Duke, A. Paton, W.K. Ford, A. Kahn and J. Carelli, Phys. Rev. B 27, 803 (1982).
4. J. Carelli and A. Kahn, Surf. Sci. 116, 380 (1982).
5. J.J Joyce, J. Anderson, M.M. Nelson and G.J. Lapeyre, Phys. Rev. B (in press); J.J. Joyce, J. Anderson, M.M. Nelson, C.Yu and G.J. Laypeyre, J. Vac. Sci. Technol. 7, 850 (1989).
6. R. Ludeke, A. Taleb-Ibrahimi, R.M. Feenstra and A.B. McLean, J. Vac. Sci. Technol. 7 (1989) 936; A.B. McLean, R.M. Feenstra, A. Taleb-Ibrahimi and R. Ludeke, Phys. Rev. B 39 (1989) 12925.
7. T. Guo, R.E. Atkinson and W.K. Ford, Phys. Rev. B (submitted)

8. C. B. Duke, D. L. Lessor, T. Guo, and W. K. Ford, J. Vac. Sci. Technol. (in press).
9. T. Guo, R. E. Atkinson and W.K. Ford, Rev. Sci. Inst. (in press)

PHOTOEMISSION CHARACTERIZATION OF THIN FILM NUCLEATION ON INERT SUBSTRATES

G. HAUGSTAD, A. RAISANEN, C. CAPRILE, X. YU AND A. FRANCIOSI
Department of Chemical Engineering and Materials Science
University of Minnesota, Minneapolis, MN 55455

ABSTRACT

Synchrotron radiation photoemission was used to characterize Sm and Mn thin film nucleation and growth on solid Xe substrates, in the 3×10^{14} - 2×10^{16} atoms/cm^2 coverage range. Film growth is well approximated by a model in which the nucleation site density remains constant and hemi-ellipsoidal particles increase in size until coalescence is achieved. Site density and average cluster size are determined from the coverage-dependence of metal and Xe photoemission intensities. Size estimates are confirmed by experimental fingerprints of coalescence.

INTRODUCTION

Many thin film nucleation experiments employ chemically inert substrates of theoretical and/or technological interest [1-5]. Recently rare gas solid substrates have been used in photoemission studies of metal particles [2-4], and in the first stages of unreactive metal-semiconductor interface synthesis [5]. Thin films grown *in situ* on solid Xe exhibit three-dimensional island growth, and the inportant parameters are therefore particle concentration, size and shape [4]. Film characterization must be non-destructive and is best performed *in situ*, to avoid changes in film morphology.

In this paper we apply synchrotron radiation photoemission to estimate Sm and Mn cluster size, shape and nucleation site density on Xe, and to measure the size-dependent valence electronic structure. We employed a simple experimental procedure to minimize work-function induced shifts of cluster and Xe photoemission features. In the particle size range explored (35-90 Å in diameter), we find that growth can be well approximated by a model which assumes a constant nucleation site density and hemi-ellipsoidal particles with varible z-aspect ratio. Site density and particle aspect ratio are nearly identical for Sm and Mn clusters on Xe.

EXPERIMENT

A bellow-mounted, closed-cycle refrigerator was used as sample manipulator in the photoelectron spectrometer, with a polished polycrystalline Cu cold finger serving as the sample substrate. A radiation shield at 60-70 K was used to minimize the thermal load. An operating base pressure $< 6 \times 10^{-11}$ Torr was routinely achieved. Metal deposition was accomplished through direct sublimation from a W coil evaporator. A water-cooled quartz thickness monitor was employed to measure metal coverage. Sm or Mn films 100 Å thick were deposited at room temperature on the polished region of the cold finger. The area was then cooled to 12-20 K for Xe condensation. By using identical metals for both the "underlayer" (below the Xe) and clusters, we avoided having the Xe film in contact with two metals of different work functions during the cluster studies. The presence of different work functions would induce Xe lineshapes shifted relative to each other and complicate the subtraction of the Xe features, which is necessary to extract the metal emission [4]. We monitored the attenuation of the underlayer emission as a function of Xe exposure to determine the total thickness of condensed Xe. The Xe films were stable in ultrahigh vacuum on the time scale of our experiment, and Auger spectroscopy showed no contaminants within experimental uncertainty.

Sm and Mn clusters were prepared by depositing the respective metal on top of Xe films 50 ± 5 Å thick. No charging effects were observed, and the detectable emission from the underlayer was less than 1% of the initial value. Synchrotron radiation photoemission measurements were performed as described in Ref. 4 at the Synchrotron Radiation Center of the University of Wisconsin-Madison.

RESULTS AND DISCUSSION

We used the coverage-dependence of the Xe and overlayer photoemission intensity to determine the growth mode of the metal film. Relative to ideal layer-by-layer growth we observe a lower attenuation rate of the Xe emission intensity and a lower rate of increase of the metal emission, as a function of metal coverage. These observations are consistent with island growth (Volmer-Weber mode). The negligible escape depth sensitivity of the results implies an average particle thickness larger than the photoelectron escape depth ($\sim$ 5 Å), in the coverage range considered here (> 3×10^{14} atoms/cm^2), and rules out metal-Xe interdiffusion [4].

The results for Sm on Xe are shown in Fig. 1. In Fig. 1a we

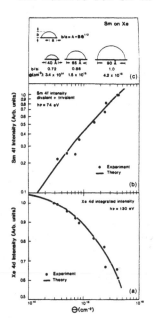

Fig. 1. a) Integrated intensity of the Xe 4d core level emission (hν = 130 eV) as a function of coverage on Xe. The solid curve marks the substrate emission expected during growth of hemi-ellipsoidal particles with a concentration of 7×10^{11} cm^{-2} and a cluster z-aspect ratio increasing from 0.5 to 1.0. b) Sum of Sm [6]H and [5]I peak intensities (hν = 74 eV) as a function of Sm coverage. c) Morphology of the hemi-ellipsoidal metal particles at selected Sm coverages. The z-aspect ratio b/a is allowed to increase according to A + B$\theta^{1/3}$, where A and B are defined in the text.

plot the integrated photoemission intensity of the Xe 4d core
levels as a function of Sm coverage. The data are normalized to
the Xe 4d emission intensity at the lowest Sm coverage explored.
In Fig. 1b we plot the sum of the peak intensities of the Sm 4f ^{6}H
and ^{5}I multiplets, proportional to the total Sm 4f photoemission
intensity [4], normalized to the intensity at the highest
coverage. The solid curves in Figs. 1a and 1b show the result of a
least squares fit of the data with a theoretical model. We
simulated three-dimensional island growth with hemi-ellipsoidal
particles (for ease of calculation) and fitted the
coverage-dependence of the Xe and metal photoemission intensities
to obtain information on particle size, growth morphology and
nucleation site density. Site density, z-aspect ratio
(height/radius) and metal escape depth were fitting parameters. We
compared single-sized particles with a Gaussian distribution of
particle sizes [1] and found the differences in the determined
parameters to be within the experimental uncertainty (10%). We
considered both a changing occupied site density and a changing
z-aspect ratio as a function of coverage, to allow for the
possibilities of particle migration and a changing degree of
substrate wetting [6]. This resulted in three independent models:
1) constant site density and aspect ratio; 2) variable site
density and constant aspect ratio; 3) constant site density and
variable aspect ratio. We obtained a good fit for case 3 (solid
lines in Fig. 1) and a fair fit for case 2, while no acceptable
fit was obtained for case 1.

Photon energies used in Figs. 1a and 1b (130 and 74 eV,
respectively) correspond to a similar Sm escape depth in the two
cases (kinetic energies of 60 and 68 eV). The model yields:

$$\frac{I_{Sm}(u)}{I_{Sm}(u_f)} = \frac{u^2}{u_f^2} \times \frac{1/2 - (1/r^2u^2)[1 - \exp(-ru)(1+ru)]}{1/2 - (1/r^2u_f^2)[1 - \exp(-ru_f)(1+ru_f)]} \quad (1)$$

and

$$I_{Xe}(u) = 1 + (N\pi D^2 u^2)\{-1 + (2/r^2u^2)[1 - (1+ru)\exp(-ru)]\} \quad (2)$$

where

$I_{Sm(Xe)}$ = intensity of Sm (Xe) emission
  u = R/D
  R = cluster radius in the plane of the substrate
  D = escape depth in Sm
  r = cluster z-aspect ratio (height/R)
  N = cluster concentration = occupied nucleation site density

and the subscript f denotes final coverage.

The fitting procedure was carried out as follows: 1) The Sm
intensity data were fitted to Eq. 1 to obtain the escape depth in
Sm, 6 Å. 2) Using this value, the Xe intensity data were fitted to
Eq. 2 to obtain the nucleation site density, 7×10^{11} sites/cm^2. 3)
The cluster aspect ratio was assumed to increase as a simple power
of coverage, $r = A + B\theta^n$, where A and n were free parameters and B
was obtained from the value of r at the highest coverage. The
results were A = 0.5, n = 1/3 and r = 1 for θ = 4.2×10^{15}
atoms/cm^2. The fit was not sensitive to the precise value of n,
and 1/4 < n < 1/2 produced similar results. Cross-sectional

diagrams for clusters at three different coverages are shown in Fig. 1c (note "b/a" = r = aspect ratio).

A Sm nucleation site density of 7×10^{11} sites/cm^2 implies particle coalescence at $\theta = 9 \times 10^{15}$ atoms/cm^2, consistent with experimental indications of coalescence. With particle growth the Sm 4f $^6H/^5I$ multiplet intensity ratio decreases, achieving the bulk value at $\theta = 1.0 \pm 0.2 \times 10^{16}$ atoms/cm^2 [4], indicating a fully coalesced Sm film. In addition, the Sm 4f ^6H and ^5I features shift to lower apparent binding energies with increasing particle size [4], and attain bulk values at $\theta \sim 1 \times 10^{16}$ atoms/cm^2.

Model 2 produced a fair fit in Fig. 1 if we allowed the nucleation site density to decrease from 1.5×10^{12} to 7×10^{11} sites/cm^2 as the 1/3 power of coverage. Nonetheless, a decreasing site density implies cluster migration and fusion, phenomena not expected for particles containing hundreds of atoms [7]. Conversely, model 3 is supported by the observation that in multilayer films wetting tends to decrease with growth [6], and by the realization that a larger fraction of adatoms are deposited directly on top of the clusters, rather than approaching via surface diffusion, with increasing particle size.

Similar fits of photoemission intensities were performed in a preliminary study of Mn clusters grown *in situ* on solid Xe, using the peak intensity of the dominant valence band feature. We determined a nucleation site density of $9 \pm 1 \times 10^{11}$ sites/cm^2 and a z-aspect ratio coverage-dependence identical to the Sm case to within the uncertainties of the parameters A and n. Knowing these values we calculated cluster diameters for coverages in the 4.1×10^{14} to 1.3×10^{16} atoms/cm^2 range, shown in Fig. 2. The open squares correspond to experimental coverages and the solid curve is included as a guide to the eye. In Fig. 3 we present energy distribution curves (EDC's) for the valence band emission from Mn particles at $h\nu = 80$ eV. Coverage increases moving upwards through the spectra and corresponds to the squares in Fig. 2. Energies are referred to the Fermi level. At top in Fig. 3 is the valence band EDC for a thick polycrystalline Mn film, while the bottom spectrum is from a Xe film of 50 Å thickness, condensed on thick Mn. The bulk Mn valence band emission contains two d-band features [8] located, respectively, near the Fermi level and at a binding energy of 2.8 eV. In Fig. 3 both features are shifted to higher binding energy and approach the bulk positions as particle size increases. Similar shifts are widely reported in photoemission studies of supported clusters and can result from both initial- and final-state effects [1,3-4]. At $\theta = 1.3 \times 10^{16}$ cm^{-2} in Fig. 3 the valence band edge is shifted by about 0.5 eV below the bulk location. This shift is similar to the final-state (charged cluster) binding energy shift observed for Sm particles of the same size [4]. The predicted coalescence coverage for Mn is $\theta = 2.2 \times 10^{16}$ cm^{-2}. At lower coverages the Mn valence band edge is deeper than expected from this mechanism [4], and a rigid shift of Xe features in the same direction is also observed. The Mn shifts may therefore contain both initial- and final-state contributions which must be deconvolved, a process we will not address here.

At low coverage in Fig. 3, the tailing of the Xe 5p can be seen above 6 eV binding energy. An accurate measurement of the total Mn valence band width thus requires the subtraction of the Xe 5p lineshape, which we have not attempted in this preliminary analysis. The width of the leading valence band edge at low

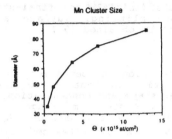

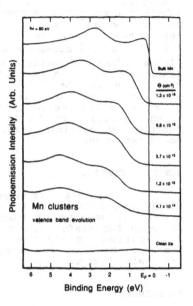

Fig. 2. Diameter of hemi-ellipsoidal Mn particles as a function of coverage, as obtained from a least-squares fit of the emission intensity to the model. The squares correspond to experimental coverages and the solid curve is included as a guide to the eye.

Fig. 3. Mn valence band emission as a function of coverage, at hv = 80 eV. The bottom-most spectrum is from a clean Xe film of 50 Å thickness. Coverages increase moving upwards through the spectra. The top spectrum is from a 100 Å - thick Mn film on polycrystalline copper, and is representative of the bulk emission. Energies are referred to the Fermi level.

binding energy, ΔE_v, decreases monotonically from 1.53 ± 0.15 eV at $\theta = 4.1 \times 10^{14}$ cm^{-2} to 0.34 ± 0.03 eV in the bulk spectrum. The splitting of the two main d-band features, ΔE_{3d}, increases monotonically from 1.9 ± 0.1 eV at $\theta = 1.2 \times 10^{15}$ cm^{-2} to 2.4 ± 0.1 eV in the bulk. We measure $\Delta E_{3d} = 2.2$ eV for $\theta = 4.1 \times 10^{14}$ cm^{-2}, inconsistent with this trend; the apparent discrepency may, however, result from a smaller signal/noise ratio at low coverage (the spectra in Fig. 3 have been obtained by averaging the actual experimental points. The data dispersion is about 20 % at the lowest coverages explored).

An increasing d-band splitting with particle growth has been observed in other transition metals [1], and is in accord with simple theories which relate band width to the square root of coordination number [1], and hence to the fraction of surface atoms. This fraction decreases from 44 % to 15 % with increasing size in the coverage range explored. The apparent narrowing of ΔE_v with particle growth is likely to be the result of the existence of a distribution of cluster sizes: core levels are known to be broader at lower coverages, due to a larger distribution of final-state binding energy shifts [1], and one expects a similar broadening of the valence band edge. In order to extract information on the particle size distribution, quantitative estimates for all of the above effects (initial- and final-state energy shifts, changes in d-band splitting, surface atom contributions to broadening) must first be obtained [9].

CONCLUSIONS

We have employed synchrotron radiation photoemission and an analytical model to analyze thin film nucleation on inert substrates and estimate particle size, shape and concentration. Sm and Mn coverages in the 3×10^{14} to 2×10^{16} atoms/cm^2 range on Xe were investigated. The similar nucleation site densities, $N_{Sm} = 7.0 \pm 0.5 \times 10^{11}$ sites/cm^2 and $N_{Mn} = 9 \pm 1 \times 10^{11}$ sites/cm^2, along with an identical particle z-aspect ratio coverage-dependence, are consistent with the lack of chemical interaction or adatom size dependence in the film morphology. The coalescence coverage predicted by the model is consistent with the observed saturation of Sm 4f effective binding energy shifts and $^6H/^5I$ multiplet intensity ratio. The Mn 3d valence band splitting increases as a function of particle size, a reflection of increasing average coordination number.

ACKNOWLEDGEMENTS

This work was supported in part by ONR under grant N00014-89-J-1407 and by the NSF Center for Interfacial Engineering at the University of Minnesota. We thank K. Wandelt and M. Strongin for useful discussion. We thank the whole staff of the Synchrotron Radiation Center of the University of Wisconsin-Madison, supported by the National Science Foundation, for their cheerful support.

REFERENCES

1. S.B. DiCenzo and G. K. Wertheim, Comments Solid State Phys. 11, 203 (1985).

2. S. Raaen, J. W. Davenport and M. Strongin, Phys. Rev. B 33, 4360 (1986)

3. S. L. Qiu, X. Pan, M. Strongin and P. H. Citrin, Phys. Rev. B 36, 1292 (1987)

4. G. Haugstad, C. Caprile, A. Franciosi, D. M. Wieliczka and C. G. Olson, Phys. Rev. B (in press)

5. G. D. Waddill, I. M. Vitromirov, C. M Aldao and J. H. Weaver, Phys. Rev. Lett. 62, 1568 (1989)

6. M. Bienfait, Surf. Sci 162, 411 (1985)

7. L. Eckertova, Physics of Thin Films (Plenum Press, New York, 1977)

8. A. Wall, S. Chang. P. Philip, C. Caprile and A. Franciosi, J. Vac. Sci. Technol. A 5, 2051 (1987)

9. G. Haugstad and A. Franciosi (to be published)

OBSERVATION OF HETEROEPITAXIALLY GROWN ORGANIC ULTRATHIN LAYERS ON INORGANIC SUBSTRATES BY IN SITU RHEED AND BY STM

MASAHIKO HARA, HIROYUKI SASABE, AKIRA YAMADA AND ANTHONY F. GARITO*
Frontier Research Program, The Institute of Physical and Chemical Research (RIKEN), Wako, Saitama 351-01, JAPAN
*Permanent Address: Department of Physics, University of Pennsylvania, Philadelphia, PA 19104, USA

ABSTRACT

An organic molecular beam epitaxy (OMBE) system has been designed and constructed with in situ reflection high-energy electron diffraction (RHEED) specifically for the deposition of organic molecular layers under ultrahigh vacuum (UHV), the order of 10^{-10} torr. The system is equipped with a portable UHV chamber which allows easy transfer of the OMBE film samples to a separate UHV scanning tunneling microscopy (STM) system. A structural investigation of heteroepitaxially grown organic ultrathin layers of copper phthalocyanine (CuPc) on inorganic substrates was carried out by the combined UHV system from less than a monolayer of CuPc.

INTRODUCTION

A detailed investigation of the initial growth structure of ultrathin films is essential to the understanding of the growth mechanism and interface formation between layers of different materials. Molecular beam epitaxy (MBE) with ultra-high vacuum (UHV) analysis techniques has long been successful in exploring such problems, primarily in the semiconductor field. However, UHV MBE techniques have not been established yet for organic molecular systems which is a prerequisite for realizing current molecular device designs in addition to fundamental interests.

The combination of diffraction methods with real space imaging of scanning tunneling microscopy (STM) allows direct studies of epitaxial layers at the atomic scale. Implementation requires either having the STM head in the MBE system, or transferring the as-grown samples under UHV atmosphere from the MBE growth chamber to the separate UHV STM. Either choice raises some practical difficulties in the UHV systems that must be solved.

Following our previous paper on the newly developed organic molecular beam epitaxy (OMBE) technique [1], here we report a detailed discussion of the initial growth condition of organic ultrathin layers by in situ reflection high-energy electron diffraction (RHEED), and also the complete UHV OMBE system, introducing a portable UHV chamber for transferring samples from the OMBE chamber to UHV STM and other analysis systems while protecting the sample from contamination.

INSTRUMENTATION

Figure 1 shows an OMBE system composed of three independent chambers connected through gate valves.

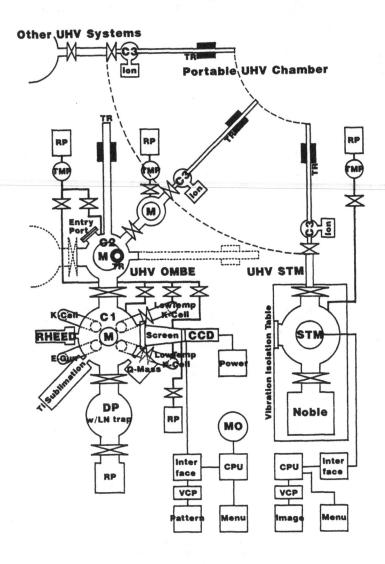

Fig. 1 Schematic diagram of the combined UHV OMBE system
C1:growth chamber, C2:entry chamber, C3:portable chamber,
TR:sample transfer rod, M:manipulator
VCP:color video copy processor

The smallest chamber (C3) can be isolated from the main body with the valve closed as a portable UHV chamber, which is maintained at UHV of less than 10^{-9} torr by an ion pump equipped with an independent power supply. The sample holder assembly is interchangeable for use with other UHV systems so that the portable chamber can combine independent UHV systems remotely located into one system.

The main difficulties in developing UHV MBE methods for organic molecular systems are their inherent high vapor pressure, thermal instability and impurity contents of other organic substances. A low-temperature Knudsen cell was specially designed for organic samples in this OMBE system. The temperature of the Knudsen cells can be controlled not only by electrical heating but also by circulating liquid nitrogen. During the baking process of the OMBE system itself, or degassing of organic molecules, the Knudsen cells are drawn outside of the growth chamber through load lock bellows, while the UHV conditions in the cells remain the same.

For RHEED analysis, a charge-coupled device (CCD) camera was positioned in front of the screen for RHEED. The diffraction patterns are stored in a 600 megabytes magneto-optical (MO) disk through a microcomputer with two-dimensional (2D) intensity profiles (16 bit) divided into 510 x 492 pixels, so that in situ RHEED patterns and 2D intensity changes can be investigated at the same time. In addition, atomic or molecular layer phase locked epitaxy (PLE) can be managed by the feedback from the results of the intensity analysis.

The UHV STM system used was based on Nanoscope II system (Digital Instruments, Inc., Santa Barbara) with a modified STM head for the UHV use. The STM chamber is mounted on a vibration isolation table and kept at a vacuum pressure of less than 10^{-9} torr during operation.

EXPERIMENTAL

Heteroepitaxial growth of organic ultrathin layers on inorganic substrates was carried out in the UHV OMBE system under a base pressure of about 2 x 10^{-10} torr.

The organic layers consists of copper phthalocyanine (CuPc) which has high thermal and chemical stabilities as well as suitable sublimation properties. The inorganic substrates were natural 2H-MoS$_2$ single crystal and KCl. 2H-MoS$_2$ is a typical layered transition metal dichalcogenide that has no dangling bonds on its cleaved face [2].

After several heat treatments to clean the substrate surface and degas the CuPc, the CuPc source was heated at 250 - 280 $^\circ$C under the order of 10^{-9} torr for deposition of the CuPc layers. The growth rate of the CuPc layers was held at under a few layers per hour. The positional and orientational order of the CuPc layers was determined during actual layer growth by in situ RHEED starting from less than a monolayer. Because of the short irradiation time, about two seconds, in taking one diffraction pattern with the CCD camera system, beam damage of the CuPc layer by the 10 - 20 keV electron beam was negligible.

The STM studies were carried out in air at room temperature. Those images were obtained in the constant current mode at 1 V (tip negative) and 0.1 nA for the CuPc films, and at 180 mV and 2.5 nA for the MoS$_2$ substrates.

RESULTS AND DISCUSSION

RHEED Patterns of Organic Ultrathin Layers

As illustrated in Fig. 2, we have confirmed the earlier in situ RHEED observations of CuPc ultrathin layers grown heteroepitaxially on a cleaved surface of MoS_2 substrate by OMBE. The narrow streaks originate from a growing CuPc layer that is molecularly flat. The CuPc layer has its own lattice constant, which is completely different from that of MoS_2 even at the initial deposition stage. This indicates that very large lattice mismatched structures can be obtained by this OMBE method. Details were reported in the previous paper [1].

Figure 3 (a) shows the RHEED pattern of the as-prepared KCl surface. The spacing between streaks in the pattern corresponds to the characteristic KCl lattice spacing. The change in the pattern after opening the shutter to the CuPc source is shown in Fig. 3 (b). Although this pattern shows rather spotty features due to diffraction from a three dimensional structure, the crossed RHEED patterns yield the growth direction, especially the b-axis orientation. When the electron beam is parallel to the [100] orientation of KCl, the deduced angle of the b-axis from the crossed RHEED pattern is 32 degrees from the normal to the substrate surface. When the substrate was rotated 45 degrees in the plane, the angle changes to 24 degrees. This indicates that the RHEED pattern can reveal the growth direction from the initial stage. These data are in very good agreement with earlier electron microscopy results on large single crystalline domains [3].

RHEED Intensity Changes during Organic Ultrathin Layer Growth

In order to investigate the possibility of RHEED intensity changes for organic molecular systems, CCD imaging of the RHEED patterns was carried out. As a preliminary observation, the intensity changes as a function of time of a RHEED pattern from CuPc layers on MoS_2 is shown in Fig. 4. Some periodic structure is apparent in the intensity profile corresponding to the layer growth. While further detailed investigations are required to reach a final conclusion, there seems to be definite indications that RHEED intensity oscillations are occurring even in the organic molecular layer growth.

STM Images of Organic Ultrathin Layer and Substrate

Figure 5 shows STM images of the surface profiles of (a) freshly cleaved MoS_2 and (b) CuPc layer on MoS_2 obtained at ambient pressure. Hexagonal packing of the sulfur atoms on the surface of MoS_2 can be clearly seen at atomic resolution. For the surface profile of the OMBE CuPc layers, the top view of individual CuPc molecules can be identified at molecular resolution. Similar measurements are currently being carried out in the UHV chamber for comparative purposes. These results will be reported separately.

Fig. 2 RHEED pattern from CuPc layers on MoS$_2$
at the MoS$_2$ [1120] azimuth

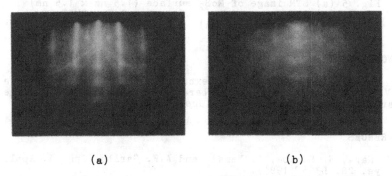

(a)   (b)

Fig. 3 (a) RHEED pattern from KCl at the [110] azimuth
(b) Crossed RHEED pattern from CuPc layers on KCl

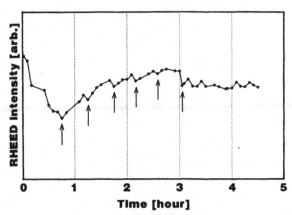

Fig. 4 RHEED intensity changes during CuPc layer growth

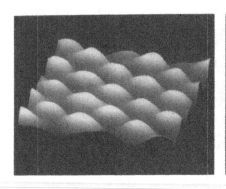 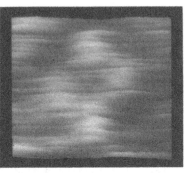

(a)   (b)

Fig. 5 (a) STM image of MoS_2 surface (1.5 nm x 1.5 nm)
       (b) STM image of one CuPc molecule on OMBE layers
       (2.3 nm x 2.3 nm)

ACKNOWLEDGMENTS

The authors would like to express their sincere gratitude to Vieetech Japan Co, Ltd. (Ibaraki, Japan) for aid in the construction of the UHV systems used in this study.

REFERENCES

1. M. Hara, H. Sasabe, A. Yamada and A.F. Garito, Jpn. J. Appl. Phys. 28, L306 (1989).
2. J.A. Wilson and A.D. Yoffe, Adv. Phys. 18, 193 (1969).
3. M. Ashida, Bull. Chem. Soc. Jpn. 39, 2632 (1966).

THE INITIAL STAGES OF MBE GROWTH OF InSb ON GaAs(100)
- A HIGH MISFIT HETEROINTERFACE.

C.J.KIELY, A.ROCKETT*, J-I. CHYI* AND H.MORKOC*
Department of Materials Science and Engineering, University of Liverpool, England.
* Coordinated Science Laboratory, University of Illinois at Urbana-Champaign, IL 61801,
USA.

ABSTRACT

The initial stages of heteroepitaxy of InSb on GaAs(100) grown by MBE have
been studied by transmission electron microscopy. Three dimensional InSb island growth
occurs in which the majority of the 14.6% misfit strain is accommodated by a square array
of $a/2<011>$ edge-type misfit dislocations. The implications of each island having a well
defined defect array before coalescence into a continuous epilayer are discussed. Some
60^0-type $a/2<101>$ interfacial defects and associated threading dislocations are also
observed in coalesced films and possible reasons for their existence are explained. A
strong asymmetrical distribution of planar defects in the InSb islands is observed and the
origin of the asymmetry is discussed. Finally some evidence for local intermixing in the
vicinity of the interface is presented.

INTRODUCTION

Recently several groups worldwide have reported successfully growing InSb
layers on GaAs (100) by molecular beam epitaxy despite a 14.6% lattice mismatch
[1-4]. The potential advantage of growing InSb layers (i.r. detectors) directly on a GaAs
substrate is that signal processing could then be integrated with the detector on a single
substrate chip, leading to significant gains in device performance.

Hall effect measurements have typically been used to assess the electrical quality
of the MBE grown InSb layers [1-4]. Reports of room temperature electron mobilities for
films between 5 and 10 μm thick lie in the range 5.7 - 6.7 x 10^4 cm^2 V s^{-1}, which are still
somewhat lower than those obtained for bulk InSb (7.6 - 8.4 x10^4 cm^2 V s^{-1}). Several
detailed microstructural characterisations of the InSb/GaAs interface have also been
reported [4-6]. A common characteristic of the epilayers produced to date seems
to be a rather high density of threading dislocations (10^6 - 10^7 cm^{-2}). Such defects are
undesirable since they act as scattering centres for electrons and hence reduce the
electron mobility of the film.

High resolution electron microscopy studies of the InSb/GaAs interface [5,6] have
revealed that most (although not all) of the misfit strain is accommodated by a square
array of pure-edge $a/2<011>$ Lomer-type interfacial misfit dislocations spaced on average
30Å apart. However, some 60^0-type $a/2<101>$ defects are also occasionally noted at the
interface, and it is these which often have epi-threading segments associated with them
[7]. RHEED studies [1,3] have indicated that the early stages of layer growth involve
three dimensional nucleation and formation of a non-pseudomorphic interface. In this
paper we present a TEM examination of the initial stages of MBE InSb growth on
GaAs(100). We discuss in some detail the formation mechanism of the Lomer-type misfit
dislocations and suggest reasons why the 60^0 type dislocations and undesirable
threading defects are also observed to form.

Previous TEM studies [4,5] have also noted the presence of planar defects such
as microtwins and stacking faults in InSb layers. We present evidence here for a
strong asymmetrical distribution of planar defects in the InSb overlayer which we believe
to be connected with differences in the mobilities of α and β-type defects in the InSb.
Finally, some observations concerning fine scale interface roughening and local
intermixing are presented.

EXPERIMENTAL

A Riber MBE system was used in this study which was equipped with As and Sb crackers to reduce As_4 and Sb_4 tetramers to As_2 and Sb_2 dimers. Semi-insulating, nominally on orientation, GaAs (100) substrates were cleaned with solvents and heated *in-situ* to remove the surface oxide. A 0.2 μm undoped GaAs buffer layer was grown at 580 °C and then cooled to 400 °C before the deposition of 50 - 200Å of InSb was commenced. Plan-view and cross-section TEM specimens were prepared by standard techniques which have been reported elsewhere [5]. All samples were examined in a Philips EM430 TEM operating between 150 and 300 kV.

RESULTS

Initial island nucleation.

A bright field axial (100) plan view moire image of InSb islands on GaAs is shown in Fig.1. Many islands before coalescence look fairly square in projection with edges running along <011> type directions. Nearest neighbour nuclei distances for this nominally on-axis substrate were typically in the range 500 - 900Å, and a nucleation density of ~200 islands/μm^2 was estimated. The (022) type moire fringes (running parallel to the island edges) are spaced on average ~16Å apart, whereas the (040) type fringes (inclined at 45° to the island edges) are on average ~11.5Å apart. The predicted parallel moire fringe spacing (assuming a 14.6% lattice mismatch) are 15.67 Å and 11.06Å for (022) and (040) reflections respectively. Hence, it is clear that nearly all the misfit strain between the InSb and the GaAs has been relieved *before* island coalescence. Furthermore, some slight relative rotations of the moire fringe directions can be observed from island to island, implying that slight deviations from perfect epitaxy are common before island coalescence

Cross-sectional images such as that shown in Fig.2 clearly demonstrate that islands tend to show one of two profiles when viewed along an (011) direction. Some islands are almost hemispherical (eg. island A) whereas others appear much flatter with distinct inclined {111} facet planes (eg. island B). This latter profile is similar to that observed for polar semiconductors grown on Si(100) substrates [8].

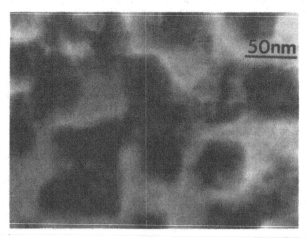

Figure 1. Axial BF Moire Fringe Image of InSb Islands on GaAs (100).

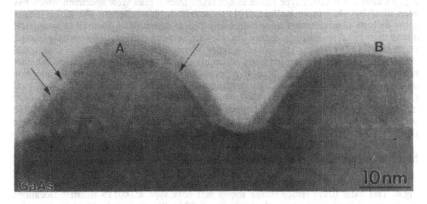

Figure 2. Cross sectional image of discrete InSb islands on GaAs.

Planar defects in islands

The {011}-type cross-sectional images such as those in Fig.2 often show stacking faults and thin microtwin lamellae on inclined {111} planes within the InSb islands. Pirouz et al [8] have analysed the stacking fault formation phenomena in prism shaped islands (similar to island B in Fig.2) of polar semiconductor islands grown on (100)Si. They proposed a very successful model based on energy and geometry arguments, which explains that such stacking faults are simply a consequence of growth accidents occurring on advancing {111} facet planes (ie. the fault plane then must lie parallel to the facet plane). Inspection of Fig.2 shows that a similar argument does not apply for the InSb islands since the fault planes usually intersect the inclined facet planes with an acute angle of 70°.

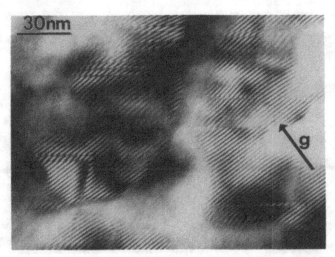

Figure 3. A $g = \{202\}$ type moire image taken from a plan view specimen which has been tilted to an inclined (101) zone axis.

Fig. 3 shows a **g** = {202} type moire image taken from a plan view specimen which has been tilted by ~45° to an inclined (101) zone axis. In this reflection the moire fringes show definite displacements on crossing any planar defect. The obvious asymmetrical distribution of these defects suggests that they are forming preferentially on only *one* set of inclined {111} planes (ie. either A-type (metal terminated) or B-type (non-metal terminated) planes). We are presently attempting to unambiguously determine whether the planar defects form on A or B-type planes using a convergent beam electron diffraction approach for polarity determination [9]. It should be noted that Fnaiech et al [10] have reported that α type dislocations are 100 times more mobile than β-type dislocations in undoped InSb. We believe that the mobility difference in α and β-type partial dislocations is intimately related to the asymmetrical planar defect distribution in InSb.

Misfit strain relief before and after island coalescence

The interface between a typical InSb island and the GaAs substrate is shown in Fig.4. This axial (011) lattice image shows that edge type a/2<011> misfit dislocations occur approximately every seventh atomic row (~30Å). The expected critical thickness for dislocation formation in this system is only one or two monolayers. Each island probably grows pseudomorphically until it reaches a critical lateral size along [011] of slightly greater than 30Å, at which point a complete a/2<011> line defect forms at the periphery of the island [5]. For isotropic islands, the second line defect forms perpendicular to the first to balance the strain relief in the island. Subsequent defects form in a similar manner at regular intervals as strain periodically accumulates at the island edges during lateral growth. The edge-type arrays are energetically more favourable than 60°-type arrays since they are more efficient at relieving misfit strain while allowing more coherent interface area to form.

One implication of each individual island having its own well defined misfit dislocation structure before coalescence is that there exists the possibility for a "dislocation array mismatch" at the boundary. Furthermore, islands on coalescence are unlikely to be completely strain free and there is likely to be a finite residual unrelaxed

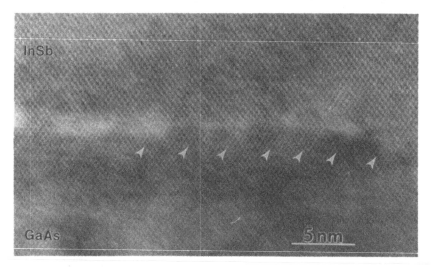

Figure 4. An axial (011) lattice image of the interface between an InSb island and the
GaAs substrate showing a/2<011> misfit dislocations.

stress associated with each boundary. This boundary stress may be either tensile (ie. extra 60° or edge-type defects required to relieve strain) or compressive (ie. two edge-type defects forced to be too close together). The combined effect of this "excess boundary stress" combined with the dislocation array mismatch and the elevated temperature (~400°C) may provide the driving force required for the a/2<011> defects to overcome the large Peierl's barrier to dislocation slip on the (100) interfacial plane. Hence some limited local rearrangement may be possible to decrease the boundary energy. However, complete rearrangement by coordinated misfit dislocation array slip to arrive at a situation where there is no dislocation array mismatch or boundary strain is very unlikely. Thus, any residual boundary strain and mismatch may generate new boundary defects and threading dislocations. Also, it is possible that loops generated near the film surface will glide down on inclined {111} planes under the influence of tensile boundary stresses in the system to form 60° segments in the vicinity of the boundary.

HREM observations of our InSb/GaAs interfaces do in fact show occasional 60° type interfacial defects. Since these 60° defects are usually associated with threading dislocation segments, it is worth discussing their origin. As noted earlier, these defects can be generated to relieve excess boundary stresses at island coalescence. In fact, any stresses generated at or after island coalescence will most likely be relieved by 60° type dislocations since these are the the most easily introduced line defects in continuous films. For instance, when InSb/GaAs interfaces are cooled from 400°C to room temperature, the thermal expansion coefficient mismatch creates an additional 0.02% lattice mismatch. This additional mismatch can be described in terms of a critical thickness of ~0.25μm for 60° dislocations. Deposition of an InSb layer thicker than ~0.25μm on top of the "just coalesced" film would result in the introduction of 60° dislocations to relieve this stress. It has also been suggested that 60° type defects may also tend to form near step edges at the substrate surface (and especially so for vicinal surfaces) [11].

Interface roughening effects

The interface plane in many regions of our InSb/GaAs samples is obviously not atomically smooth (Fig.5), but does still show a/2<011> type misfit dislocation arrays. It is uncertain whether this roughening occurs during growth or whether the GaAs buffer layer surface was unduly rough prior to InSb deposition. However, an interesting point to note is that where gross undulations occur, vertical moire fringes (ie. perpendicular to the

Figure 5. A cross sectional image of a region of InSb/GaAs interface which exhibits considerable roughening.

interface) occur on cross-section images which are due to the mismatch between $g^{InSb}=\{022\}$ and $g^{GaAs}=\{022\}$. The measured average spacing of 16.7Å is somewhat larger than the 15.7Å spacing expected. This discrepancy could be explained by some local strain driven interdiffusion near the interface. For instance, the moire fringe spacing expected for the difference between $g^{InGaSb}=\{022\}$ and $g^{GaAs}=\{022\}$, (where 5% Ga has been introduced on In sites), would be 16.5Å (assuming Vegard's Law). Such intermixing would not be inconsistent with previous SIMS depth profile studies on this interface [3].

CONCLUSIONS

The initial stages of growth of InSb on GaAs(100) proceeds by a 3D nucleation mechanism and most of the 14.6% lattice mismatch in each island is accommodated by a square array of $a/2<011>$ misfit dislocations. Unrelieved boundary stresses and misfit dislocation array mismatches upon island coalescence can lead to the introduction of 60^0-type defects and epi-threading dislocation segments (as can the thermal expansion coefficient mismatch between InSb and GaAs). For low dislocation density films it is likely that incorporation of an InSb/GaSb superlattice in the InSb layer may be required as a barrier to threading dislocation propagation through the epilayer.

A strong asymmetrical distribution of planar defects such as stacking faults and microtwins have been observed in the InSb islands. We believe that this situation may arise due to a large difference in the mobility of α and β dislocations in InSb. Finally, some interface roughening was observed in our samples and although the precise origin of this effect is unknown, it was noted from moire fringe spacing measurements that some localised intermixing of InSb and GaAs had possibly occurred.

ACKNOWLEDGEMENTS

This work was supported by the US Department of Energy under contract DE-AC-02-76ER 01198, the U.S.Air Force Office of Scientific Research under contract 86-0111 and IBM. The microscopy was carried out in the Center for Microanalysis of Materials, University of Illinois which is supported by the DOE.

REFERENCES

[1] A.J.Norieka, M.H.Francombe and C.E.Wood, J.Vac. Sci.Technol., A 1, 558, (1983).

[2] J-I.Chyi, S.Kalem, N.S.Kumar, C.W.Litton and H.Morkoc, Appl.Phys.Lett., 53, 1092, (1988).

[3] G.M.Williams, C.R.Whitehouse, C.F.McConville, A.G.Cullis, T.Ashley, S.J.Courtney and C.T.Elliot, Appl.Phys.Lett., 53, 1189, (1988).

[4] R.L.Williams, R.Droopid, R.A.Stradling , K.W.Barnham, S.N.Holmes, J.Laventy, C.C.Philips, E.Skuras, R.Thomas, X.Zhang, A.Staton-Bevan and D.W.Pashley, Semicond. Sci. and Technol., 4(8), 663, (1989).

[5] C.J.Kiely, J-I.Chyi, A.Rockett and H.Morkoc, Phil.Mag.A., 60(3), 321, (1989).

[6] C.F.McConville, C.R.Whitehouse, G.M.Williams, A.G.Cullis, T.Ashley, M.S.Skolnick, G.T.Brown and S.J.Courtney, Journ.Cryst.Growth, 95, 228, (1989).

[7] J.Matthews, J.Vac.Sci.Technol, 12, 126, (1975).

[8] P.Pirouz, F.Ernst and T.T.Cheng, Mat.Res.Soc.Symp.Proc, 116, 57, (1988).

[9] P.Spellward and A.R.Preston, Inst.Phys.Con.Ser., 93, 29, (1988).

[10] M.Fnaiech, F.Reynaud, A.Couret and D.Caillard, Phil.Mag.A., 55, 405, (1987).

[11] D.J.Eaglesham, M.Aindow and R.C.Pond, Mat.Res.Soc.Symp.Proc., 116, 267, (1988).

NANOLAYER REACTIONS IN ALUMINUM-METAL INTERFACES

E. V. BARRERA, M. W. RUCKMAN, AND S. M. HEALD
Division of Materials Science, Department of Applied Science, Brookhaven
National Laboratory, Upton, NY 11973

ABSTRACT

Surface extended x-ray absorption fine structure (SEXAFS) measurements
on the nanometer level were made for Al/M interfaces where M was Cu or Ni.
The samples were studied immediately after deposition and after heat treatments. Significant differences in interface reactions were observed
depending on deposition direction (Cu on Al or Al on Cu) and the amount of
mixing was also related to whether M was Cu or Ni. The SEXAFS measurements
revealed that there were no detectable amounts of C or O present. One percent Zn was observed to be in the Al layers. The results obtained from the
as-deposited interfaces correlate well with data obtained from buried interfaces of like element combinations.

INTRODUCTION

This research gives preliminary results from studies of Al-Cu and Al-Ni
interfaces which demonstrates the usefulness of surface extended x-ray
absorption fine structure (SEXAFS) to study interface reactions. Previous
research has shown that buried interfaces of Al on Cu deposited at room
temperature react to form 30 Å of predominately $CuAl_2$. Further reaction
occurred after heating to 120° C for 5 min., as observed by Heald et al.
[1-3] using glancing angle EXAFS and x-ray reflectivity. Research, using
x-ray photoemission, showed that for Al-Cu and Al-Ni, the interfaces reacted
at room temperature whether the films were Al/metal or metal/Al [4,5]. The
before mentioned results are for short-range reactions where the sensitivity
of the combined EXAFS and x-ray reflectivity technique is ~10 Å [2] and the
of the x-ray photoemission studies have monolayer sensitivity on surfaces.
To investigate the Al-metal reaction at a buried interface, where the probe
sensitivity is as sensitive as the photoemission, surface extended x-ray
absorption fine structure (SEXAFS) is used.

SEXAFS is a technique which generally uses total or partial electron
yield detection methods to monitor the x-ray absorption [6]. Simply put,
an atom excited by x-rays with energy greater than its absorption edge,
releases a photoelectron which undergoes scattering by the surrounding
atoms. The modification of the final state of the photoelectron produces
oscillations in the extended absorption spectra with increasing energy.
This fine structure provides short-range atomic structure information from
the vantage point of the excited atoms. Since SEXAFS is normally conducted
in an ultrahigh vacuum system, detection can be made of partial monolayers
on surfaces and shallow buried interfaces. Approximate probing depths are
~500 Å for total yield and ~30 Å for partial yield.

EXPERIMENTAL PROCEDURE

Thin polycrystalline films of Al-metal were deposited on a thin tanta-
lum foil at room temperature. The samples were either metal on Al bilayers
or Al-metal-Al trilayers where the metal was either Cu or Ni. The Al thicknesses were 120-240 Å. Thicknesses were measured by an oscillating crystal
thickness monitor in the SEXAFS vacuum chamber. The vacuum system, after
baking, reached levels of 8×10^{-10} - 2×10^{-9} Torr with LN_2 cold trapping. The
vacuum remained in the 10^{-9} Torr range during depositions where times
between depositions were less than 15 min. Cu and Ni were evaporated using

99.999% pure source materials from Al_2O_3 coated W boats. Al was deposited using 99.999% pure source material from an effusion cell. Deposition rates between 10 and 20 Å/min were achieved. Rates remained constant during the depositions. The quality of the films is considered high since samples made in another vacuum system, using the same procedures, showed O and C free-metal layers when analyzed by Auger electron spectroscopy [7]. Zinc contamination from the Al MBE source was detected by the SEXAFS in later runs and was determined to be ~1 % of the Al layers. The samples investigated in the study are summarized in Table. I.

Table I.
Al-M Interface Samples in Bilayers (B) or Trilayers (T)

Designation	Sample Description
Cu20B	20 Å Cu/49 min, 10^{-9} Torr/240 Å Al/Ta
Cu22B	22 Å Cu/14 min, 10^{-9} Torr/136 Å Al/Ta
Cu40B	20 Å Cu/Cu20B
Cu120B*	80 Å Cu/Cu40B
Cu48T	72 Å Al/15 min, 10^{-9} Torr/48 Å Cu/16 min, 10^{-9} Torr/120 Å Al
Cu45T	45 Å Al/11 min, 10^{-9} Torr/45 Å Cu/13 min, 10^{-9} Torr/146 Å Al
Ni20B	20 Å Ni/9 min, 10^{-9} Torr/130 Å Al
Ni40B	20 Å Ni/Ni20B
Ni60B	20 Å Ni/Ni40B
Ni100B*	40 Å Ni/Ni60B
Ni40T	50 Å Al/15 min, 10^{-9} Torr/40 Å Ni/12 min, 10^{-9} Torr/146 Å Al

*Designates samples that were annealed and further analyzed.

The SEXAFS experiments were conducted at NSLS Beam Line X-11A. Ni and Cu k-edge data were measured by total yield either from a grid or a channeltron. Experiments were run on two different occasions. In the first run both grid and channeltron modes were used. The data from the channeltron had a low noise level, but had glitches throughout the spectrum due to the non-linearity of the channeltron. The data from the grid had more noise but less of a problem with glitches. The Chi data reported for the Cu SEXAFS was taken using the channeltron. The Cu edge data and the Ni Chi data were taken using the grid. The experiments for Cu were repeated during the second run where better vacuum was achieved with shorter times between depositions. Vacuum conditions during the first run were $2-5 \times 10^{-9}$ Torr and were $8 \times 10^{-10} - 3 \times 10^{-9}$ Torr during the second run.

For comparison, EXAFS was measured for CuAl, $CuAl_2$, NiAl, $NiAl_3$, Cu, and Ni. This data serves as standards for comparing Chi and the edge data directly.

Resistive sample heating was also conducted, although the temperature was not measured accurately. Annealing temperatures exceeded 500°C as measured by an optical pyrometer. Heating time intervals were for 1 min. and only the thick Cu on Al (Cu120B) and Ni on Al (Ni100B) samples were annealed.

RESULTS OF AL-METAL INTERFACES

Interfaces of Al and Cu

To study the Al on Cu interfaces, Cu was deposited on Al(Cu20B) and followed by further Cu depositions with SEXAFS analysis between the depositions. The Cu20B sample had a long time between the Al and Cu depositions so it was necessary to repeat the experiment (Cu22B). Figure 1(a) shows the Chi data for Cu20B, Cu40B, Cu120B, and the annealed Cu120B as compared to the trilayer (Cu48T), CuAl, CuAl$_2$ and Cu. The Cu bilayer data compare well with Cu and appear to represent unreacted Cu on Al. The smaller amplitude seen for the deposited Cu suggests that the coordination number for the Cu layers is lower than the expected 12 for Cu. This may possibly be explained by a high grain boundary density in the thin films. As the sample thickness increases, the matrix volume increases and the grain boundary volume fraction decreases until the expected coordination is reached. The amplitude for Cu120B does not equal that of Cu and may again be due to the small grain size as compared to that in the standard Cu foil. The trilayer, Cu48T compared favorably with CuAl and CuAl$_2$, indicating that Al on Cu has reacted upon deposition. The drop in the amplitude of the Chi data for Cu48T is partially due to the Cu/Al interface. The resemblance to CuAl and CuAl$_2$ at low k suggests that either CuAl$_2$ or CuAl has formed in the the Al/Cu interface.

The SEXAFS analysis of the Al-Cu samples was repeated during the second run. Figure 2 shows the Cu k-edge data for the bilayers and trilayers. Even though the vacuum conditions and times between depositions were better for Cu22B and Cu45T, the results remain that Cu/Al did not react while Al/Cu did.

Interfaces of Al and Ni

The results of the Al-Ni interfaces are given in the same way as that for Al-Cu. Figure 1(b) shows the Chi data for Ni20B, Ni40B, Ni100B and the annealed Ni100T samples. The trilayer, Ni40T is also given followed by NiAl, NiAl$_3$ and Ni. Once again the bilayer does not show a reaction and the amplitude is again low as compared to that for Ni. The amplitude increases as the Ni layer thickness increases, however for Ni/Al, the amplitude decreases with annealing. This may mean that the Ni layer is changed by heating. One possible explanation is that limited grain boundary diffusion occurs which is postulated to cause a drop in the amplitude [7].

The trilayer, Ni40T does not appear to have reacted. It compares favorably with Ni and again the lower amplitude may be due in part to the Ni/Al interface. The Ni edge data is not given here because it is not as informative as the Cu edge data. The Al-Ni experiments were conducted during the second run with the same experimental conditions as the Cu22B and Cu45T samples, therefore these results do not differ from that for the Al-Cu interface due to different experimental conditions.

DISCUSSION

It appears that there is a difference in the structure of Al/Cu and Cu/Al where the Al/Cu interface is reacted upon deposition at room temperature and the Cu/Al interface is not. The Al/Ni interfaces, on the other hand, did not react in either case. The results from buried Al/Cu interfaces also showed an initially reacted interface and this research seems to confirm that result. However, the unreacted Cu/Al interface does not agree with that seen by x-ray photoemission [4] for reasons which may be experimental since detection sensitivities of the two techniques are the same. Current experiments on Al/Ni buried interfaces also showed that the Al/Ni interface did not react upon deposition [7]. And again these results do not agree with the photoemission studies [5].

Zn contamination was present in the Al layers at a level of ~1% yet it is unilkely that this played a role in the interface reactions since only 1% Zn would be present in the interface of these samples. Oxygen in the interface may be a consideration as to why these results differed from the photoemission studies. For those experiments the vacuum was always below 7x10^{-10}

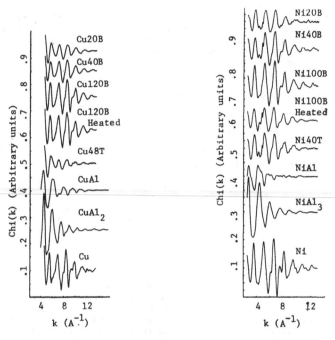

Fig. 1. (a) The Chi (k) data after pre-edge and background substraction for Cu20B, Cu40B, Cu120B, annealed Cu120B, Cu48T, CuAl, CuAl$_2$, and Cu. (b) The Chi (k) data after pre-edge and background subtraction for Ni20B, Ni40B, Ni100B, annealed Ni100B, Ni40T, NiAl, NiAl$_3$, and Ni.

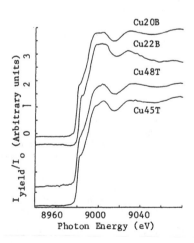

Fig. 2. Cu k-edge data for Cu20B, Cu22B, Cu48T and Cu45T.

Torr where for the present experiments, vacuum conditions were mid 10-9 Torr during depositions. At this level, for 15 min. intervals between depositions one might expect one Langmuir of gases to interact with the surface. But since the chamber was baked, it is likely that most of the residual gas is hydrogen. Still it should be considered that more oxygen could find its way to the Al surface compared to the metal before the overlayer is deposited. This would explain the differences seen between the Al/Cu and Cu/Al interfaces but does not explain the results for the Al-Ni interfaces. The repeated Al-Cu experiments also discount the oxygen consideration since the results obtained were the same as that for the first run of experiments.

CONCLUSION

The conditions of room temperature formed Al-Cu and Al-Ni thin films were studied using SEXAFS. The sensitivity of SEXAFS is equal to that of x-ray photoemission yet buried interfaces can more easily be studied. Cu Chi and edge data revealed that Cu/Al did not react while Cu/Al did react when the interface was formed at room temperature. The observed Al/Cu reaction confirmed results seen for buried Al/Cu bilayers using EXAFS and x-ray reflectivity. The Al-Ni samples did not show a reacted interface whether Al/Ni or Ni/Al were formed. The unreacted Al/Ni interface is in agreement with that seen for buried Al/Ni bilayers as seen using EXAFS and x-ray reflectivity. Still there are significant deviations in these results from what has been observed by x-ray photoemission. This may at this time be attributed to different levels of vacuum at which the experiments were conducted. Further SEXAFS analysis of these and other Al/metal thin films are planned with 10^{-10} Torr vacuum conditions to be used.

ACKNOWLEDGMENTS

The authors would like to thank J. K. D. Jayanetti for his help in conducting these experiments. The support for this research was provided by the U.S. Department of Energy, Division of Materials Sciences under Contract Nos. DE-AS05-80ER10742 and DE-AC02-76CH00016.

REFERENCES

1. S. M. Heald, H. Chen, and J. M. Tranquada, Glancing-angle extended x-ray absorption fine structure and reflectivity studies of interfacial regions, Phys. Rev. B 38, 1016-26 (1988).

2. H. Chen, Studies of Cu-Al interfaces using glancing angle x-ray reflectivity and EXAFS, Ph.D. Dissertation, The City University of New York, 1989.

3. H. Chen and S. M. Heald, Glancing angle EXAFS studies of interfacial reactions: An application to Cu-Al thin films, Solid State Ionics 32/33, 924-9 (1989).

4. D. DiMarzio, H. Chen, M. W. Ruckman, and S. M. Heald, Photoemission and glancing-angle extended x-ray absorption fine structure studies of vacuum-deposited Al/Cu bilayers, J. Vac. Sci. Technol. A7 (3), 1549-53 (1989).

5. M. W. Ruckman, L. Jiang, and Myron Strongin, Room temperature reaction between polycrystalline Ni/Al bilayers deposited in ultrahigh vacuum, J. Vac. Sci. Technol. A, in press.

6. P. H. Citrin, An overview of SEXAFS during the past decade, J. de Phys. C8 (12), 437 (1986).

7. S. M. Heald and E. V. Barrera, X-ray reflectivity and absorption study of Ni/Al reactions: The role of O impurities, submitted to March 1990 Meeting of the American Physical Society.

PART II

Heterointerface Structure

MEASUREMENT OF THE RELATIVE POSITION OF ADJACENT CRYSTALS AND THE MODELLING OF INTERFACIAL STRUCTURE

M.H.I. EL-ERAKI, C.J. KIELY AND R.C. POND
Department of Materials Science and Engineering, The University of Liverpool, P.O. Box 147, Liverpool L69 3BX, U.K.

ABSTRACT

Various techniques using transmission electron microscopy for determining the relative position of two crystals separated by an interface are reviewed briefly. Measurements obtained by these means can be used to identify unrelaxed interfacial configurations. The modelling of relaxed configurations based on these unrelaxed candidate structures requires additional information from experimental techniques such as electron diffraction and high resolution microscopy, or theoretical investigation by computer simulation. This procedure is illustrated for a selection of grain and interphase boundaries. $\{211\}$ grain boundaries between twin related crystals of Al and Ge, and interfaces in $NiSi_2$:Si, $CoSi_2$:Si and Al:GaAs epitaxial systems, are discussed.

1. INTRODUCTION

Experimental techniques are now being developed which can be used to determine directly the atomic structure of interfaces. For example, developments in high resolution electron microscopy (hrem), and the use of synchrotron radiation, are likely to prove increasingly valuable in this respect. However, as with all techniques, these new methods require time for proper development and are also subject to certain limitations. On the other hand, electron microscopists have devised other techniques which individually provide incomplete information about the structure of interfaces, but which have been used in combination to reveal the structure of several grain boundaries and interphase interfaces. One parameter in particular, the relative position of the adjacent crystals, can be established using a variety of techniques. Moreover, theoretical calculations have demonstrated that interfacial energy is profoundly influenced by this parameter, and stable configurations are therefore expected to be characterised by specific values. The present article is a brief review of the experimental methods used to determine relative positions, and an illustration of the way in which measured values form the basis for the subsequent modelling of relaxed structures.

2. EXPERIMENTAL METHODS FOR MEASURING RELATIVE POSITION

α-Fringes

The relative position of adjacent crystals can be determined in certain cases by the formation of displacement, or α-fringes. This method relies upon the excitation of Bloch waves in the bulk of the two crystals which are identical except possibly for their phase. Therefore the technique is restricted to grain boundaries with special relative orientations such that several crystal planes of the same form are parallel in the adjacent crystals. If parallel sets of planes are offset by virtue of a relative displacement of the two crystals, interference will result, leading to the formation of characteristic stacking-fault like contrast in two-beam or systematic diffraction conditions when the interfacial plane is inclined to the specimen surface. Since the fringe contrast in a given micrograph is sensitive to the magnitude and sign of the phase difference, $2\pi g_c.p$ where g_c is the common reciprocal space vector used to form the image and p is a vector defining the relative position of the two crystals, the value of $g_c.p$ can therefore be determined by comparing the observed contrast to simulations. Thus, if images are obtained using at least three separate and non-coplanar common reflections, p can be determined [1,2]. Three special requirements are necessary in order for this technique to be successful. Firstly, the planes oriented for diffraction must be exactly parallel in the two crystals in order to avoid additional contrast effects such as moiré fringes or dislocation images. Secondly, reflections due to non parallel planes must not be excited, since these would couple to the selected beam and lead to thickness-fringe-like contrast contributing to the final image. Thirdly, imaging should be confined to relatively low index reflections since the intensity of fringes associated with high index reflections is generally low. In addition to these experimental requirements, ambiguities in the determination of unrelaxed interfacial structures can arise, particularly if high index common reflections are used. For example, there may be a multiplicity of structures for which all the common planes used in the experiment are aligned. On the other hand, accurate determinations of p can be made under favourable circumstances, because the fringe contrast is very sensitive to the value of $g_c.p$. For example, offsets of planes with magnitudes equal to about 2% of the interplanar spacing can be detected.

Despite the constraints described above, the α-fringe technique has been used in several careful studies of grain boundaries having appropriate crystallography, i.e. so-called 'coincidence' boundaries. The first reported investigations were concerned with measurements of the relative position of twin related crystals ($\Sigma=3$) of Al where the interfacial planes were (111), (151) and ($\bar{2}$11) [1,3,4]. Interfaces in semiconductor materials, particularly Si and Ge, have also been characterised. The $\Sigma=3$ ($\bar{2}$11) interface has been studied by several researchers [5,6,7,8,9], and will be discussed in more detail later. In addition, the $\Sigma=9$ (1$\bar{2}$2) and $\Sigma=11$ (2$\bar{3}$3) interfaces in Ge have been studied by Papon et al.[10], and the $\Sigma=5$ (1$\bar{3}$0) by Bacmann et al. [11].

Moiré Fringes

Moiré fringe patterns, formed in images of one crystal superimposed on another, have been used to measure lattice displacement across coincidence grain boundaries. Matthews and Stobbs [12] studied {112} lateral twin boundaries in Au deposited on MoS_2 substrates using this method. The relative position of the Au crystals was deduced from the offset between moiré fringes in the adjacent crystals. No displacement was found for fringes oriented perpendicular to the boundary, but inclined sets of fringes exhibited some displacement on crossing the boundary, and this was related to a rigid body shift (of typically 0.06 nm) perpendicular to the boundary. The moiré fringe method for determining rigid body shifts is potentially very accurate and has the advantage of not requiring detailed image simulations. However, in practice, it is of limited applicability due to the rather specific specimen geometry required. Also, it is not clear whether the interaction between the deposited crystals and the substrate influences the relative position of the former.

Fresnel Fringes

Fresnel fringe contrast, which is observed parallel and in close proximity to an interface formed between two crystals when viewed approximately edge-on, has also been used successfully to monitor the rigid body shift between adjacent crystals. In particular $\Sigma 3$ (111) twin boundaries in Cu have been studied carefully using this method by Wood et al. [13] and Boothroyd et al. [14]. Fresnel fringes have their origin in the elastic diffuse scatter located around strong beams and the fringe intensities have been shown to be altered by the relative position of the adjacent crystals. However, the precise fringe structure and contrast is also known to be a function of other parameters such as image defocus, beam convergence and sample thickness. Hence, the approach which has to be adopted is to simulate the fringe profile, using multislice calculations (which take all the above parameters into account), and to compare these with the experimental images. Displacements perpendicular to the boundary of ~0.06 d_{111} ((0.01 nm) were detected for the $\Sigma 3$ (111) Cu case using this method. It should also be noted that this method is sensitive to distortions caused by impurity segregation at the interface and, in general, it is a problem to separate the contrast effects due directly to displacements from those arising indirectly from the presence of a segregant.

High Resolution Electron Microscopy

The principle of this method is to use axial hrem imaging on very thin cross-sectional samples in which the interface is oriented exactly 'edge-on'. Multislice image simulations should be carried out and compared to the experimental images. This is necessary because the hrem image is sensitive to many instrumental parameters (e.g. image defocus, spherical aberration coefficient, beam convergence, beam tilt) as well as specimen parameters such as sample thickness. Furthermore if a heterointerface is being studied the two component crystals will probably exhibit different variations in image contrast as a function of thickness. Hence, the different absorption parameters and extinction distances for the component crystals must also be taken into account. Provided it is clear whether the dark spots or bright spots on each side of the interface correspond to either atom columns or holes in the structure, the rigid body shift across an interface can be measured to better than 0.03 nm using an instrument with a point to point resolution of about 0.3 nm [15]. Furthermore it is advisable to measure the rigid body shifts between the two crystals at a reasonable distance away from the boundary plane (> 1.5 nm) to avoid 'interfacial distortions' and Fresnel fringe effects.

The relative positions of adjacent crystals have been measured, or at least those components perpendicular to the beam direction, in a wide range of metallic, ceramic and semiconducting materials. As the resolution of microscopes improves, these measurements are becoming subsumed into what is now approaching direct determination of interfacial structure. Therefore, we shall not refer to particular studies here as examples where relative position only has been established, but will describe some specific instances later when interfacial modelling is considered.

Convergent Beam Electron Diffraction

Convergent beam electron diffraction (cbed) has been used to study both 'plan view' and 'edge-on' specimens. Using this method it should be possible to determine the symmetry of a bicrystalline specimen, and hence to identify relative positions consistent with the observed symmetry. In cases where a multiplicity of relative positions is consistent with the observed symmetry, it is possible in principle to distinguish between the various possibilities by considering the fine structure in holz reflections. This latter procedure would involve comparison of experimental observations with Bloch wave calculations to simulate relevant branch structures. Eaglesham et al. [16] have estimated the sensitivity of cbed patterns to small changes of relative position by rocking curve calculations, and they deduced that displacements greater than about 0.02 nm should be detectable. These authors also noted that the electron probe should be positioned on regions of interface not too close to interfacial dislocations, since the displacement fields of these can modify the symmetry of cbed patterns.

The cbed method has been used successfully to determine the relative position of interphase boundaries using 'plan-view' specimens. In particular, Eaglesham et al. [16] have investigated (001) and (111) interfaces between $NiSi_2$ and Si, and also (001) interfaces between Al and GaAs [17]. On the other hand, the study of grain boundary structures, using both 'plan view' and 'edge-on' specimens, has revealed additional complications compared to the studies mentioned above. This arises because bicrystals where both crystals have the same structure can exhibit symmetries which inter-relate the two crystals (antisymmetry), e.g. mirror planes and diad axes parallel to the interface. Such bicrystals have been investigated systematically by Schapink et al. [18,19,20,21] and El-Eraki [22], and both research groups have reached similar conclusions. In the case of 'plan-view' specimens, these antimirror planes and antidiad axes are precisely those symmetry elements which would be expected to give rise to diffraction effects for which reciprocity needs to be invoked. However, such effects were never observed, probably because these symmetries tend to be broken by specimen preparation; for example, if the boundary plane is not exactly central in a 'plan-view' specimen, antimirror and antidiad symmetry is not exhibited by the bicrystal. In the case of 'edge-on' specimens, a different complication arises associated with the size of the probe. When the probe size is relatively large, the resultant cbed pattern tends towards being an incoherent superposition of the patterns obtained from each of the two crystals separately [22]. For smaller probe sizes, the patterns tend to show symmetry intermediate between this and the true bicrystal symmetry. Caron and Schapink [21] have demonstrated that double rocking wide angle zone axis patterns, obtained using a small probe focussed on the boundary plane, leads to more reliable determination of bicrystal symmetry.

Interfacial Dislocations

Interfacial dislocations are line-defects which separate regions of interfacial structure which are either i) identical, ii) equivalent, i.e. related by a symmetry operator, and hence energetically degenerate, or iii) distinct. The Burgers vectors of the discontinuities in the first two classes are constrained by the symmetry of the adjacent crystals, whereas those of the latter are not [23]. When an array of dislocations is present in an interface in order to accommodate misfit and/or misorientation, and where the Burgers vectors and defect spacings have been determined using some physical technique, it is possible in certain circumstances to make deductions concerning the relative position of the adjacent crystals. If the network comprises only dislocations of the first type, no conclusion can be reached. On the other hand, if the relative position is such that the bicrystal's symmetry is lower than the highest possible for that interface, then an array of dislocations of the above type can decompose into dislocations with smaller Burgers vectors which separate regions of equivalent structure. Because of the duality relationship between dislocation arrays and relative position of the crystals, which reflects the symmetry of the bicrystal in question [24,25], it is possible to 'read' from the experimental observation information about the relative position. If dislocations of the third type are present, regions of interface with different specific free energies will be present, and hence the areas of distinct interfacial structure may no longer be equal, but overall symmetry is expected to be maintained [25]. One instance where such considerations have been used is the investigation of grain boundary structure in Mg0 by Sun and Balluffi [26].

3. MODELLING INTERFACIAL STRUCTURES

Once the relative position of two adjacent crystals has been established using one of the techniques described above, and assuming there are no ambiguities, the next step is to identify the set of unrelaxed structures which are consistent with the experimental measurement. Such a multiplicity may exist, particularly in the cases of crystals exhibiting multi atomic bases. This arises because, for a given relative position of the crystals, the interface may be located at different levels. A completely general crystallographic method has been devised for obtaining the multiplicity of unrelaxed structures, and the interested reader is referred to the original papers [4,27]. In addition, this method also incorporates the consequences of bicrystal symmetry. For example, it can be used to obtain the set of crystallographically equivalent interfacial configurations which arise when the relative position leads to symmetry breaking.

In the remaining part of the paper we discuss the modelling of the relaxed structures of a selection of interfaces. Once the candidate unrelaxed configurations have been obtained, it is necessary to assess further experimental evidence obtained from additional techniques which are sensitive to atomic relaxations. The techniques used to provide this further data are primarily hrem and electron diffraction, as will be illustrated below. Another important tool in the investigation of relaxed structures is atomistic simulation; by these means, the energy of alternative configurations can be calculated and used to support experimental observations.

$\Sigma=3$ {1$\bar{2}$1} Grain Boundaries in Al

The relative positions of the adjacent crystals in this case were measured by Pond and Vitek [3] and Pond [4]. Using the α-fringe method, it was found that the crystals are relatively displaced parallel to [111], and that a small expansion parallel to [1$\bar{2}$1], and designated $\underline{e}$, is present. The resulting bicrystal symmetry is plml, and hence two symmetry related configurations are expected, and these were observed experimentally. Taking the experimental measurement of relative position, three distinct unrelaxed structures can be identified as candidates for the actual relaxed configurations, corresponding to locating the interface plane at one of three sequential positions differing by (2$\bar{4}$2) lattice plane spacings. Hrem has not yet been used to study this particular interface, and electron diffraction studies did not reveal additional reflections indicating significant relaxation. However, the interface has been studied by computer simulation [3], and the relaxed structure obtained by this means was found to be in excellent agreement with one of the unrelaxed possibilities. Fig. 1 is a [10$\bar{1}$] projection of the relaxed structure, and it can be seen that in this metallic structure very little local atomic relaxation has occurred, and the relative position of the crystals is such that compact groupings of 'hard sphere' atoms arise at the interface. It was shown by Pond [2] that when symmetry related domains coexist in an interface, separated by a line-defect [23], it is possible to make accurate measurements of $\underline{e}$ by the α-fringe method. For example, the relative positions of the adjacent crystals corresponding to the symmetry related structures can be expressed as $\underline{p}_1 = q\,[111] + \underline{e}$ and $\underline{p}_2 = -q\,[111] + \underline{e}$. When a common reflection $\underline{g}_C$ is used to form an α-fringe image, the amplitude of the resultant fringes depends on the magnitude of $\underline{g}_C.\underline{p}_1$ and $\underline{g}_C.\underline{p}_2$. Clearly, if $|\underline{e}| = 0$, complementary contrast would always be expected for the two structures. However, when $|\underline{e}| \neq 0$, complementary behaviour is expected when reflections are used for which $\underline{g}_C.e = 0$; an example is shown in fig. 2(a) which is an image formed using $\underline{g}_C = [111]$. Therefore, using reflections where $\underline{g}_C.\underline{e} \neq 0$, the complementarity of fringe contrast is lost, as is illustrated in fig. 2(b) which is an image obtained using $\underline{g}_C = \bar{3}1\bar{1}$. If the values of $\underline{g}_C.\underline{p}_1$ and $\underline{g}_C.\underline{p}_2$ are subsequently determined separately, it follows that $\underline{g}_C.\underline{p}_1 + \underline{g}_C.\underline{p}_2 = 2\underline{g}_C.\underline{e}$, and hence $\underline{e}$ can be found with enhanced accuracy. In the present case $\underline{e}$ was found to be 0.02 $[\bar{1}2\bar{1}]$, which has a magnitude of 0.02 nm.

$\Sigma=3$ ($\bar{1}1\bar{2}$) Boundaries in Ge

($\bar{1}1\bar{2}$) boundaries in Si and Ge have been studied using the α-fringe technique by several researchers [5,6,8,9], and also by means of hrem [28,29]. The spread of experimental measurements is reasurringly small [30], and the values obtained are very similar for Si and Ge. Similarly to the case for Al, the coincident mirror symmetry parallel to (202) is conserved, and two symmetry related structures can arise, but, as far as the authors are aware, these two structures have not so far been found coexisting in the same specimen. Atomic relaxation at ($\bar{1}1\bar{2}$) grain boundaries in Si and Ge has been studied extensively by hrem, electron diffraction and computer simulation, and the comparison with the structure of the boundary in Al discussed above is most interesting. As was the case for Al, three candidate unrelaxed structures can be identified, corresponding to different locations of the interfacial plane, which are consistent with the relative position of the crystals as determined by the α-fringe technique, and these have been illustrated elsewhere [5]. Electron diffraction and hrem studies by Bourret and Bacmann [31] and Bourret et al. [29] have demonstrated that reconstruction occurs in Ge boundaries leading to a centred 2 x 2 structure with spacegroup clml. Moreover, atomistic calculations [32] have confirmed that the energy of this proposed structure is lower than that of alternative models considered. The reconstructed structure is illustrated in fig.3, which is taken from the work of Paxton and Sutton. No dangling bonds are present in the model, and we note the interesting reconstruction parallel to [110], i.e. the bonding between pairs of atoms such as numbers 1, 1', 11, 11' and the ones directly below or above them in the figure. These bonds also exhibit the maximum bond stretching that occurs, and hence the highest bond energies calculated in the study.

A selected area electron diffraction pattern taken from a 'plan-view' specimen containing a ($\bar{1}1\bar{2}$) Ge boundary and obtained in our laboratories is shown in fig.4. The beam direction is perpendicular to the interface, and the observed reflections are illustrated schematically in fig.5. Crystal reflections are coincident in the zero layer for this section of reciprocal space, as shown in fig. 5(a), and reflections that can arise by double diffraction are depicted in fig.5(b). Reflections arising due to the centred 2 x 2 reconstructed interfacial structure are illustrated in fig.5(c), and it can be seen that some of these are not coincident with doubly diffracted beams. Thus, for

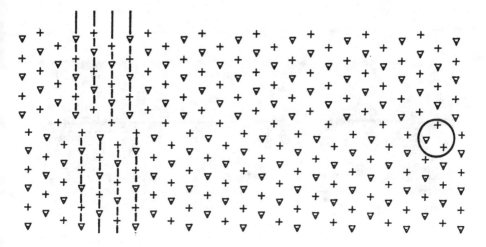

← [111]

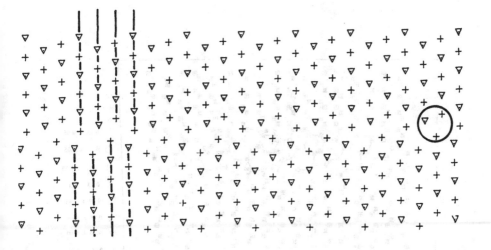

Fig.1: [01$\bar{1}$] projection of fully relaxed computed structures for (1$\bar{2}$1) $\Sigma = 3$ interface in Al [3]; the symbols represent the ...ABAB... stacking along [01$\bar{1}$]. Upper and lower configurations are symmetry related (energetically degenerate) interfaces characterised by opposite relative displacements parallel to [111]. The offset of the (111) planes has been indicated, and a group of closely packed atoms at the interface has been circled.

84

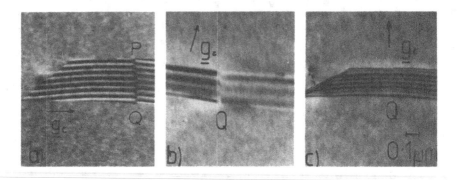

Fig.2: Bright field electron micrographs showing α-fringes formed by a Σ = 3 (T2T) interface in Al; symmetry related structures are separated by the defect PQ. (a) g_C = TTT, w = 0.12, (b) g_C = 31T, w = 0.81, (c) g_C = 220, w = 1.06. Note that the α-fringes for the two structures are complementary in (a) where $g_C.\underline{e}$ = 0, identical in (c) where $g_C.q[111]$ = 0, but not complementary in (b) where $g_C.\underline{e} \neq 0$ (see text).

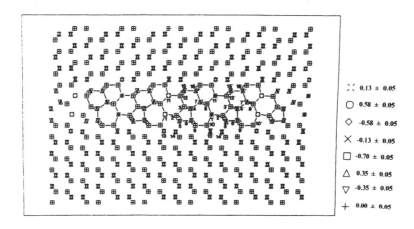

Fig.3: Computer simulated Σ = 3 (T1Z) interface in Ge, taken from the work of Paxton and Sutton [32]. The structure is viewed along [110], and the centred 2x2 reconstruction is apparent.

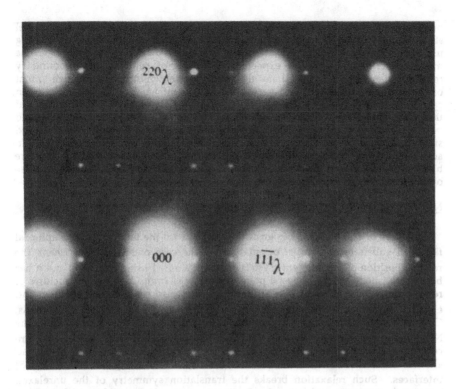

Fig.4: Selected area diffraction pattern taken from a plan view bicrystal specimen of Ge containing a Σ = 3 (T12) interface.

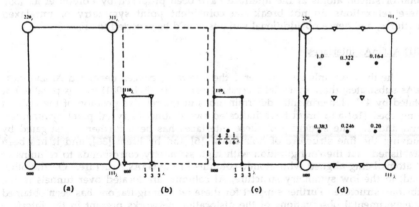

Fig.5: Schematic illustration of the reflections in the selected area diffraction pattern shown in fig.4; (a) crystal reflections, (b) double diffraction reflections, (c) grain boundary reflections, and (d) calculated intensities of some grain boundary reflections.

purposes of comparison between experimental observations and the model structure, it is advisable to concentrate attention on the centring reflections. Making use of the kinematical approximation, and taking the atomic coordinates from the model described above, the expected intensities of centring grain boundary reflections have been calculated and the values are shown in fig.5(d) relative to that of the strongest reflection, and very good agreement has been obtained between calculated and observed values [22]. Thus, there is extensive experimental and theoretical support for the structure proposed by Papon et al. [7]. However, it should be noted that Bourret et al. [29] found small discrepancies between observed and simulated hrem micrographs in the regions of the highest energy bonds. In addition, although Paxton and Sutton [32] confirmed that the centred 2 x 2 structure had lower energy than alternative structures considered, they note that one of these other structures has very similar energy.

NiSi$_2$: (001) Si and CoSi$_2$: (001) Si Interfaces

Layers of NiSi$_2$ and CoSi$_2$ grow on (001) Si with the unit cells of the epitaxial film and substrate in parallel orientation. In the case of NiSi$_2$Si specimens the relative position of the adjacent crystals has been studied using cbed [16], and it has been established that the bicrystal point symmetry is 2mm. The atomic structures of relaxed (001) NiSi$_2$:Si and CoSi$_2$:Si interfaces have also been studied by hrem and electron diffraction. By matching cross-sectional images to contrast simulations, Cherns et al. [33] suggested that the interfacial structure for NiSi$_2$:Si is such that the Ni atoms at the interface have six fold coordination, and Si atoms are everywhere coordinated tetrahedrally. However, further observations using both hrem and electron diffraction have revealed 2 x 1 relaxations at both NiSi$_2$ [34] and CoSi$_2$ [35] interfaces. Such relaxation breaks the translation symmetry of the unrelaxed interfacial structures, and hence 1 x 2 relaxations are also possible, and have been detected experimentally [35]. Further work is necessary in order to establish the details of these reconstructed interfacial structures, and possible dimerisation of metal or silicon atoms at the interface have been proposed by Loretto et al. [35]. These relaxations do not break the coincident point symmetry of unrelaxed structures, and hence no additional variants would be expected to arise.

(001) Al:GaAs Interfaces

The final example we consider is the interface between deposited Al and (001) GaAs substrates; these epitaxial bicrystals form with their (001) planes parallel but rotated by 45°. Experimental determinations of the relative position of the crystals using cbed [16] and hrem have indicated two distinct bicrystal point symmetries, 2mm and 1. The structure of these interfaces has been further investigated by studying the fine structure of holz lines [16] and by hrem [36], and it has been established that the configuration with high symmetry corresponds to columns of Al atoms being aligned with atomic columns in the GaAs structure. On the other hand, in the low symmetry structure, Al columns are located over tunnels in the substrate structure. Further support for these two configurations has been obtained by experimental observations of the dislocation networks present in the interface which accommodate the misfit. In the case of the high symmetry interface, an orthogonal array of misfit dislocations with $1/2<110>$ Burgers vectors would be expected. However, the relative position of the crystals in the low symmetry configuration breaks all the bicrystal symmetry, and hence four crystallographically equivalent forms arise. The dislocation network described above would then be expected to decompose into dislocations with Burgers vectors equal to $1/4<110>$ and separated by half the above spacing [36]. This is completely consistent with the analysis presented in section 2, and, in the present case, leads to a significant reduction in the elastic energy of the misfit dislocation array.

ACKNOWLEDGEMENTS

The authors are very grateful to Dr. J.J. Bacmann for providing Ge bicrystals, and his encouragement.

REFERENCES

1.   R.C. Pond and D.A. Smith, Canad. Metall. Q. 13, 33 (1974).

2.   R.C. Pond, J. Micros. 116, 105 (1979).

3.   R.C. Pond and V. Vitek, Proc. R. Soc. Lond. A357, 453 (1977).

4.   R.C. Pond, Proc. R. Soc. Lond. A357, 471 (1977).

5.   D. Vlachavas and R.C. Pond, Inst. Phys. Conf. Ser. No. 60, 159 (1981).

6.   C. Fontaine and D.A. Smith, Appl. Phys. Lett. 40, 153 (1982).

7.   A.M. Papon and M. Petit, Scripta Met. 19 391 (1985).

8.   A. Rocher and M. Labidi, Revue Phys. Appl. 21, 201 (1986).

9.   Ph. Komninou, Th. Karakostas and P. Delavignette, J. Mat. Sci. 21, 3817 (1980).

10.   A.M. Papon, M. Petit and J.J. Bacmann, Phil. Mag. A49, 573 (1984).

11.   J.J. Bacmann, A.M. Papon, M. Petit and G. Silvestre, Phil. Mag. A51, 697 (1985).

12.   J.W. Matthews and W.M. Stobbs, Phil. Mag. A36, 373 (1977).

13.   G.J. Wood, W.M. Stobbs and D.J. Smith, Phil. Mag. A50, 375 (1984).

14.   C.B. Boothroyd, A.P. Crawley and W.M. Stobbs, Phil. Mag. A54, 663 (1986).

15.   J.M. Gibson, Ultramicroscopy 14, 11, (1984).

16.   D.J. Eaglesham, C.J. Kiely, D. Cherns and M. Missous, Phil. Mag. A60, 161 (1989).

17.   D. Cherns, C.J. Kiely and D.J. Eaglesham, Proc. Mat. Res. Soc. Symp., 75 321 (1985).

18.   F.W. Schapink and F.J.M. Mertens, Scripta Met. 15, 611 (1981).

19.   R.P. Caron and F.W. Schapink, Ultramicroscopy, 17, 383 (1985).

20.   F.W. Schapink, S.K.E. Forghany and B.F. Buxton, Acta Cryst. A39, 805 (1983).

21. F.W. Schapink, S.K.E. Forghany and R.P. Caron, Phil. Mag. A53, 717 1986.

22. M.H.I. El-Eraki, Ph.D. Thesis, University of Liverpool (1989).

23. R.C. Pond, in Dislocations in Solids, Vol. 8, edited by F.R.N. Nabarro (North-Holland, Anmsterdam, 1989) p.1.

24. W. Bollmann, Crystal Defects and Crystalline Interfaces (Springer-Verlag, Berlin, 1970).

25. R.C. Pond, in Grain Boundary Structure and Kinetics, edited by R.W. Balluffi (ASM, Columbus 1980).

26. C.P. Sun and R.W. Balluffi, Phi. Mag. A46, 63 (1982).

27. R.C. Pond, D.J. Bacon, A. Serra and A.P. Sutton, accepted for publication in Metall. Trans. (1990).

28. A. Bourret and J.J. Bacmann, Inst. Phys. Conf. Ser. No. 78, 337 (1985).

29. A. Bourret, L. Billard and M. Petit, Inst. Phys. Conf. Ser. No. 76, 23 (1985).

30. M. Cheikh, A. Hairie, F. Hairie, G. Nouet and E. Paumier, accepted for publication in J. Phys. Colloq. (1990).

31. A. Bourret and J.J. Bacmann, Proc. J.I.M.I.S. 4, Minakami Spu (1985).

32. A.T. Paxton and A.P. Sutton, J. Phys. C 21, L481 (1988).

33. D. Cherns, G.R. Anstis, J.L. Hutchison and J.C.H. Spence, Phil. Mag. A 46, 849 (1982).

34. J.L. Batstone, J.M. Gibson, R.T. Tung and D. Loretto, EMSA Proceedings, (1989).

35. D. Loretto, J.M. Gibson, A.E. White, K.T. Short, R.T. Tung, S.M. Yahsave and J.L. Batstone, Appl. Phys. Letts. - in press (1989).

36. C.J. Kiely and D. Cherns, Phil. Mag. A. 59 (1), 1, (1989).

HIGH-RESOLUTION IMAGING OF CoGa/GaAs AND ErAs/GaAs INTERFACES

Jane G. Zhu, Stuart McKernan, Chris J. Palmstrøm* and C. Barry Carter
Department of Materials Science and Engineering, Bard Hall, Cornell University,
Ithaca, NY 14853;
*Bellcore, 331 Newman Springs Road, Red Bank, NJ 07701

ABSTRACT

CoGa/GaAs and ErAs/GaAs grown by molecular-beam epitaxy have been studied using high-resolution transmission electron microscopy (HRTEM). The epitactic interfaces have been shown to be abrupt on the atomic scale. Computer simulations of the HRTEM images have been obtained for different interface structures under various specimen and image conditions. Practical problems in the comparison between the simulated and experimental images are discussed.

INTRODUCTION

The epitactic growth of metal compounds, such as NiAl [1], CoGa [2] and NiGa [3], and rare-earth arsenides, such as ErAs [4,5], LuAs [5] and YbAs [6], on GaAs has recently been reported. These materials have potential applications for both stable contacts on compound semiconductors and three-dimensional device structures. The layers on each side of the heterojunction of these newly developed systems have different crystal structures, unlike the extensively studied semiconductor-semiconductor systems [7]. Transition metal-gallium (or aluminium) compounds, e.g., CoGa, CoAl, FeAl, NiGa and NiAl, have the CsCl (B2) structure and can be epitactically grown on GaAs with about 2% lattice mismatch. Misfit dislocations at the interface of CoGa and GaAs have recently been investigated [8]. Most of the rare earth mono-arsenides have the NaCl structure [9]. The lattice mismatch between ErAs and GaAs is about 1.6%. In this paper, the study of CoGa/GaAs and ErAs/GaAs structures by high-resolution transmission electron microscopy (HRTEM) is reported.

Atomic-resolution characterization of the interface configuration is very important since the interface can dominate the electronic properties of the devices. A determination of the interface location by the difference in the background intensity in conventional transmission electron microscopy is not reliable since the images at the interface region could be modulated by artifacts of sample preparation. HRTEM studies are necessary to precisely locate the interface position. The detailed interpretation of micrographs, however, requires the comparison of simulated and experimental images.

EXPERIMENTAL

The CoGa/GaAs and ErAs/GaAs layers used in this study were grown in a VG-V80H MBE system. Co was evaporated by an e⁻-gun. Er, Ga and As₄ were evaporated from conventional effusion cells. 50 nm of CoGa was grown at 450°C after the growth of a 500 nm GaAs buffer layer at 600°C. Ga was deposited to obtain a Ga-terminated surface prior to the growth of CoGa. ErAs was grown at 400°C after the growth of a 500nm-GaAs buffer layer at 600°C. The As₄ shutter was open during the period when the substrate temperature was lowered from 600°C to the CoGa or ErAs growth temperature. Details of the MBE growth can be found elsewhere [2,5].

The samples were characterized by cross-sectional HRTEM using a JEOL 4000EX operated at 400 keV. {010} cross sections are chosen instead of {011} cross sections since the line directions of misfit dislocations at these interfaces are close to <010> directions [8,11]. The {010} cross sections are particularly suitable for the CoGa/GaAs case because it is difficult to produce cross-grating lattice images from {011} projection.

Image simulations were performed using the TEMPAS multislice program [10]. This program allows the user to construct a foil consisting of different layers. Each layer corresponds to a supercell in dimensions of $11.3 \times 0.565 \times 0.565$ nm. An interface can be incorporated simply by putting GaAs on one side and the epilayer on the other side of the supercell. A very long dimension perpendicular to the interface (11.3 nm) is necessary to avoid spurious interference due to the periodic extension used in the multislice calculations. To simplify the simulation process, the misfit and strain in the epilayers have been ignored, which is not too far from the real situation if the areas considered are between the misfit dislocations in a strain-relaxed layer.

CoGa/GaAs INTERFACE

When CoGa is grown on GaAs, how the first layer of CoGa is deposited on the GaAs surface, or what the CoGa/GaAs interface atomic configuration is, has not been well understood. Different interface atomic structures are possible. The GaAs structure has alternating Ga and As (100) planes. Similarly, the CoGa structure has alternating Co and Ga (100) planes. When the atoms at the interface are arranged in the minimum energy configuration, it is possible that the interface contains more than one element, i.e., the interface could be a mixture between CoGa and GaAs. As a simple approximation, CoGa is rigidly placed on the GaAs with the interface perfectly flat. Images have been simulated for four different interface configurations, as shown in Fig. 1, to study the influence of the interface atomic configurations on the HRTEM images. The Ga sublattice is discontinuous in "1" and "2", but continuous in "3" and "4", although the sublattice is fcc in GaAs and cubic in CoGa.

{010)} cross-section images of CoGa/GaAs interfaces were simulated for a range of specimen thicknesses and defocus values using the TEMPAS multislice program. Selected simulated images, with the defocus value $\Delta f = -60$ nm and the thicknesses $t = 5, 10, 15, 20$ nm, are shown in Fig. 2. When the sample thickness is thinner than 20 nm, the images for the different interface structures look similar. Substantially different images are obtained for the four different interface structures when the sample is 20 nm thick.

An experimental HRTEM image of CoGa/GaAs is shown in Fig. 3. The image is different from area to area, which indicates that the TEM specimen is not uniform in thickness. The GaAs side of the sample appears thinner than 20 nm. The comparison between simulated and experimental images shows that the images simulated for structure "2" tentatively have the best match with the experimental images. However, all of the interfaces simulated in this paper have simple structures and hence the previous results do not rule out the possibility that the interface may have a more complicated structure. A determination of the exact interface atomic configuration by comparing the simulated and experimental images is difficult due to the following reasons. First, the simulated images are sensitive to the interface structure only for relatively thick specimens. However, the thickness is difficult to control when making TEM samples. Second, the TEM samples used in this study are thinned to perforation by ion milling, and the preferential ion milling can alter the interface contrast. Third, the strain due to misfit is expected to affect the interface contrast but this has been neglected in the image simulations. Fourth, there are probably dislocations lying in the interface parallel to the specimen foil for a specimen thicker than 20 nm [8]. These dislocations can change the image contrast more than 10 nm away from

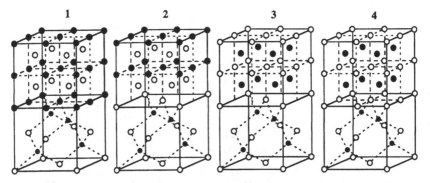

Fig. 1. The schematic diagram of the four different CoGa/GaAs interface structures used for the image simulation, ● Co, O Ga, ● As.

the interface [11]. These issues should be considered in the comparison between simulated and experimental images.

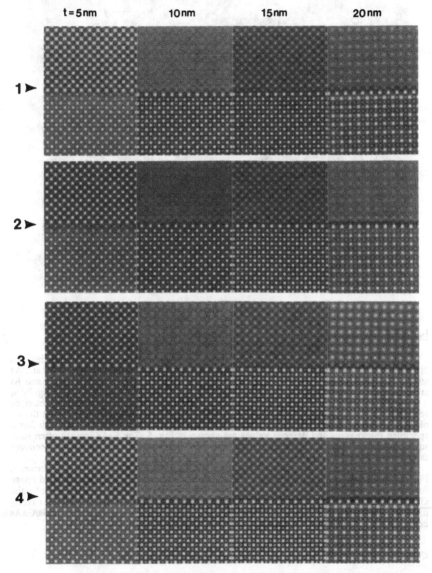

Fig. 2. Computer simulated images of CoGa/GaAs for four different interface structures (t = 5, 10, 15, 20 nm and Δf = -60 nm).

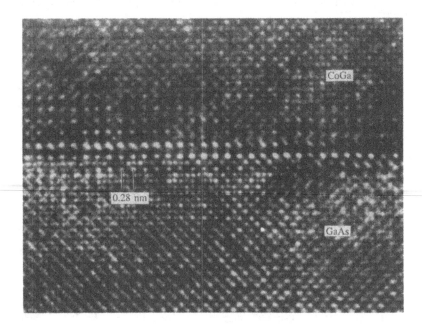

Fig. 3. (010) cross-section HRTEM image of CoGa/GaAs specimen.

ErAs/GaAs INTERFACE

An analysis of high-resolution images of ErAs/GaAs shows both similarities and differences with CoGa/GaAs. The {100} surfaces of ErAs contain equal numbers of Er and As atoms. The {010} cross-section images do not change when the atomic positions of Er and As are exchanged. In a similar manner to the CoGa/GaAs simulation, ErAs is put on GaAs rigidly as a first-order approximation to study the variation of interface contrast with different interface structures. Two structures shown in Fig. 4 have been simulated, where ErAs ends on the As plane in structure "1" and on the Ga plane in structure "2". The simulated images for different specimen thicknesses and defocus values are shown in Fig. 5. The ErAs images often have square patterns. The simulated images for these two different structures are similar for specimens thicker than 15 nm.

An experimental image from a (010) cross-section of an ErAs/GaAs sample is shown in Fig. 6. The ErAs/GaAs interface is abrupt within one or two monolayers. The simulated image for 5nm-thick specimen is overlaid on the experimental image. Both the simulated ErAs and GaAs images are closely matched individually with the experimental images but not exactly matched across the interface. All of the practical problems discussed for the CoGa/GaAs heterostructure also need to be considered for the ErAs/GaAs system.

CONCLUSIONS

Images of the CoGa/GaAs and ErAs/GaAs interfaces have been simulated for a range of specimen and image conditions using the TEMPAS multislice program. These images have been compared to the experimentally obtained images. The interfaces can be located in both the experimental and simulated images within a monolayer. The practical problems in comparing the

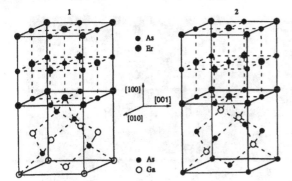

Fig. 4. Schematic diagrams of the two ErAs/GaAs structures simulated.

simulated and experimental images have been outlined. The HRTEM images show that the MBE grown ErAs/GaAs and CoGa/GaAs interfaces are abrupt within one or two monolayers. The ErAs/GaAs interface is very flat on the atomic scale.

ACKNOWLEDGEMENTS

The authors would like to thank Dr. D. R. Rasmussen for helpful discussion on computer simulation and R. Coles for maintaining the microscope and M. Fabrizio for assisting on photographic work. Jane G. Zhu acknowledges support by the Semiconductor Research Corporation Center for Microscience and Technology under grant No. 89-SC-069. The Materials Science Center Facility for Electron Microscopy at Cornell is supported, in part, by NSF.

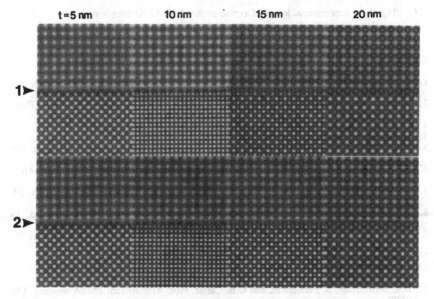

Fig. 5. Simulated images of ErAs/GaAs (t = 5, 10, 15, 20 nm, Δf = -50 nm) for the two structures shown in Fig. 4.

94

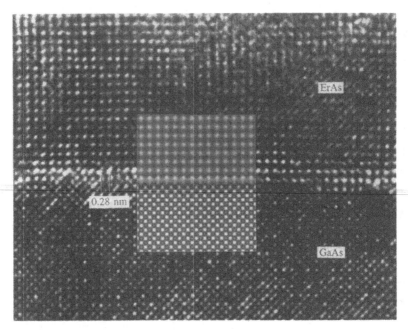

Fig. 6. An experimental image of a cross-section ErAs/GaAs sample with a
simulated image (t = 5 nm, Δf = -50 nm) overlaid on it.

REFERENCES

1. T. Sands, J.P. Harbison, W.K. Chan, S.A. Schwarz, C.C. Chang, C.J. Palmstrøm, and
   V.G. Keramidas, Appl. Phys. Lett. **52**, 1216 (1988).
2. C.J. Palmstrøm, B.-O. Fimland, T. Sands, K.C. Garrison, and R. Bartynski, J. Appl. Phys.
   65, 4753 (1989).
3. A. Guivarc'h, R. Guérin, and M. Secoué, Electron. Lett. **23**, 1004 (1987).
4. C.J. Palmstrøm, N. Tabatabaie, and S.J. Allen, Jr., Appl. Phys. Lett. **53**, 2608 (1988).
5. C.J. Palmstrøm, K.C. Garrison, S. Mounier, T. Sands, C.L. Schwartz, N. Tabatabaie, S.J.
   Allen, Jr., H.L. Gilchrist, and P.F. Miceli, J. Vac. Sci. Technol. **B 7**, 747 (1989).
6. H.J. Richter, R.S. Smith, N. Herres, M. Seelmann-Eggebert, and P. Wennekers, Appl.
   Phys. Lett. **53**, 99 (1988).
7. See, for example, T. Nakamura, M. Ikeda, S. Muto, and I. Umebu, Appl. Phys. Lett. **53**,
   379 (1988).
8. J.G. Zhu, C.B. Carter, C.J. Palmstrøm, and K.C. Garrison, Appl. Phys. Lett. **55**, 39
   (1989).
9. P. Villars and L.D. Calvert, in *Pearson's Handbook of Crystallographic Data for Intermetallic
   Phases* (American Society for Metals, Metals Park, Ohio, 1985).
10. R. Kilaas, Proc. **45**th Annual EMSA meeting, 66 (1987).
11. J. G. Zhu, C. J. Palmstrøm, and C. B. Carter, to be published in the EMAG-MICRO
    proceedings in the Institute of Physics Conference Series, London 1989.
12. D.R. Rasmussen, S. McKernan, and C.B. Carter, Proc. **47**th Annual EMSA meeting, 130
    (1989).

PLAN-VIEW CBED STUDIES OF NiO-ZrO$_2$(CaO) INTERFACES

V.P. Dravid*, M.R. Notis*, C.E. Lyman* and A. Revcolevschi**
* Department of Materials Science & Engg., Lehigh University, Bethlehem, PA 18015 USA
** Laboratoire de Chimie des Solides, Universite de Paris-Sud, 91405 Orsay Cedex, FRANCE

ABSTRACT

Low energy lamellar interfaces in the directionally solidified eutectic (DSE) NiO-ZrO$_2$(CaO) have been investigated using transmission electron diffraction and imaging. The symmetry of this bicrystal and an aspect of interfacial relaxations in the form of symmetry lowering in-plane rigid body translation (RBT) have been explored by performing convergent beam electron diffraction (CBED) experiments of plan-view bicrystals. Edge-on interfaces have also been studied by conventional and high resolution transmission electron microscopy (CTEM and HRTEM respectively), and electron diffraction fine structure analysis. Despite certain experimental difficulties due to interfacial defects and strain, plan-view CBED patterns offered valuable information concerning bicrystal symmetry and indicated no symmetry lowering RBT in this bicrystal. The suitability of plan-view CBED is briefly discussed in view of its potential as a technique to determine bicrystal symmetry and RBT.

INTRODUCTION

Interphase interfaces form a special class of general crystalline interfaces. Unlike grain boundaries, interphase interfaces can be equilibrium defects in a crystalline solid. Low energy interphase interfaces, in particular, have a marked influence on nucleation and growth processes in solid-state phase transformations. The structure of crystalline interfaces is an intensely pursued subject because it is believed that the generic structure-property relation can be extended to bicrystal and polycrystal interfaces. Among the available geometric approaches [1] to describe a given bicrystal and interface, *Bicrystallography* [2] has recently emerged as an important and useful approach. Unlike the CSL and O-lattice type concepts [3], which utilize only the translational symmetry of the crystals, bicrystallography embodies complete symmetries (translational, point and combination, i.e. space group symmetries) of the adjoining crystals. For given orientation of the two crystals and the interface plane (for planar interfaces), it is always possible to assign a point group and space group (if there is at least one dimensional translation symmetry) to the bicrystal. If the interface in question is a grain boundary, additional symmetry elements such as color reversing operations may also arise. There are many advantages in describing crystalline interfaces in the framework of bicrystallography. It can be anticipated that just as single crystal properties are expected to follow crystal symmetry, bicrystal properties should also conform to bicrystal symmetry. However, more importantly, bicrystallography provides a unique framework for analyzing interfacial defects in a rigorous and unified manner. Symmetry breaking while creating the bicrystal may lead to interfacial defects and symmetry compensation is important in determining morphology of embeded crystals. Therefore, both theoretical and experimental aspects of bicrystallography deserve detail attention.

Pond & Bollman [4], and Pond & Vlachavas [5] have discussed the theoretical aspects of bicrystallography. However, there have been only a few experimental investigations of bicrystal symmetry and interface relaxation. Symmetry determination is one popular application of convergent beam electron diffraction (CBED) technique. Recently CBED technique has been applied to the study of

Mat. Res. Soc. Symp. Proc. Vol. 159. ©1990 Materials Research Society

grain boundaries [6] and interphase interfaces [7]. Just as CBED pattern symmetries from single crystals are related to the true crystal symmetry, it can be assumed that CBED pattern symmetries from bicrystals may also be related to the bicrystal symmetry. However, the motivation behind CBED studies has been not just the investigation of bicrystal symmetry but also, indirectly, some aspects of interface relaxation such as in-plane rigid body translation. It can be shown that for a periodic interface, in-plane RBT can reduce the point group symmetry of the bicrystal. This reduction in symmetry can be investigated by CBED. It may perhaps be worthwhile to illustrate this point schematically with reference to Figure 1. The symmetry of the square depicted in Fig. 1 A (with respect to its center) is 4mm. If we superimpose another identical square on top of the original one, without lateral translation, the symmetry of the composite remains the same, i.e. 4mm. However, if the second square is superimposed with lateral translation (analogous to RBT) as shown in Fig. 1C, the overall symmetry of the composite is now reduced to 2mm. Therefore intuitively one can imagine that symmetry lowering RBT may manifest itself as reduced symmetry (with respect to maximum symmetry bicrystal) in CBED patterns.

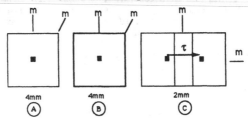

Fig. 1: Schematic illustration of reduction in composite symmetry due to lateral translation (RBT). **(A)** A square exhibiting 4mm symmetry, **(B)** Superposition of an identical square without translation preserves original 4mm symmetry of **(A)** , but lateral translation reduces it to 2mm as in **(C)**.

Schapink and co-workers [6] investigated some aspects of the theory and practice of CBED from grain boundaries and established the relationship between the diffraction groups and the dichromatic point groups. More recently, Eaglesham et al. [7] studied electron diffraction from epitaxial bicrystals using Bloch wave approach and also presented the detailed analysis of experimental CBED patterns, including HOLZ effects, from $NiSi_2/Si$ and Al/GaAs bicrystals. These authors showed that symmetry effects alone can deduce the possible projected forms of the interface and higher order Laue zone (HOLZ) diffraction effects can also be utilized in determining the extent of RBT. We have recently successfully applied this technique to a model system of GaAs/InP epitaxial bicrystal [8], which is readily amenable to plan-view studies. The aim of this paper is to apply the technique of plan-view CBED to crystallographically and structurally more complicated interfaces as in $NiO-ZrO_2(CaO)$ DSE (hereafter referred to as NZ, N for NiO and Z for ZrO_2). The crystallography of this DSE has been previously established [9]. Briefly, the orientation relationship follows:

$$[1\bar{1}0] \text{ N } // [001] \text{ Z}$$
$$[\bar{1}\bar{1}2] \text{ N } // [010] \text{ Z, and}$$
$$[111] \text{ N } // [100] \text{ Z } // \text{ interface normal}$$

It has also been previously shown that (111) N // (100) Z // interface has two dimensional periodicity and corresponds to a definite minimum in interface energy [10]. The objective of this study is to explore the bicrystal symmetry by plan-view CBED and also to determine the extent of in-plane RBT (if any) in low energy interfaces in NZ bicrystal.

EXPERIMENTAL

The conditions and parameters for directional solidification of NZ eutectic have been discussed elsewhere [9]. Both transverse (for edge-on (111) N // (100) Z interface) and longitudinal (for plan-view) sections were prepared for TEM imaging and diffraction. It should be pointed out that requirements for plan-view specimens are quite stringent [6] and it was difficult to prepare such sections from NZ DSE because of the microscopic deviations in lamellar regularities. Despite these problems some specimens fulfilled the requirements for plan-view studies. Electron microscopy studies were performed using a Philips 400T at 120 kV for CBED (in both microprobe and STEM modes) and a Philips 430T at 100-300 kV for electron diffraction fine structure analysis. HRTEM imaging was done on the Philips 430T operated at 250 kV and image simulations were performed using the MacTempas simulation programs.

RESULTS & DISCUSSION

Fig. 2A and 2B are the representative CBED patterns from the NZ bicrystal in plan-view mode. Fig. 2A is the zero order Laue zone (ZOLZ) pattern which represents the projected diffraction symmetry (PDS) of the bicrystal. Fig. 2B is the whole pattern (WP), which corresponds to the 'true' 3-dimensional (3-D) symmetry. For the given orientation of the bicrystal (and the interface), and for zero in-plane RBT (i.e. maximum symmetry bicrystal) one expects ZOLZ symmetry (i.e. PDS) to be the intersection symmetry of the PDS of NiO and that of ZrO_2. That is, **6mm** (of [111] NiO)$\cap$**4mm** (of [100] ZrO2) = **2mm** as shown schematically in Fig. 3A. In other words, only two aligned mirrors and one common two-fold rotation axis should survive. WP symmetry of this composite, however, is given by **3m** (of [111] NiO)$\cap$ **4mm** (of [100] ZrO2) = **m**, which is the only common symmetry element (see Fig. 3B). The observed symmetries of ZOLZ and WP, despite extensive double diffraction effects, as may be seen from Figs. 2A and B are 2mm and m respectively. These observations indicate that there is no symmetry lowering in-plane RBT present in the NZ bicrystal. Further observations were also made to confirm the presence of single true mirror plane in the WP and also the absence of the other mirror plane (and two-fold rotation axis). Many CBED patterns were obtained by tilting the bicrystal along both the ZOLZ mirrors. Fig. 4A shows an off-axis [111] N // [100]Z CBED WP exhibiting one mirror plane, which is the same WP mirror as in on-axis pattern. The same orientation from only NiO has been shown for comparison in Fig. 4B. It can be readily appreciated that the bicrystal CBED pattern indeed has one mirror plane, not only in WP but also in the BF disc. No other additional point group symmetry elements could be found. The 'm' symmetry of the NZ bicrystal along [111] N // [100] Z is the maximum symmetry expected. Therefore no in-plane RBT, which otherwise would have reduced the point symmetry, is present in the NZ bicrystal interface. This result is consistent with our HRTEM observations [11] of edge-on NZ interfaces and their simulations. The third component of RBT which is perpendicular to the interface, was investigated using edge-on HRTEM imaging and electron diffraction fine structure analysis [11] and found to be zero within the accuracy of the techniques.

At this point it is worth mentioning that symmetry-breaking and conservation in CBED is a fiercely debated topic in recent years. How symmetric is symmetric, and how genuine is the broken symmetry are two of the outstanding problems in CBED [12]. Extreme caution must be exercised in interpreting CBED pattern symmetries when ambiguities are expected. Plan-view CBED is also prone to (specimen related) artefacts of CBED because of interfacial strain and defects which may distort the pattern symmetry. However, by careful choice of probe size, location and by repeated experiments in search of high symmetry patterns, it is possible to remove certain ambiguities. Higher symmetry patterns should be

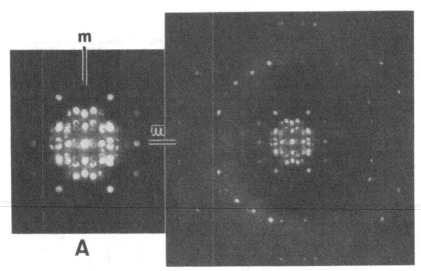

Fig. 2 : CBED patterns from plan-view (111)N//(100)Z bicrystal.
(A) ZOLZ, exhibiting 2mm projection symmetry and,(B) WP showing m symmetry

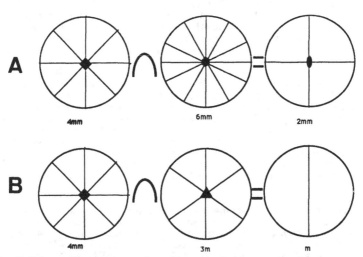

Fig. 3: Schematic representation of intersection symmetry due to superposition. (A) 4mm PDS of [100] Z intersecting 6mm PDS of [111] N results in 2mm composite symmetry. (B) 4mm WP (3-D) symmetry of [100] Z intersecting 3m WP symmetry of [111] N resulting in survival of only single coincident mirror.

considered as representative over lower symmetry ones because lowering of symmetry may not be genuine and may be due to specimen artefacts. The CBED patterns presented in this paper appear to be minimally affected by specimen-probe artefacts and there is good confidence that symmetry exhibited by these patterns represents expected true symmetry of the NZ bicrystal.

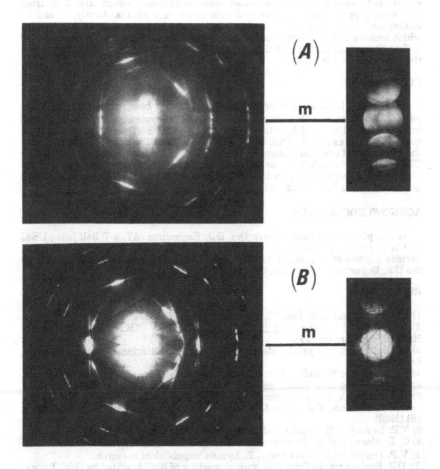

Fig. 4 : (A) Off-axis plan-view CBED pattern from NZ bicrystal exhibiting clear survival of single WP mirror. **(B)** Off-axis CBED pattern from NiO exhibiting m WP and BF symmetry, for comparison with (A).

Traditionally CTEM (α-fringe contrast [13]) and HRTEM techniques have been utilized for determining the presence and extent of RBT. HRTEM imaging has stringent requirements which include suitable specimen thickness and orientation, to mention a few. Furthermore, HRTEM can indicate only the projected component(s) of the RBT. CTEM technique involving α-fringe technique has been successfully utilized for coincidence related grain boundaries [13]. Its utility as a powerful experimental technique for determining RBT has been demonstrated for grain boundaries. However, its applicability to interphase interfaces, where exact coincidence virtually never exists, can be limited. Plan-view CBED, in retrospect, does have certain advantages. It is particularly useful for analyzing semiconductor bi- or multilayers which usually are readily available in plan-view. Plan-view offers larger interface area for imaging and diffraction than the traditional cross-sectional views. Therefore statistically significant results can be obtained from specimens which are free from specimen preparation artefacts and thin foil strain effects. Lastly, it can be anticipated that with better understanding of electron-specimen interaction which include HOLZ effects, quantification of intensities of CBED reflections and diffuse scattering; CBED techniques may offer much more than just symmetry of the bicrystals and interfaces.

CONCLUSIONS

(111) NiO // (100) ZrO_2 interfaces in NZ DSE have been investigated by plan-view CBED technique. Simple arguments based on symmetry superposition and intersection are utilized in interpreting CBED pattern symmetries. It has been shown that the expected ZOLZ (PDS) and WP (and BF) pattern symmetries from the NZ bicrystal conform to zero in-plane RBT. It can be concluded that plan-view CBED is a viable technique for the determination of bicrystal symmetry and in-plane RBT. However, just as in single crystal CBED studies, aspects of CBED artefacts need to be investigated more thoroughly.

ACKNOWLEDGEMENTS

It is a pleasure to acknowledge Drs. D.J. Eaglesham (AT & T Bell labs., USA) and R.C. Pond (University of Liverpool, U.K.) for stimulating discussions on various aspects of bicrystal symmetry and CBED. This research is supported by the U.S. Department of Energy, Grant No. DE-FG02-84ER45150.

REFERENCES

1) A.P. Sutton and R.W. Balluffi, Acta Met., 35, 2177 (1987).
2) R.C. Pond, in Dislocation in Solids, 8, edited by F.R.N. Nabarro (North-Holland, Amsterdam, 1989).
3) W. Bollman,in Crystal Defects and Crystalline Interfaces (Springer-Verlag, Berlin 1970)
4) R.C. Pond and W. Bollman, Phil. Trans. Royal Soc. London, 292, 449 (1979) .
5) R.C. Pond and D.S. Vlachavas, Proc. Royal Soc. London, A386, 95 (1983).
6) F.W. Schapink, S.K.E. Forghany and B.F. Buxton, Acta Cryst., A39, 805 (1983).
7) D.J. Eaglesham, C.J. Kiely, D. Cherns and M. Missous, Phil. Mag., A60, No.2, 161 (1989).
8) V.P. Dravid, C.E. Lyman and M.R. Notis, unpublished research.
9) G. Dhalenne and A. Revcolevschi, J. Cryst. Growth, 69, 616 (1984).
11) V.P. Dravid, M.R. Notis and C.E. Lyman unpublished research.
12) D.J. Eaglesham, in Proc. 47th annual meeting of EMSA, edited by G.W. Bailey (San Francisco Press, CA) 480 (1989).
13) R.C. Pond and V. Vitek, Proc. R. Soc., A357, 453 (1977)

DOUBLE CRYSTAL X-RAY DIFFRACTION MEASUREMENT OF A TRICLINICLY DISTORTED AND TILTED $Al_xGa_{1-x}As$ UNIT CELL PRODUCED BY GROWTH ON OFFCUT GaAs SUBSTRATES

A. LEIBERICH AND J. LEVKOFF
AT&T Bell Laboratories, Engineering Research Center, Princeton, N.J. 08540

ABSTRACT

Corrections are required for double crystal X-ray diffraction characterization of epitaxial $Al_xGa_{1-x}As$ layers grown on offcut GaAs (100) substrates. Double crystal X-ray diffraction measurements show that the cubic film unit cell defined by Vegard's law is triclinicly distorted and tilted with respect to the substrate unit cell. The distortion and tilt angles oppose each other defining a crystal geometry where the substrate and film <100> axes remain approximately coplanar with the surface normal. This film/substrate crystal geometry leads to formulation of a model describing heteroepitaxy on offcut (100) substrates. When film atoms are bonded to an offcut substrate, the already tetragonaly distorted film unit cell is subjected to additional cell distortions. The magnitude of this additional strain depends on where the film atoms are positioned on a substrate terrace. The first few layers of film atoms establish strain grades across individual substrate terraces. Constrained by the geometry of this interface region and driven by strain relaxation in the net growth direction, subsequent heteroepitaxy forms the measured film/substrate crystal geometry.

INTRODUCTION

Within given temperature and growth regimes, MOCVD homoepitaxy on GaAs (100) substrates, offcut towards the [110][1] [2] and [100] axes[3] results in formation of ordered terrace steps, which on the average, are aligned with the (100) projection of the surface normal. In device technology, homoepitaxy in most cases precedes growth of heteroepitaxial films. Auvray et al.[2] have suggested that $Al_xGa_{1-x}As$ growth onto a conceptually ordered stepped surface, produces a correlation between the substrate surface morphology and the film/substrate crystal geometry. The present study is aimed at investigating the crystal geometry of MOCVD $Al_xGa_{1-x}As$ grown on GaAs (100) substrates, offcut such that the (100) projection of the surface normal falls between the [110] and [100] axes. Conceptually, this general offcut angle alignment produces a substrate surface constructed of terraces terminated by steps and kinks. Previous work has shown that the epitaxial film unit cell grown on such offcut substrates is subject to shear strain[4].

MOCVD SYNTHESIS

A schematic of the offcut GaAs (100) substrate wafers is shown in Fig. 1. The definition of the crystallographic axes conforms with the Semiconductor Equipment and Materials Institute's (SEMI) convention[5]. The wafer's surface normal $\hat{n}_s$ is inclined by $\Theta_o = 2.5° \pm 0.1°$ from the <100> axis. The projection of the surface normal onto the (100) plane was oriented with an angle of $\zeta_o = 65° \pm 3°$ spanning counter clockwise from the <01$\bar{1}$> axis. Film synthesis entailed MOCVD growth of a ≈0.2 μm thick GaAs buffer layer with subsequent deposition of $Al_xGa_{1-x}As$ films, ranging in thickness from 0.5 to 0.8 μm. The reactants were trimethylgallium, trimethylaluminum and arsine, mixed with hydrogen carrier gas. The growth temperature and pressure were set at 680 °C and 60 torr, respectively. As a final deposition step, all samples received a ≈100Å GaAs capping layer. The sample layer thicknesses and concentrations were measured with Rutherford Backscattering Spectrometry (RBS) and are listed in Table I. More details on the MOCVD synthesis are given in a previous publication[6].

TABLE I.   Layer concentrations and thicknesses measured by RBS.

Sample	$Al_xGa_{1-x}As$ concentration $x(RBS)$ (± 0.04)	$Al_xGa_{1-x}As$ thickness [Å]	GaAs cap thickness [Å]
1	1.00	5,150	80
2	0.71	6,000	90
3	0.69	7,600	140

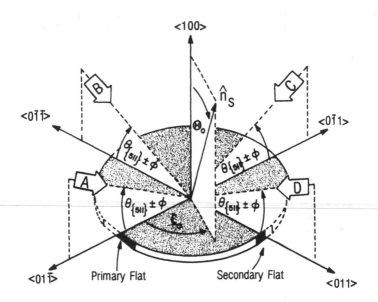

FIG. 1. A schematic of the offcut GaAs (100) substrate wafer. The crystallographic axes conform with the SEMI convention. The surface normal $\hat{n}_s$ is inclined by $\Theta_o = 2.5°$ from the $<100>$ axis. The (100) projection of the surface normal spans an angle $\zeta_o = 65°$ measured counter clockwise from the wafer's major flat, marked by the $<01\bar{1}>$ axis. The scanning directions X = A, B, C and D produce four pairs of line splittings (ω_b^X, ω_a^X). The subscripts "b" and "a" imply incident angles $\theta_{\{511\}} - \phi$ and $\theta_{\{511\}} + \phi$, respectively.

X-RAY DIFFRACTION WITH CONVENTIONAL ANALYSIS FORMALISM

Double crystal X-ray diffraction (DCXRD)[7], allows the measurement of crystal lattice strains, provided the incident X-ray wavelength and a crystal lattice constant for a volume fraction within the samples' near surface is known. Diffraction spectra, referenced to the GaAs lattice constant $a_{GaAs} = 0.56536$ nm[8], were measured with a two-crystal diffractometer equipped with a Cu K_α X-ray source. Depending on the choice of asymmetric or symmetric diffraction geometry, the first crystal was aligned for {511}a[8] or (400) reflections, respectively. Collimating slits were centered about the $Cu\ K_{\alpha 1}$ line, thus defining an incident X-ray wavelength of 0.1540562 nm[9]. The present analysis assumes the parametrization of the $Al_xGa_{1-x}As\ /\ GaAs$ crystal geometry as defined by Estop et al.[10], describing synthesis of strained epitaxial $Al_xGa_{1-x}As$ in terms of elastic theory and Vegard's law[11]. For a tetragonally distorted $Al_xGa_{1-x}As$ unit cell, the perpendicular strain measured by DCXRD defines the aluminum concentration x such that[10]:

$$(\Delta a/a)_\perp = C_x \times x \quad with \ C_x = 2.821 \times 10^{-3}. \tag{1}$$

The constant C_x agrees with DCXRD and RBS characterization of $Al_xGa_{1-x}As$ films grown for the present work on on-axis cut GaAs (100) substrates using the above described MOCVD synthesis parameters. The value of C_x given in Eq. (1) further agrees with characterization by DCXRD and Nuclear Reaction Analysis[12].

Using conventional formalism which assumes a tetragonal distortion of the $Al_xGa_{1-x}As$ unit cell, DCXRD yields the measurement of the perpendicular and parallel lattice strains. From line splittings ω_b and ω_a, resulting from two separate asymmetric diffraction geometries, labeled {511}b and {511}a, the perpendicular and parallel lattice strains can be calculated such that[8]:

$$(\Delta a/a)_\perp = \frac{a_\perp - a_{GaAs}}{a_{GaAs}} = \left[\frac{\sin(\theta_{\{511\}})}{\sin\left[\theta_{\{511\}} + (\omega_b + \omega_a)/2\right]} \times \frac{\cos(\phi)}{\cos\left[\phi + (\omega_b - \omega_a)/2\right]} \right] - 1 \qquad (2)$$

and

$$(\Delta a/a)_{//} = \frac{a_{//} - a_{GaAs}}{a_{GaAs}} = \left[\frac{\sin(\theta_{\{511\}})}{\sin\left[\theta_{\{511\}} + (\omega_b + \omega_a)/2\right]} \times \frac{\sin(\phi)}{\sin\left[\phi + (\omega_b - \omega_a)/2\right]} \right] - 1 . \qquad (3)$$

The perpendicular and parallel lattice constants of the tetragonal unit cell are given by $a_\perp$ and $a_{//}$. The angles $\theta_{\{511\}} = 45.0688°$ and $\phi = 15.7932°$ are the Bragg angle and the angle spanning the (100) and {511} diffraction planes of the substrate crystal[13], respectively.

X-RAY DIFFRACTION FROM Al$_x$Ga$_{1-x}$As GROWN ON OFFCUT GaAs (100)

A (100) cut GaAs crystal, rotated about the two [110] axes which are coplanar with the (100) plane, yields eight accessible {511} asymmetric reflection geometries defined by four {511} planes. The line splittings associated with these reflections were used to measure the symmetry of the film/substrate crystal geometry. Figure 1 defines the required scanning directions of the incident X-ray beam, marked by arrows labeled A, B, C and D. Taking scanning direction A for instance, allows access to two {511} planes. Positioning the incident X-ray beam at angles $\theta_{\{511\}} - \phi$ and $\theta_{\{511\}} + \phi$ with respect to the (100) plane, produces line splittings labeled here by ω_b^A and ω_a^A, respectively. The same definition applies to scanning directions B, C and D producing pairs of line splittings (ω_b^B, ω_a^B), (ω_b^C, ω_a^C) and (ω_b^D, ω_a^D). To establish the scanning directions, the sample wafer is rotated on the goniometer by a fixed angle ζ, defined such that the scanning directions A, B, C and D imply $\zeta = 0°$, $90°$, $180°$ and $270°$, respectively. The Miller indices of the diffracting crystal planes associated with each scanning direction are given in Table II. The pairs of {511} diffraction line splittings (ω_b^X, ω_a^X) with X standing for A, B, C or D are listed in Table II. The differences in splittings $\omega_b^X - \omega_a^X$ for diffraction conditions A versus C and B versus D, conform to a symmetry that cannot be explained by mere tilting of a tetragonal Al$_x$Ga$_{1-x}$As unit cell, implying a non-tetragonal unit cell distortion. The detailed analysis given in a separate publication[14] interprets the data in Table II to reflect a triclinic distortion and separate unit cell tilt of the epitaxial film unit cell.

TABLE II.   Line splittings from {511} reflections.

Sample	(51̄1)b ω_b^A [arcsec] (±0.5)	(5̄1̄1)a ω_a^A [arcsec] (±0.5)	(51̄1)b ω_b^B [arcsec] (±0.5)	(511)a ω_a^B [arcsec] (±0.5)	(51̄1)b ω_b^C [arcsec] (±0.5)	(51̄1)a ω_a^C [arcsec] (±0.5)	(511)b ω_b^D [arcsec] (±0.5)	(5̄1̄1)a ω_a^D [arcsec] (±0.5)
1	-672.7	-365.6	-696.8	-406.2	-688.3	-394.7	-666.7	-355.3
2	-495.6	-267.5	-510.3	-297.1	-506.3	-290.3	-492.2	-261.1
3	-447.3	-241.8	-458.9	-268.4	-455.3	-261.2	-440.6	-235.0

Figure 2 summarizes schematicly the steps leading to interpretation of the X-ray diffraction line splittings listed in Table II. Figures 2(a) and 2(b) define the conventional tetragonal distortion. A cubic film unit cell with lattice constants $a_x = a_{GaAs} + (a_{GaAs} - a_{AlAs})x$ is fitted to the cubic substrate unit cell lattice constant $a_0 = a_{GaAs}$. The construction in Fig. 2(b) implies commensurate growth such that a_0 equals $a_{//}$, with $a_\perp$ given by Eq. (1). For the present analysis the conventional tetragonal distortion is assumed to form an intermediate geometrical construction leading to the final film/substrate unit cell geometry with further unit cell distortion and tilt.

To calculate the film composition, the line averages $\overline{\omega}_b$ and $\overline{\omega}_a$ given in Table III, are inserted into Eqs. (2) and (3), accommodating the film lattice dilatation. The film concentration x, given by Eq.(1) is listed as x(DCXRD) in Table III, along with the perpendicular and parallel strains, $(\Delta a/a)_\perp$ and $(\Delta a/a)_{//}$, respectively. The conceptual, intermediate tetragonal distortion defined by $\overline{\omega}_b$ and $\overline{\omega}_a$ implies "commensurate" growth as shown by the measured parallel strains given in Table III. The quantitative measurements of the epitaxial film concentration by RBS, labeled x(RBS) in Table I, are in reasonable agreement with the concentrations x(DCXRD) given in Table III.

TABLE III. Lattice strains and concentrations measured by DCXRD.

Sample	$\overline{\omega}_b$ [arcsec] (±0.5)	$\overline{\omega}_a$ [arcsec] (±0.5)	$(\Delta a/a)_\perp$ $[10^{-4}]$ (±0.01)	$(\Delta a/a)_{//}$ $[10^{-4}]$ (±0.04)	x(DCXRD) (±4.4%)
1	-681.1	-380.5	27.84	0.01	0.986
2	-501.1	-279.0	20.44	-0.11	0.724
3	-450.5	-251.6	18.39	-0.03	0.651

Figures 2(c) and 2(d) sketch the remaining unit cell distortions. Figure 2(c) depicts a triclinic distortion of the tetragonal unit cell by an angle $\Delta\delta$. The angle $\Delta\delta$, with components $\Delta\alpha$ and $\Delta\beta$, defines the triclinic unit cell distortion, with the triclinic cell being defined by angles α, β and γ[13]. Throughout the analysis the angle γ is assumed to remain fixed at 90°. Figure 2(d) completes the geometric construction and produces the final film/substrate unit cell geometry with an external unit cell tilt $\Delta\tau$, defined in three dimensions. The components of $\Delta\tau$, $\Delta\tau_\alpha$ and $\Delta\tau_\beta$, are colinear with, but are defined to point in opposing directions to $\Delta\alpha$ and $\Delta\beta$, respectively.

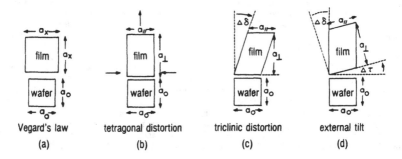

FIG. 2. Schematic of the observed film unit cell distortion. Figure 2(a) and 2(b) display a conventional tetragonal unit cell distortion for the case of commensurate growth. Figure 2(c) and 2(d) depict the triclinic distortion and the separate unit cell tilt, respectively. The step in Fig. 2(d) rotates the film unit cell with respect to the substrate crystal.

CALCULATION OF THE FILM UNIT CELL DISTORTION AND TILT ANGLES

The components of the tilt angle $\Delta\tau$ and the triclinic distortion angle $\Delta\delta$ are calculated using the data given in Tables II and III[14]. The results, superimposed on a (100) projection of the substrate wafer, are shown in Fig. 3. The angles ζ_δ and ζ_τ, defined analogous to ζ_o, result from polar transform of the angular components $\Delta\alpha$, $\Delta\beta$, $\Delta\tau_\alpha$ and $\Delta\tau_\beta$. The upper right quadrant of Fig. 3 defined by the axes $\Delta\alpha$ and $\Delta\beta$, displays the (100) projection of the offcut angle Θ_o. The calculated triclinic distortion angles are shown to align within $\approx 10°$ of the projection of the offcut angle, such that $\zeta_\delta \approx \zeta_o$. The calculated tilt angles, given by components, $\Delta\tau_\alpha$ and $\Delta\tau_\beta$, define the lower left quadrant. The projections of the tilt angles, pointing in opposite direction to $\Delta\alpha$ and $\Delta\beta$, respectively, are approximately colinear with the projections of the distortion angles and the offcut angle, such that $\zeta_\tau \approx \zeta_o + \pi$. The numerical values of the parameters plotted in Fig. 3 are listed in Table IV.

TABLE IV. Crystallographic parameters calculated from {511} diffraction geometry.

Sample	$\Delta\alpha$ [arcsec] (±1)	$\Delta\beta$ [arcsec] (±1)	$\Delta\delta$ [arcsec] (±1)	ζ_δ [degrees] (±2)	$\Delta\tau_\alpha$ [arcsec] (±1)	$\Delta\tau_\beta$ [arcsec] (±1)	$\Delta\tau$ [arcsec] (±1)	ζ_τ [degrees] (±2)
1	27.4	5.9	28.0	57.2	23.9	6.8	24.9	241
2	24.5	4.7	24.9	55.9	17.0	3.9	17.4	238
3	22.5	3.2	22.7	53.1	15.3	4.5	16.0	241

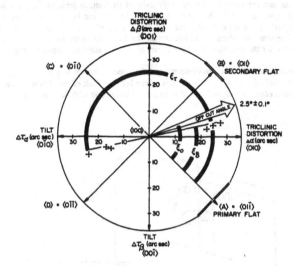

FIG. 3. Measured and calculated crystallographic parameters superimposed on a (100) projection of the substrate wafer. Shown is the alignment of the (100) projection of the surface normal labeled "OFF CUT ANGLE". Plotted are the calculated components of the triclinic distortion angle ($\Delta\alpha$, $\Delta\beta$) and the external unit cell tilt components ($\Delta\tau_\alpha$, $\Delta\tau_\beta$). The data points represent the values given in Table IV.

The data interpretation implies that a model given by Auvray et al.[2] for offcut angles inclined towards a [110] axis, can be extended to describe the film/substrate crystal geometry of wafers cut with general alignment of the offcut angle. For general offcut angle alignment, the substrate and film <100> axes remain coplanar with the surface normal, implying that[6]

$$\Delta\tau = \tan^{-1}\left[(\Delta a/a)_\perp \, (a_o \, / \, l_t)\right] = \tan^{-1}\left[(\Delta a/a)_\perp \tan(\Theta_o)\right].\qquad (4)$$

Here, the parameter l_t is defined as the average length of a substrate terrace, measured parallel to the (100) projection of the surface normal $\hat{n}_s$. For convenience the substrate terrace step height is chosen to equal the substrate unit cell lattice constant. The described geometry does accommodate substrate step bunching[14], allowing for arbitrary choice of the geometric building blocks. Plausible arguments given with regard to a model based on atomic size terrace steps were presented previously[6].

The unit cell tilt angles calculated from Eq. (4) with the perpendicular strains given in Table III, are labeled $\Delta\tau(Eq. (4))$ and listed in Table V. Also given in Table V are the values of $\Delta\tau$ which were calculated from analysis of the {511} diffraction line positions and which were listed in Table IV. For the case of the AlAs / GaAs sample, Table V gives an experimental value of the unit cell tilt, labeled $\Delta\tau(400)$, obtained through measurements and analysis with symmetric (400) X-ray reflections[14]. The measured and calculated unit cell tilts shown in Table V are in good agreement.

TABLE V.   Unit cell tilts evaluated by various methods.

Sample	$\Delta\tau$ [arcsec] (± 1.0)	$\Delta\tau(Eq. (4))$ [arcsec]	$\Delta\tau(400)$ [arcsec] (± 1.0)
1	24.9	25.1	25.1
2	17.4	18.4	-
3	16.0	16.6	-

106

DATA INTERPRETATION AND GROWTH MODEL

For the purpose of illustration, an ordered substrate surface, with some of its features arbitrarily chosen, is sketched in Fig. 4. The GaAs (100) substrate used for the present experiments was offcut in a general direction, such that the (100) projection of the surface normal is not aligned with one of the major axes. Conceptually, an ordered substrate surface consists of regular step edges and kinks. Again, for convenience the film and substrate unit cells are chosen as building blocks, for the described geometry is not affected by substrate step bunching.

As defined by Eq. (4), the magnitude of the offcut angle determines the average terrace length l_t, measured parallel to the (100) projection of the surface normal $\hat{n}_s$. The relative phase between kink edges of two adjacent terraces was chosen arbitrarily; the kink edges are aligned such that they remain coplanar with the dashed rectan-

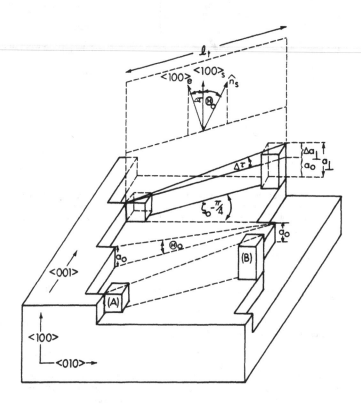

FIG. 4. Schematic of an ordered substrate surface displaying the initial stages of heteroepitaxy. The surface normal $\hat{n}_s$ is inclined by the offcut angle Θ_o from the substrate <100> axis, labeled $<100>_s$. The (100) projection of the surface normal points between the <010> and <011> substrate axes. The <100> axis of the epitaxial film, labeled $<100>_e$, is tilted by an angle $\Delta\tau$ from $<100>_s$. After film unit cell distortion and tilt, the film <100> axis remains coplanar with the substrate <100> axis and the surface normal. Locations (A) and (B) on the substrate surface define step and edge sites, respectively. The phase between the kink sites is chosen arbitrarily. The described geometry allows for substrate step bunching, thus the choice of unit cells as building blocks is justified.

gle containing the surface normal $\hat{n}_s$ and the <100> axes of the epitaxial film and substrate, labeled $<100>_e$ and $<100>_s$, respectively.

The present X-ray diffraction analysis suggests a model for the initial stages of heteroepitaxy on offcut substrates. As shown in Fig. 4, an epitaxial film unit cell nucleated at a step site on the substrate terrace at location (A), conceptually experiences triaxial boundary conditions during elastic deformation, implying a non-tetragonal distortion. Conversely, at the opposite side of the same terrace on an edge site at location (B) that same epitaxial film unit cell would be constrained by bi-axial boundary conditions, resulting in tetragonal distortion. Thus a continuous chain of epitaxial film unit cells spanning locations (A) to (B), forms a strain grade giving rise to a tilt $\Delta\tau$ spanning the epitaxial and substrate (100) planes. Thus initial heteroepitaxial growth forms an interface region, where the number of film/substrate atom bonds and thus the number of growth constraints imposed onto the film atoms, changes as function of position on a terrace. The strain grade formed in direction of the (100) projection of the surface normal produces maximum lateral strain relaxation per unit length. Subsequent heteroepitaxial growth onto this strained interface produces the measured film/substrate crystal geometry. The described geometry implies that the rows of <100> atoms project from the substrate through the epitaxial layer without changing direction at the film/substrate interface. On the other hand, the rows of atoms aligned in direction of the substrate <010> and <001> axes do change direction upon crossing the interface.

In summary, $Al_xGa_{1-x}As$ heteroepitaxy on offcut GaAs (100) substrates, seems to be governed by the following growth dynamics.

(1) For specific synthesis parameters, homoepitaxy forms an ordered substrate surface, defined by the magnitude of the offcut angle and the alignment of the (100) projection of the surface normal. An average terrace length is defined to span colinear to the (100) projection of the substrate surface normal.
(2) Initial heteroepitaxy establishes an interface region, forming strain grades such that the *step* and *edge* sites along individual terraces are characterized by $(\Delta a/a)_{\perp} = 0$ and $(\Delta a/a)_{\perp} = C_x \times x$, respectively.
(3) Maximum strain relaxation per unit length is obtained colinear to the direction of the average terrace length. Thus the formation of strain grades produced colinear (100) projection of the surface normal, results in a tilt angle $\Delta\tau \approx (\Delta a/a)_{\perp} \tan(\Theta_o)$ spanning the film and substrate (100) planes.
(4) As in the case of conventional tetragonal unit cell distortion, most of the unit cell strain is produced by increased separation between the epitaxial film atoms in the net growth direction. This effect explains the compensation of the triclinic distortion and external unit cell tilt angles, implying $(\Delta\delta - \Delta\tau) \approx 0$.

CONCLUSIONS

A procedure has been developed for the measurement of important crystallographic parameters with DCXRD. The presented formalism produces significant corrections to conventional formalism which postulates a tetragonal distortion of the $Al_xGa_{1-x}As$ unit cell when grown on $GaAs$ substrates. The formalism given in the present study, allows for correct measurement of film concentrations, while additionally providing detailed information on the film/substrate crystal geometry. Although the present study deals with $Al_xGa_{1-x}As$ grown on offcut $GaAs$ substrates, the discussed X-ray diffraction formalism is applicable for any cubic semiconductor system. Multicrystal X-ray diffraction might prove to be the only tool with which the above materials parameters can be extracted in a precise, cost-effective and timely manner.

When $Al_xGa_{1-x}As$ is grown on offcut GaAs (100) substrates, the cubic film unit cell defined by Vegard's law is tetragonally, then triclinicly distorted and finally tilted with respect to the substrate unit cell. This unit cell distortion/tilt action remains in plane with the substrate <100> axis and the surface normal. The film/substrate unit cell geometry is correlated to an ordered substrate surface morphology, produced by homoepitaxy preceding film deposition. An average terrace length is defined by both the magnitude of the offcut angle and the (100) projection of the surface normal. The measured film/substrate unit cell distortion and tilt is quantitatively accounted for by a model describing heteroepitaxial growth onto an interface region, formed by initial film growth and producing strain grades spanning across the substrate terraces. The measured film/substrate crystal geometry allows "defect" free strained heteroepitaxial growth on offcut substrates.

Displaying a triclinic distortion with compensating external unit cell tilt, the strained heteroepitaxial film's formal commensuracy is not readily defined. For a discommensurate film the three major orthogonal crystallographic axes remain parallel for the film and substrate unit cells, while the continuity along individual strings of atoms may be interrupted by dislocations spanning across the film/substrate interface. For a heteroepitaxial film displaying a triclinic distortion with compensating external unit cell tilt, only strings of film and substrate atoms aligned with the <100> axis remain both parallel and continuous across the film/substrate's interface. The strings of film and substrate atoms aligned with the <010> and <001> axes remain continuous, but not parallel across the film/substrate's interface. Such strained heteroepitaxial film is characterized by a coherent unit cell geometry, but is neither formally commensurate nor discommensurate.

ACKNOWLEDGEMENTS

The authors acknowledge R.F. Roberts and M.J. Yuen for their support and guidance. We are grateful to J.R. Moore for his assistance during MOCVD synthesis. We thank W.L. Brown and K.S. Short for advice with the RBS analysis.

REFERENCES

[1] T. Fukui and H. Saito, J. Vac. Sci. Technol. B6(4), 1373 (1988).

[2] P. Auvray, M. Baudet and A. Regreny, *Workbook of the Fifth Int'l Conf. on MBE* (1988), p. 403; J. Cryst. Growth, 95, 288 (1989).

[3] K. Wada, A. Kozen, H. Fushimi and N. Inoue, J. Cryst. Growth 93, 935 (1988).

[4] W.J. Bartels and W. Nijman, J. Cryst. Growth 44, 518 (1978).

[5] *Book of SEMI Standards*, Vol. 3 (Semiconductor Equipment and Materials Institute, Mountain View, CA, 1986).

[6] J. Levkoff and A. Leiberich, J. Appl. Phys., submitted for publication.

[7] A.H. Compton and S.K. Allison, *X-rays in Theory and Experiment* (New York, D. Van Nostrand, 1935).

[8] A.T. Macrander, G.P. Schwartz and G.J. Gualtieri, J. of the Electrochem. Soc. 134 (9), 578 C (1987).

[9] J.A. Ibers and W.C. Hamilton, eds., *Int'l Tables for X-ray Crystallography*, Vol. IV (Kynoch, Birmingham, England, 1974).

[10] E. Estop, A. Izrael and M. Sauvage, Acta. Cryst. A32, 627 (1976).

[11] G.A. Rozgonyi, P.M. Petroff and M.B. Panish, J. Cryst. Growth 27, 106 (1974).

[12] F. Xiong, T.A. Tombrello, H.Z. Chen, H. Morkoc and A. Yarin, J. Vac. Sci. Technol. B6, 2 (1988).

[13] B.D. Cullity, *Elements of X-ray Diffraction*, 2nd ed. (Addison-Wesley, Reading, MA, 1978).

[14] A. Leiberich and J. Levkoff, J. Vac. Sci. Technol., submitted for publication.

INTERFACE STRAIN AND THE VALENCE BAND OFFSET AT THE LATTICE MATCHED $In_{0.53}Ga_{0.47}As/InP$ (001) INTERFACE

MARK S. HYBERTSEN
AT&T Bell Laboratories, 600 Mountain Avenue, Murray Hill, NJ 07974

ABSTRACT

Total energy minimization calculations show that the interface bonds are strained in nominally lattice matched $In_{0.53}Ga_{0.47}As/InP$ (001) heterostructures, in agreement with recent X-ray measurements. Anion intermixing relieves the interface strain. The calculated valence band offset varies with the interface bond lengths so the minimum energy structure must be used for a given composition. Then the calculated offset is independent of composition and is in good agreement with experiment. A simple model exhibits the qualitative features revealed by these calculations.

I. INTRODUCTION

Measurement, prediction and control of semiconductor heterojunction band offsets continues to be a challenging problem [1]. The technologically important $In_{0.53}Ga_{0.47}As/InP$ (001) system offers a heretofore unexplored parameter which influences the band offsets: the interface bond lengths or interface strains. Figure 1a illustrates a sequence of planes for the $In_{0.53}Ga_{0.47}As/InP$ (001) heterostructure. Two chemically distinct interfaces result requiring InAs bonds on the left and MP bonds on the right ($M = In_{0.53}Ga_{0.47}$). Consideration of bulk bond lengths [2] suggests that $d_{InAs} > d_{InP} = d_{MAs} > d_{MP}$ giving a positive interface strain on the left side of the barrier (InP) with a compensating negative strain on the right side. The interfaces are structurally as well as chemically different. Anion intermixing at the interfaces may alter the local strain: e.g. $As_{0.5}P_{0.5}(=A)$ at each interface (Fig. 1b) reduces the strain at the interfaces to essentially zero. Since $d_{AIn} > d_{AM}$, there is an internal strain coordinate ϵ'. Thus local strains are to be expected at the interfaces with the magnitude "tunable" from zero to several percent depending on interface composition. X-ray diffraction scans of samples grown with the intention of exhibiting the asymmetric interfaces illustrated in Fig. 1a gave the first experimental evidence that significant strains exist near the interfaces [3].

The calculations reported here show that this ideal heterojunction (Fig. 1a) has a large (6%) interface strain. Chemical intermixing (Fig. 1b) is shown to relieve the interface strain. The calculated valence band offset at the abrupt interface varies by 150 meV for $\epsilon=0$ - 8%. For the total energy minimum at $\epsilon=0.06$, the present first principles calculation yields $\Delta E_v = 410 \pm 50$ meV, in very good agreement with experiment [4-6]. Furthermore, if the local strain is relieved by chemical intermixing, the calculated valence band offset is *unchanged*, for the minimum energy structure which is now $\epsilon'=0.03$. The qualitative behavior of the band offset is captured by a simple model.

II. THEORETICAL APPROACH

The local density functional approach (LDA) [7] with the pseudopotential technique [8] is used to calculate the selfconsistent ground state charge density and total energy for a series of MAs/InP(001) superlattices with different interface configurations. The alloy (M) is treated with the virtual crystal approximation. A planewave basis set is used with a 9 Ry energy cutoff for the superlattice calculations. For a given interface configuration, the valence band offset is calculated from the change in potential across the interface, $\Delta < V_{loc} >^{int}$:

$$\Delta E_v = (E_v - <V_{loc}>)_{MAs}^{bulk} - (E_v - <V_{loc}>)_{InP}^{bulk} - \Delta <V_{loc}>^{int}. \quad (1)$$

The position of the bulk valence band edge E_v is required relative to the average local potential $<V_{loc}>$ in the bulk separately for each constituent material, taken from highly converged (24 Ry cutoff) calculations [9]. There is a further correction due to the fact that the LDA valence band edge is not rigorously the electron removal energy [10]:

$$\Delta E_v = \Delta E_v^{LDA} + \delta_v, \quad (2)$$

where δ_v is a self energy correction which derives from proper treatment of the many-body interaction effects in the constituent materials. It is estimated that $\delta_v \approx 110 \pm 40$ meV for the MAs/InP system [11].

III. CALCULATED INTERFACE PROPERTIES

The interface calculations are done using the theoretical lattice constant of InP (d_0, within 0.5% of the calculated lattice constant for $In_{0.53}Ga_{0.47}As$) and a $(MAs)_3(InP)_3$ superlattice. The total energy as a function of interface strain parameter ϵ is shown in Fig. 2a. The minimum energy configuration corresponds to $\epsilon=0.06$.

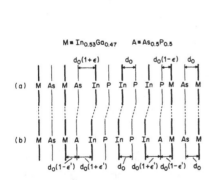

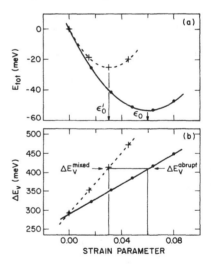

Figure 1. (a) A sequence of planes in the $In_{0.53}Ga_{0.47}As/InP$ (001) heterojunction system illustrating local interface strains. (b) Chemical intermixing which relieves the net local strain in the interface molecular layer.

Figure 2. (a) Superlattice total energy as a function of the local interface strain parameter. (b) Calculated valence band offset as a function of local strain parameter. Solid line: abrupt case (Fig. 1a). Dashed line: chemically intermixed case (Fig. 1b).

Consideration of bulk bond lengths [2] would suggest $\epsilon \approx 0.03$. However, InAs (for example) is biaxially compressed on InP, so the perpendicular lattice parameter should be altered according to elastic compliance coefficients. This would suggest $\epsilon \approx 0.065$ which corresponds to strong preservation of local bulk bond lengths. As the interface strain parameter is varied, the interface dipole decreases markedly. The variation of the dipole with local bond length translates to a significant increase in the valence band offset with ϵ (Fig. 2b). The minimum energy value of $\epsilon = 0.06$ yields $\Delta E_v = 410$ meV, 120 meV larger than would be found under the assumption of unstrained interface bonds.

Variations in interface composition might relieve the interface strain and hence alter the valence band offset. The particular change in interface stoichiometry illustrated in Fig. 1b almost completely relieves the interface strain: the molecular layer spacing is uniform ($2d_0$). The anion layer A is here modeled with the virtual crystal approximation. The total energy is a minimum for $\epsilon' = 0.03$ (Fig. 2a). The offset depends strongly on ϵ' (Fig. 2b) with twice the slope found in the abrupt case. However, it is the same as the abrupt case for $\epsilon = \epsilon' = 0$ and, importantly, for the locally minimum energy structures ($\epsilon = 0.06$, $\epsilon' = 0.03$).

The validity of the simple one-parameter structural models illustrated in Fig. 1 have been carefully examined. Calculations in which all the internal coordinates are allowed to vary in order to further lower the total energy show insignificant changes.

IV. COMPARISON TO EXPERIMENT

The interface structure for the $In_{0.53}Ga_{0.47}As/InP$ (001) system has been accurately probed with X-ray diffraction techniques [3]. The fit to the X-ray scan included interface strains modeled over a molecular layer. The data show clear evidence of interface strain which translates to $\epsilon = 0.038$ (InAs interface) and $\epsilon = -0.034$ (MP interface). This supports the simple model in Fig. 1a and is in reasonable agreement with the present microscopic calculations. The X-ray experiments are insensitive to interdiffusion at the scale illustrated in Fig. 1b. Furthermore, the beam sequence in the growth of the sample is consistent with having As/P mixture at the interface. This may alter the measured strain from the value calculated here. In fact, the present calculations show that the strain may have any value from zero to 0.06 depending on interface composition (chemical abruptness).

Precise measurements for the band offsets at the $In_{0.53}Ga_{0.47}As/InP$ (001) heterojunction based on transport [4,5] and internal photoemission [6] yield the valence band offset (T=0) in the range 340-400 meV. These are based on the measured ΔE_c, extrapolated as necessary to $T=0$, and combined with the measured $\Delta E_g(T \rightarrow 0) = 613$ meV. The experimental data is to be compared to the theoretical result of 410 ± 50 meV. If the interface strain had been neglected, the theory would have given 290 meV, considerably smaller than the experimental range. Recent optical measurements [12] on $In_{0.53}Ga_{0.47}As/InP$ (001) quantum wells were accurately accounted for using $\Delta E_v = 380$ meV.

V. SIMPLE MODEL FOR INTERFACE SCREENING

It has long been recognized, based on simple and general arguments, that an unreconstructed polar-nonpolar interface (e.g. Si/GaAs) gives rise to long range electric fields [13]. It has not been previously recognized that the *same* argument applies to polar-polar interfaces such as those studied here. The *difference* in charge transfer in the material on the left and on the right sides of the interface must be compensated by a finite screening charge at the interface in order to eliminate long range electric fields. These ideas may be extended to form a very simple model of the interface screening which captures many qualitative features of interface dipoles (and hence band offsets) observed in microscopic calculations.

The semiconductors are reduced to charge sheets alternating $+q$ (anion) and $-q$ (cation) electronic charge (Fig. 3). In total energy calculations, these charges are taken as the volume integral of the total charge between planes which stand at the midpoints between the (001) (for example) cation and anion planes. (This definition of q is arbitrary, but an obvious choice.) Then when crossing the interface from MAs to InP, there is a change in charge transfer Δq. Deep in the bulk to left and right, the charge on the planes must approach q and $q+\Delta q$ respectively, but near the interface there may be screening charges. These are restricted to the simplest possibility: a change of charge on the interface anion and cation planes only (Fig. 3). Now consider the case when there is no net screening charge. The boundary condition on the left is zero *average* electric field. Then, as shown in Fig. 3, a long range electric field results on the right from integrating Poisson's equation: $<E> = \Delta E/2 = 2\pi e\Delta q/A$ where A is the area of the unit cell in the (001) plane. The condition required for no average electric field on the right is that $\Delta q_{As} + \Delta q_{In} = -\Delta q/2$. In this case a well defined dipole results which is illustrated in Fig. 3 and may be straightforwardly evaluated to be (zero interface strain)

$$\Delta V_{dip} = -\frac{4\pi e^2 d_0}{A}\left[\Delta q_{As} + \frac{1}{4}\Delta q\right]. \tag{3}$$

In the total energy calculations described above, the screening charge is indeed nearly restricted to the interface planes and is precisely $-\Delta q/2$. Also, the dipole calculated from Eq. (3) agrees with the dipole found above within about a factor of two. Clearly the real screening charge is not restricted to planes and form factors enter the full calculation.

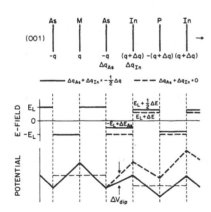

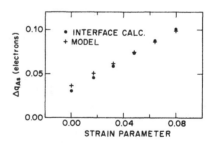

Figure 3. A sheet charge model for the (001) interface studied. The corresponding electric fields and potentials are shown (including the appropriate sign for electronic charge).

Figure 4. Comparison of the selfconsistent anion screening charge to the simple model described in the text.

The boundary conditions are not sufficient to specify fully the dipole in Eq. (3). Simple energetic considerations and reference to other bulk calculations complete the model. Assign a general "site energy" ϵ_{As} to the As (anion) plane and similarly for the M (cation) plane. Then, in the bulk, the charge q must minimize the total energy in this model:

$$E_{tot} = (\epsilon_M - \epsilon_{As})q_{MAs} + \pi e^2 d_0 q_{MAs}^2/A, \tag{4a}$$

$$q_{MAs} = (\epsilon_{As} - \epsilon_M)A/2\pi e^2 d_0, \tag{4b}$$

where the second term (4a) is the energy due to the electric field between the planes and q $= q_{MAs}$. Similar expressions follow for $q + \Delta q = q_{InP}$. The same type of energy expression may be written down for the interface energy. When minimized it yields

$$\Delta q_{As} = (q_{MAs} - q_{InAs})/2 \tag{5}$$

where q_{InAs} is the charge transfer for a bulk material composed of the interface anion and cation. The simplicity and power of this model is that the unknown site energy difference $\epsilon_{As} - \epsilon_{In}$, which must contain the information about the interface dipole, is transformed to a charge transfer which may by calculated for a bulk material. Furthermore, q_{InAs} contains all the information about the interface strain. This simple model for the interface screening charge may be tested directly. A series of calculations for bulk InAs with in plane lattice constant d_0 tetragonally distorted according to $d_\perp = (1 + \epsilon)d_0$ have been done. This then yields the interface screening charge Δq_{As} according to Eq. (5) for each interface strain. The model is compared to the full calculation in Fig. 4 which illustrates that this simple model is rather accurate for the magnitude of the screening charge.

This model can account for many qualitative features of band offsets:

(1) When extended to (110) interfaces, the orientation independence of the band offset observed for GaAs/AlAs interfaces is reproduced.

(2) The bond length dependence is explained because the charge transfer q_{InAs} decreases with increasing separation (ϵ).

(3) When applied to the intermixed interface discussed above, the dipole is found to be the same as the abrupt case for zero strain and to lowest order, the dependence on ϵ' is linear with twice the slope found for the abrupt case. This naturally leads to a band offset that is independent of measured interface strain (degree of intermixing) when combined with the local preservation of "bulk" bond lengths.

Thus, this model provides a simple picture for interface screening which faithfully reproduces the features of complex calculations.

ACKNOWLEDGEMENTS

I thank R.C. Fulton for programming assistance and X.-J. Zhu and Prof. S.G. Louie for making unpublished calculations available.

REFERENCES

[1] See, for example, *Heterojunction Band Discontinuities: Physics and Applications* edited by F. Capasso and G. Margaritondo (North-Holland, New York, 1987).

[2] One finds $d_{InAs}=1.514$ Å, $d_{InP}=1.467$ Å, and $d_{MP}=1.418$ Å resulting in $\epsilon=0.032$ based on data from *Zahlenwerte und Funktionen aus Naturwissenshaften und Technik*, Vol III of *Landolt-Bornstein*, edited by K.H. Hellwege (Springer, New York, 1982), Vol. 17a.

[3] J.M. Vandenberg, *et al.*, Appl. Phys. Lett. **53**, 1920 (1988).

[4] S.R. Forrest, *et al.*, Appl. Phys. Lett. **45**, 1199 (1984). S.R. Forrest in Ref. 1.

[5] R.E. Cavicchi, *et al.*, Appl. Phys. Lett. **54**, 739 (1989); D.V. Lang in Ref. 1.

[6] M.A. Haase, N. Pan, and G.E. Stillman, Appl. Phys. Lett. **54**, 1457 (1989).

[7] P. Hohenberg and W. Kohn, Phys. Rev. **136**, B864 (1964); W. Kohn and L.J. Sham, Phys. Rev. **140**, A1133 (1965). Correlation data: D.M. Ceperley and B.I. Alder, Phys. Rev. Lett. **45**, 566 (1980) as parameterized in J.P. Perdew and A. Zunger Phys. Rev. B **23**, 5048 (1981).

[8] G.P. Kerker, J. Phys. C **13**, L189 (1980). S.G. Louie, S. Froyen and M.L. Cohen, Phys. Rev. B **26**, 1738 (1982).

[9] The spin-orbit splitting, estimated from experiment (Ref. 2), contributes 84 meV to the band offset.

[10] S.B. Zhang, *et al.*, Solid State Commun. **66**, 585 (1988).

[11] The bare exchange contribution is calculated directly while the correlation contribution is estimated from detailed calculations [X.-J. Zhu and S.G. Louie, private communication] for GaAs, InAs, and InP.

[12] D. Gershoni, *et al.*, Phys. Rev. B **39**, 5531 (1989).

[13] W.A. Harrison, *et al.*, Phys. Rev. B **18**, 4402 (1978)

ATOMIC STRUCTURE OF CaSi$_2$/Si INTERFACES

CHRIS G. VAN DE WALLE
Philips Laboratories, North American Philips Corporation, Briarcliff Manor, NY 10510

ABSTRACT

The CaSi$_2$/Si interface is studied with state-of-the-art first-principles calculations. Various models for the interfacial structure are examined, in which the Ca atoms at the interface exhibit 5-, 6-, 7-, or 8-fold coordination. The structures with sevenfold coordination (as in bulk CaSi$_2$) have the lowest energy. However, the sixfold- and eightfold-coordinated structures are only ~0.1 eV higher in energy. Schottky barrier heights are briefly discussed.

INTRODUCTION

Epitaxial silicide/silicon interfaces may have great technological impact, because of their potential use in Schottky barriers, Ohmic contacts, or metal-base transistors. The CaSi$_2$/Si interface, recently proposed by Morar and Wittmer[1], is of particular interest because the absence of d-electrons in Ca makes this system qualitatively different from transition-metal silicides such as NiSi$_2$ and CoSi$_2$. Morar and Wittmer[1], using transmission-electron microscopy (TEM), showed that CaSi$_2$, when grown epitaxially on Si(111), assumes the trigonal-rhombohedral phase[2] (a=3.855 Å, c=30.6 Å), which is nearly lattice matched to Si ($a_{Si} = 5.43$ Å, $\sqrt{2} \times a_{Si} \approx 2a$). The interfaces were atomically abrupt and step-free.

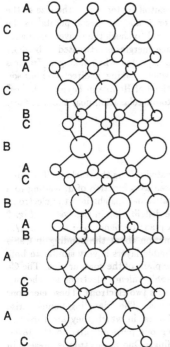

In this paper I examine the properties of bulk CaSi$_2$ and the atomic and electronic structure of the CaSi$_2$/Si interface with state-of-the-art theoretical techniques, based on pseudopotential-density-functional theory in a superlattice geometry. These methods are eminently suited for such an investigation, based on their established reliability both in studies of semiconductor interfaces[3] and of metals.[4]

The hexagonal unit cell of CaSi$_2$, space group D_{3d}^5-$R\bar{3}m$, contains six CaSi$_2$ molecules and has a 30.6 Å repeat distance along the c axis. As illustrated in Fig.1, the structure can be pictured as a stacking in the [111] direction of double layers of Si atoms, which are shifted and rotated by 180°, with Ca atoms in between.[2] For stacking in the [111] direction, there are three possible positions for each (Si or Ca) atom, which can be labeled by A, B, C. If we label each triple layer, which consists of one CaSi$_2$ formula unit, with one letter (corresponding to the position of the Ca atom in the triple layer), the overall stacking sequence can be described as AABBCC.

Figure 1: *Schematic representation of the trigonal-rhombohedral CaSi$_2$ structure in a projected view. Large spheres represent Ca atoms, small spheres Si. Two planes of atoms are shown; projected "short" bonds connect atoms in the plane of the figure to those in a plane below.*

Each Ca atom sits between two Si double layers, and has a different coordination with respect to each of these two. With respect to one Si double layer, the Ca sits in a so-called T_4 site (a threefold site on top of a second-layer Si atom), at a distance of 3.03 Å from three Si atoms in the first layer, and 3.06 Å from the Si atom in the second layer. With respect to the other Si double layer, the Ca sits in a so-called H_3 (threefold hollow) site, at a distance of 3.03 Å from its three Si neighbors in the first layer. In total, the Ca is therefore sevenfold-coordinated. The Si atoms have three Si neighbors at 2.45 Å and three Ca neighbors at 3.03 Å. In addition, some Si atoms have another Ca neighbor at 3.06 Å.

METHODS

The calculations are based on local-density-functional theory[5] and *ab initio* nonlocal normconserving pseudopotentials.[6] A supercell geometry is used,[3] in which layers of Si and $CaSi_2$ are periodically repeated, forming a superlattice $(CaSi_2)_m(Si_2)_n$. In principle, the full $CaSi_2$ bulk unit cell can only be represented if $m \geq 6$. However, it can be expected (and was verified) that the details of the stacking sequence in the bulk which involve changes more than six atoms (or 10 Å) away from the interface have no effect on the interfacial structure. The small lattice mismatch (0.4%) between $CaSi_2$ and Si is neglected; all calculations are carried out using the Si lattice constant. Results for an isolated interface can be obtained provided the interfaces are sufficiently well separated. The interface energy can be determined by taking the supercell energy and subtracting the energies of the corresponding slabs of bulk material. The latter are calculated in the same geometry and using the same convergence parameters as the supercell, to minimize systematic errors.

In order to be able to investigate a large number of different interface structures, most calculations were carried out with a 6 Ry energy cutoff (determining the size of the plane-wave basis set), a supercell with m=2 and n=2 or 3, and 39 special points[7] in the irreducible part of the Brillouin zone. Tests were performed to check the convergence with respect to all parameters. The total error bar on interface energies is estimated to be ±0.1 eV. However, some cancellation of systematic errors is expected when taking *differences* of interface energies, so that the accuracy is higher when making a direct comparison between different interface structures. Finally, calculations of Hellmann-Feynman forces[8] were carried out for selected interfaces to investigate atomic relaxation near the interface.

RESULTS

Bulk CaSi_2

First, the structure of bulk $CaSi_2$ itself was investigated. Using the stacking illustrated in Fig. 1 and a fixed c/a ratio, the energy was calculated as a function of a, leading to a lattice constant which was within 1% of its experimental value. Insight into the electronic structure can be obtained from inspection of the charge distribution in the system. Fig. 2 shows a contour plot of the valence charge density in a plane perpendicular to a $CaSi_2$/Si interface. The upper part of the figure provides information about the bonding in $CaSi_2$ itself. Fig. 2 shows that the charge density in the Si double layers is very similar to bulk Si, as evidenced by examination of the Si-Si bonds in the plane of the contour plot. The Ca atoms are characterized by a very low charge density (only valence electrons are shown), which should not be surprising, since Ca easily gives up its two electrons. These electrons go into dangling-bond-like states on each of the Si atoms. The "dangling bonds" arise as follows: the Si atoms are arranged in corrugated layers, in which they are threefold coordinated. Each Si therefore has an orbital sticking out perpendicular to the layer, in a direction in which no atoms are present for bonding. One Si electron is present in

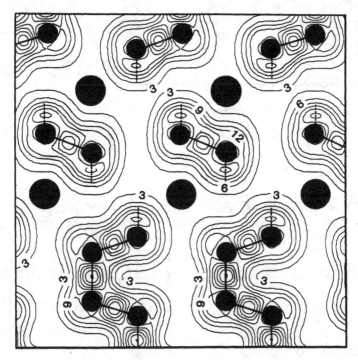

Figure 2: *Contour plot of the valence charge density in a plane perpendicular to the CaSi₂/Si interface with the 7B structure. The contour spacing is 3; units are electrons per unit cell, scaled to a unit cell containing 2 Si atoms (8 electrons). Small spheres represent Si atoms, large spheres Ca. Si-Si bonds are drawn in solid lines; dangling-bond states are indicated by dotted lines. Note the stacking fault on the Si side.*

this orbital, while the Ca contributes a second, making this "dangling bond" negatively charged. The electrostatic interaction between the negative charges on the Si layers and the positively charged Ca holds the layers together. One can therefore consider $CaSi_2$ to be a compound with mixed covalent (in the Si layers) and ionic (between Si and Ca) bonding.

It can be concluded that the electronic structure of $CaSi_2$ is very different from the transition-metal silicides $CoSi_2$ and $NiSi_2$. The latter are characterized by strong covalent interactions between the metal d states and the Si sp³ hybrids.[9] In contrast, little covalent bonding is evident between Ca and Si atoms.

Structure of the interface

Many different possibilities for the atomic arrangements at the interface were investigated. They can be classified according to the coordination of the Ca atoms at the interface, which can be 5-, 6-, 7-, or 8-fold, and according to the relative orientation of the $CaSi_2$ and the Si crystals. In a type-A structure the relative orientation is the same, while in a type-B structure the $CaSi_2$ is rotated by 180° with respect to the Si. Some of the structures that were examined in this study are illustrated in Fig. 3.

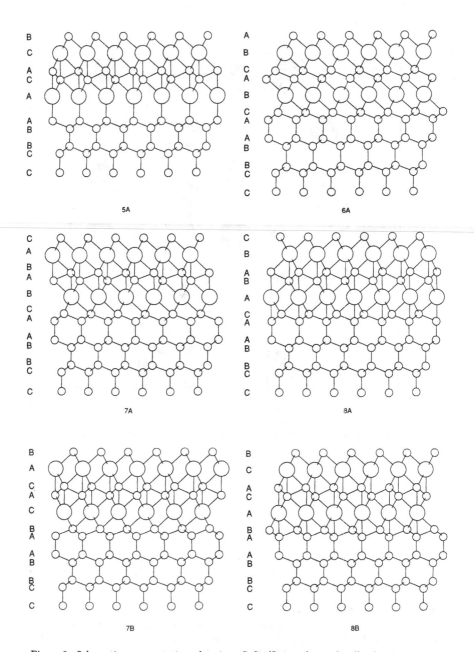

Figure 3: *Schematic representation of various CaSi$_2$/Si interfaces. Small spheres represent Si atoms, large spheres Ca. The 5B and 6B interfaces, which were also examined in the calculations, are not shown here.*

In principle, there is a choice between terminating the $CaSi_2$ with a Ca layer which is threefold coordinated to the bulk $CaSi_2$ (i.e. Ca in an H_3 site), or fourfold (in a T_4 site). This corresponds to having either a full period at the interface (i.e. AABB...) vs. only one-half period (i.e. ABB...). The first possibility occurs in the 6A and 6B structures, while the second case occurs for all five-, seven- and eightfold coordinated systems studied here. In the sevenfold-coordinated structure, Ca sits in an H_3 site with respect to the Si substrate, while in the eightfold structure it sits in a T_4 site. In a fivefold-coordinated structure, the $CaSi_2$ has a half period at the interface, with the Ca atoms sitting on top of first-layer Si atoms. Note that all the shown structures have a double layer of Si at the interface. Having only a single Si layer at the interface is energetically extremely unfavorable. Also note that for the six- and higher-fold structures, the "B" orientation (in which $CaSi_2$ is rotated by 180° with respect to the Si substrate) implies that the double layer of Si at the interface exhibits a stacking fault when viewed with respect to the Si substrate.

The interface energy of the various models is given in Table 1. The values in this Table were calculated assuming that all structures are "ideal", i.e. no relaxations were included. The atomic positions on the $CaSi_2$ side were taken to be those in the bulk up to the last Ca atom (but at a lattice constant $a=5.43$ Å$/\sqrt{2}$). The bond lengths in the double layer of Si at the interface were taken to be those of bulk Si. This actually corresponds to the optimum geometry for the sevenfold structures. For the fivefold structures, the Ca-Si distance at the interface was taken to be equal to the smaller one of those in bulk $CaSi_2$. The effect of relaxations was examined for selected interfaces, using calculated Hellmann-Feynman forces. Relaxations are only important for the 8-fold interfaces, where the outer Si-Si double layer has to expand in order to generate an acceptable distance between Ca and the second-layer Si atoms. Even there, the results indicate that the bond lengths are changed by less than 2%, shifting the energy by less than 0.05 eV.

Table 1: *Interface energies for various $CaSi_2/Si$ interfaces, characterized by the coordination of interfacial Ca (5-, 6-, 7-, or 8-fold) and the relative orientation of $CaSi_2$ with respect to Si (A or B). All values in eV per interface unit cell.*

	5	6	7	8
A	0.94	0.05	-0.03	0.08
B	0.88	0.05	-0.03	0.09

From Table 1, we conclude that the fivefold structures are high in energy. Among the others, the sevenfold interfaces are lowest, but the sixfold and eightfold structures are only slightly higher. The energy difference between A and B orientations is very small.

Schottky barrier

The (p-type) Schottky barrier height is defined as $\Phi = E_F - E_v$ where E_F is the metal Fermi level and E_v is the valence-band maximum of the semiconductor. The self-consistent interface calculations allow the derivation of this quantity in a manner analogous to the technique used for semiconductor heterojunctions, as described in Ref. 3. However, there are numerical complications which make the calculation of Schottky barrier heights much harder than that of heterojunction band offsets. In addition, LDA errors may be significant, as discussed by Das et al.[10] I therefore put an error bar of 0.5 eV on the calculated values of Φ; however, the trends between different interface structures are much more reliable. The following values were obtained for Φ: 1.02 eV for the 7A and 7B interfaces, and 0.95

eV for the 8A and 8B interfaces. The metal Fermi level is thus located in the upper part of the Si band gap. No experimental values are available to date. I also examined whether the Schottky barrier would be sensitive to the specific atomic positions near the interface. Increasing the Si-Si distance in the Si double layer at the interface by 4% leads to an increase in Φ of less than 0.1 eV.

DISCUSSION

The sevenfold-coordinated structure which I find to be lowest in energy is also the one that was observed in the experiments of Morar and Wittmer.[1] Their TEM results showed that the break in the bulk $CaSi_2$ structure falls between Ca layers of the same type. They also found that the relative orientation of the epitaxial $CaSi_2$ and bulk Si in their sample was of type-B. The Si double layer at the interface can be regarded as a continuation of the bulk $CaSi_2$ structure; considered with respect to the bulk Si lattice, it can be viewed as a stacking fault characterized by rotation of 180° about the c axis. In the present calculations, no energy difference is found between the 7A and 7B orientations.

It should actually come as no great surprise that the sevenfold structure is most stable, since it exhibits the same coordination for the interfacial Ca as in the bulk. More intriguing is the question why the sixfold and eightfold structures are only slightly higher in energy. The answer can be found in the nature of the bonding between Ca and Si atoms, as discussed above. No covalent bond is formed, and the cohesion results from electrostatic attraction between Ca^{++} and negative Si "dangling bonds". For all structures with sixfold or higher coordination, the stacking is such that Ca is surrounded by a "cage" of six of these dangling bonds. The interaction with a seventh or eighth Si atoms occurs via the charge density in the backbond region of a Si atom directly above or below the Ca, at a slightly larger distance than the other Si (see Fig. 2). The change in energy due to this additional interaction is quite minor, on the order of 0.1 eV. This was also found from a comparison of the energy of different stacking sequences in bulk $CaSi_2$. Various interface structures are therefore expected to have competing interface energies, as seen in Table 1. It should therefore be anticipated that experimentally more than one of these structures can be obtained, making this a challenging system for growth studies.

ACKNOWLEDGMENTS

Thanks are due to J. F. Morar for encouraging me to investigate this problem. I am also grateful to P. E. Blöchl, J. Tersoff, and M. Wittmer for helpful discussions.

REFERENCES

[1] J. F. Morar and M. Wittmer, Phys. Rev. B **37**, 2618 (1988).

[2] J. Evers, J. Solid State Chem. **28**, 369 (1979).

[3] C. G. Van de Walle and R. M. Martin, Phys. Rev. B **35**, 8154 (1987).

[4] R. J. Needs, Phys. Rev. Lett. **58**, 53 (1987).

[5] P. Hohenberg and W. Kohn, Phys. Rev. **136**, B864 (1964); W. Kohn and L. J. Sham, *ibid.* **140**, A1133 (1965).

[6] G. B. Bachelet, D. R. Hamann, and M. Schlüter, Phys. Rev. B **24**, 4199 (1982);

[7] A. Baldereschi, Phys. Rev. B **7**, 5212 (1973); D. J. Chadi and M. L. Cohen, Phys. Rev. B **8**, 5747 (1973); H. J. Monkhorst and J. D. Pack, Phys. Rev. B **13**, 5188 (1976).

[8] H. Hellmann, *Einführung in der Quanten Theorie* (Deuticke, Leipzig, 1937), p. 285; R. P. Feynman, Phys. Rev. **56**, 340 (1939).

[9] J. Tersoff and D. R. Hamann, Phys. Rev. B. **28**, 1168 (1983).

[10] G. P. Das, P. Blöchl, N. E. Christensen, and O. K. Andersen, Phys. Rev. Lett. **63**, 1168 (1989).

ATOMIC STRUCTURE OF DISLOCATIONS AND INTERFACES IN SEMICONDUCTOR HETEROSTRUCTURES

A. S. Nandedkar, S. Sharan and J. Narayan
North Carolina State University,
Dept. of Materials Science and Engineering,
Raleigh,N.C. 27695-7916.

ABSTRACT

We have studied and analyzed critical thickness required for the formation of misfit dislocations in semiconductor heterostructures. The present analysis has been carried out assuming that the nucleation of a misfit dislocation is controlled by the activation energy. The energy of the coherent interface and the misfit dislocation configuration has been evaluated using a discrete dislocation analysis. Further, atomic structures of coherent and semicoherent interfaces containing misfit dislocations of the type $a/2<110>\{111\}\uparrow<110>$ (60° dislocation), $a/2<110>\{001\}\uparrow<110>$ (90° dislocation) and $a/6<112>\{111\}\uparrow<110>$ (90° partial dislocation) in Ge/Si system were simulated. The total energy, consisting of both core and elastic energy components, was calculated using Stillinger-Weber interatomic potentials. The results show that the energy is a strong function of the nature of dislocations. A 60° dislocation is found to exist in undissociated form in shuffle configuration, but the core contains a dangling bond at the interface. The core of a 90° dislocation reconstructs at the Ge/Si interface with no dangling bonds. The calculated atomic configurations of dislocations and interfaces are found to be in good agreement with high resolution transmission electron microscopy observations.

INTRODUCTION

The formation of films on lattice mismatched semiconductor substrates is of considerable importance in the fabrication of advanced semiconductor devices. Defects such as dislocations play an important role in the development of semiconductor heterostructres.The growth of a lattice mismatched film on a substrate leads to the formation of a coherently strained layer only if the lattice mismatch or the thickness is sufficiently small. As the thickness of the film increases, the strain energy increases to a point where it is energetically favorable to releive stress by creating misfit dislocations and accomodating them near the interface. This value of film thickness is defined as critical thickness.The dislocation and defect densities in semiconducting films are too high at present ($>10^6$ /cm^2). In order to reduce the dislocation density to acceptable levels ($<10^4$ /cm^2), it is necessary to understand the origin, nucleation and propagation of dislocations. It is also important to understand the atomic structures of dislocations and interfaces.

Atomic structures of defects and interfaces are required to evaluate the physical and electronic properties of materials. Here we have calculated atomic structure of $a/2<110>\{111\}\uparrow<110>$ (60° dislocation) and $a/2<110>\{001\}\uparrow<110>$ (90° dislocation) at Ge / Si interface. The dislocations were simulated on a computer and the atomic structure was obtained by energy minimization techniques [1]. The Stillinger-Weber [2] interatomic potential was used to calculate the energy of atomic configuration of dislocations. A perfect 60° dislocation is simulated in shuffle configuration where widely spaced {111} planes are sheared off against each other. Geometrical modelling of this particular configuration was done by Hornstra [3]. We have carried out high resolution transmission electron microscopy observations of interface and dislocations in Ge/Si systems. The dislocations observed at the Ge/Si interface are of undissociated 60° type. The experimental results are compared with the calculated atomic structure of dislocations and interfaces. A perfect 90° dislocation ($a/2<110>\{001\}\uparrow<110>$ is also simulated at Ge/Si interface. The energy of this configuration was compared with that of coherent Ge/Si interface. Stillinger-Weber potential was modified by Ding-Anderson [4] to suit germanium. A Stillinger-Weber potential is found to be adequate for both the properties of molten silicon and the bulk properties of solid silicon, including the properties of defects in solid silicon [5].

RESULTS AND DISCUSSION

The thickness of the epilayer at which misfit dislocations will be nucleated to relieve the homogeneous strain energy corresponds to the critical thickness for the system. The existence of the critical thickness was first considered by Van der Merwe [6,7] and later by numerous authors including Matthews et. al. [8-10] and People and Bean [11,12]. Atomistic calculations using Molecular Dynamics [13,14] and Monte Carlo [15] simulation techniques have also been used to evaluate the critical thickness.

In the present work, the critical thickness is determined after considering the activation energy required to nucleate a lattice misfit dislocation at the free surface and for it to glide under the stress field associated with a coherent interface. The self and interaction energy terms associated with the dislocation configuration in the finite epilayer are obtained by satisfying the free surface boundary conditions using the surface dislocation analysis. The formation of a misfit dislocation follows the mechanism proposed by Narayan et. al. [16] and Kyam et. al. [17], wherein dislocations with Burgers vector a/2 <101> and sense vector along the <110> direction are generated at the free surface and subsequently glide on the {111} planes towards the interface. Dislocations with Burgers vector a/2 [101] make a 60°angle with both the [1T0] and [110] directions , indicating that the dislocations are of a mixed character. In our analysis, a misfit dislocation with Burgers vector a/2 [101] lying along the [1T0] direction with the slip plane (111) is considered to be nucleated at a distance of one lattice parameter from the free surface in the presence of the stress field of the coherent interface. The lattice misfit dislocation is nucleated near the free surface when the strain energy density of the coherent film exceeds the total misfit dislocation configuration energy. This becomes the necessary condition for the formation of a misfit dislocation, whereas a sufficient condition is that the glide of the misfit dislocation towards the interface should be accompanied by a decrease in the energy of the system.

The critical thickness has been numerically determined for the $Ge_{1-x} Si_x$ /Si system at different values of x, giving rise to different values of the misfit parameter, f, between the epilayer and substrate. The energy of the misfit dislocation is zero when it is outside the crystal, since both the work done and the self and interaction energy terms are zero. The work done in moving the misfit dislocation into the epilayer increases linearly with the distance, d, from the free surface while the self and interaction energy terms increase more slowly. The variation of these two energy terms gives rise to the resultant activation energy barrier. A misfit dislocation nucleated at distances greater than that at which the activation energy becomes maximum move towards the interface since this is accompanied by a lowering in energy of the configuration. Thus, if the coherent strain energy is larger than the activation energy, the misfit dislocation will move towards the interface. The strain energy of the epilayer in the absence of a misfit dislocation and that of the misfit dislocation configuration is calculated as a function of the epilayer thickness. The critical thickness is reached when the activation energy is overcome by the coherent strain energy. The strain energy of the coherent film is smaller than the activation energy of the misfit dislocation for epilayer thicknesses below the critical value, hence the film remains coherent. The strain energy associated with the coherent film increases more rapidly than the misfit dislocation energy with increasing epilayer thickness. Therefore, the activation energy required to nucleate a misfit dislocation is overcome with increasing values of the epilayer thickness. The magnitude of the critical thickness obtained for various values of the misfit parameter, f, i.e. for different values of x is shown in Fig. 1 along with the results obtained by previous workers.

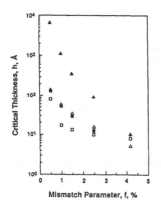

Figure 1

Figure 1 : Comparison of critical thickness predictions as a function of the misfit parameter obtained from the theories of Matthews[10](■), Van der Merwe[7] (□) and People and Bean[11,12](▲), along with the results obtained from the analysis described in the text (Δ).

Figure 2-a shows a calculated atomic structure of a 90° (a/2<110>{001}) dislocation at Ge/Si interface. The extra half plane is in silicon substrate and it terminates at the interface as shown by an arrow. The core of the dislocation consists of a pentaring in the compressional side of the dislocation (silicon substrate), and a septaring in the dilatational side (Ge film). This is a typical core configuration for this dislocation, similar to that observed in bulk silicon [18]. For this configuration the self energy of the core, E_c, is 0.56 eV / Å (Table 1) compared to 0.49 in the bulk [18]. The lengths of bonds are in the range -3% to +3% of the ideal bond length and there are no unsatisfied bonds present in the configuration. This fact suggests that there are no dangling bonds present in the core of this dislocation. When a bond stretches beyond 8% of its ideal value, it is considered to be a dangling bond [19]. The total energy of this dislocation for a radius of 28 Å was calculated to be 2.78 eV / Å.

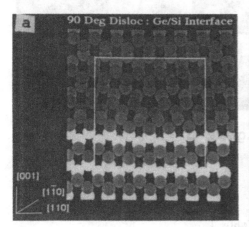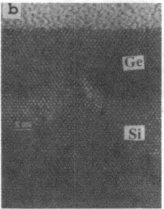

Figure 2

Figure 2-b shows a high resolution micrograph of a 90° dislocation at Ge / Si interface [20]. It is difficult to see a pentaring and a septaring which are characteristic of a core of this dislocation as shown in Figure 2-a. In order to resolve individual rows of atoms in <110> direction, a point to point resolution of 1.34 Å is needed, which is not available even in presently available ultra-high resolution electron microscope. With the resolution limit of our microscope (2.6 Å), two rows of <110> chains separated by 1.34 Å appear as a single dot. If each dot represents two rows of atoms, we find that the core structure is consistant with a pentaring and a septaring of atoms. Extra half plane of atoms is present in silicon substrate. The plane terminates at an arrow shown in Figure 2-b, and appears as an extra row of light dots at an angle of about 54° to the interface. A Burgers circuit is completed around this dislocation (Figure 2-b) and the corresponding Burgers vector is shown.

Figure 3-a shows a calculated atomic structure of a 60° dislocation (a/2<110>{111} | <110>). This is an undissociated dislocation in shuffle configuration with a calculated value of core energy 1.13 eV / Å for a radius of 5 Å. The core of a 60° dislocation has one octaring of atoms istead of a pair of pentaring and septarings contained in the core of a 90° dislocation. Extra half plane of atoms is still present in silicon substrate although the arrangement of atoms is different than that observed for a 90° dislocation. The calculated value of total energy of a 60° dislocation with a radius of 28 Å is 5.13 eV / Å.

A high resolution transmission electron micrograph of a perfect 60° dislocation at Ge / Si interface is shown in Figure 3-b. It is difficult to judge if a 60° dislocation is in shuffle or glide configuration in the high resolution micrographs, because the resolution is not high enough to reveal individual chains of atoms. Table 1 summarizes the energies associated with the two dislocations at Ge / Si interface [20]. The difference between these two dislocations at Ge / Si interface becomes more evident when a Burger's circuit is completed around each dislocation. In case of 90° dislocation, Burger's vector is perpendicular to the dislocation line and in case of 60° dislocation it is at an angle to the line. The respective Burger's vectors are shown in Figure 2-b and Figure 3-b.

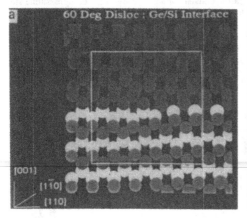 

Figure 3

Table I

Energies of Dislocations in $eV/\mathring{A}$		
Ge/Si Interface : SW Potential		
Dislocation Type	Core $(R \approx 5\mathring{A})$	Total $(R \approx 28\mathring{A})$
$a/2[110](001) \uparrow [1\bar{1}0]$	0.56	2.78
$a/2[101](\bar{1}11) \uparrow [110]$	1.13	5.13

But the coherent strain in the film is not relieved completely by the presence of one dislocation alone. Periodic occurrence of dislocations is necessary in both x and z directions to completely relieve the strain in the film. Here x direction is shown horizontally and z direction is perpendicular to the plane of the paper. For a 90° dislocation, a dislocation should be present at every twentyfifth atom (in Ge film) in x and z direction for a complete relaxation of coherent strain in the film. Here we are mainly studying atomic structure of dislocation at the Ge / Si interface. Presence of other dislocations is not going to alter the core structure or core energy, although it will have significant effect on the elastic region of this dislocation.

The energy of the coherent Ge film grown on Si substrate is plotted in Figure 4-a. When these energy values are compared with the energy of the film with a 90° dislocation (Figure 4-b), it is clear that the energies of a configuration with the dislocation is lower after 27 Å thickness of the film. Hence Figure 4 demonstrates that coherent strain energy of the film is reduced due to fromation of the dislocation.

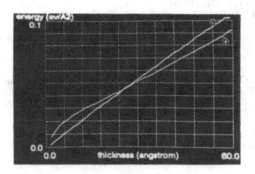

Figure 4

It is difficult to assess the status of a dangling bond in semiconductors. We have had to reconstruct the core of a 60° (undissociated shuffle configuration) dislocation. In case of a 60° dislocation, there was only one unsatisfied bond in the computational cell, and it was not possible to reconstruct without stretching the other bonds to an unacceptable high level. Hence it is deduced that this dislocation has one dangling bond per 3.84 Å length of the dislocation line.

SUMMARY

The critical thickness for the formation of misfit dislocations has been evaluated as a function of the lattice miamatch for an epilayer growing on a substrate which is free of threading dislocations. The critical thickness of the epilayer is obtained by assuming that the misfit dislocation nucleated near the free surface of the epilayer glides down on the slip plane to the interface. The misfit dislocation is nucleated when the strain energy of the coherent medium exceeds the energy of the misfit dislocation configuration.

We have calculated atomic structures of a 90° and 60° dislocation at Ge / Si interface, using Stillinger-Weber interatomic potentials. These structures were found to be in good agreement with high resolution transmission electron microscopy observations of the dislocations. Their relative energies suggest that under equilibrium conditions, the Ge / Si interface has lower energy in the presence of a 90° dislocation. It is determined from the calculations and direct observations of an undissociated 60° dislocation that it exists at Ge / Si interface in shuffle configuration.

REFERENCES
1 A.S.Nandedkar and J.Narayan, Phil.Mag. A,56(5)(1987),625.
2. F.H.Stillinger and T.A.Weber, Phys. Rev. B, Vol.31-8,5262,1985.
3 J. Hornstra, Phys.Chem.Solids,5(1958),129.
4. K.Ding and H.C. Andersen, Phys.Rev. B, vol. 34-10,6987,1986.
5. B.W. Dodson, Phys. Rev. B, vol. 33⁻10,7361,1986.
6. J. H. Van der Merwe, J. Appl. Phys., 34, 123, (1962).
7. J. H. Van der Merwe, in Single Crystal Films, Eds. M. H. Francombe and H. Sato, Pergamon Press, Oxford, p. 139, (1964).

8. J. W. Matthews, S. Mader, and T. B. Light, J. Appl. Phys., 41, 3800, (1970).
9. J. W. Matthews, A. E. Blakeslee, and S. Mader, Thin Solid Films, 33, 253, (1976).
10. J. W. Matthews in Dislocations in Solids, Ed. F. R. N. Nabarro, North Holland Publishing Company, p. 470, (1979).
11. R. People and J. C. Bean, Appl. Phys. Lett., 47, 322, (1985).
12. R. People and J. C. Bean, Appl. Phys. Lett., 49, 229, (1986).
13. M. H. Grabow and G. H. Gilmer, Initial Stages of Epitaxial Growth, Eds. R. Hull, J. M. Gibson and D. A. Smith, MRS Proc., 94, 15, (1987).
14. M. H. Grabow and G. H. Gilmer, MRS Fall Meeting (1987).
15. B. W. Dodson and P. A. Taylor, Appl. Phys. Lett., 49, 642, (1986).
16. J. Narayan, B. C. Larson and W. H. Christie, Laser - Solid Interactions and Laser Processing, Eds. S. D. Ferris, H. J. Leamy and J. M. Poate, MRS Boston, p. 440 (1978).
17. E. P. Kvam, D. J. Eaglesham, D. M. Maher, C. J. Humphreys, J. C. Bean, G. S. Green and B. K. Tanner, Proc. Mater. Res. Soc., 104, 623 (1987).
18. A.S. Nandedkar and J. Narayan, Phil. Mag. A, In Press.
19. R. Jones, Phil. Mag. B, 42(1980),213.
20. A.S. Nandedkar and J. Narayan, Mat. Sci. Eng., A113, 51

GROWTH MODE AT THE Ge/(1$\bar{1}$02) SAPPHIRE INTERFACE

GEOFFREY P. MALAFSKY

Naval Research Laboratory, Wash, DC 20375-5000

ABSTRACT

The growth mode of Ge on the (1$\bar{1}$02) surface of sapphire is explored with X-ray photoelectron spectroscopy. Ge exists in two bonding states at the interface, Ge-Ge and Ge-sapphire. Ge forms islands at submonolayer coverage for deposition temperatures of 25°C and 625°C. The formation of the islands is revealed by the rapid increase in the relative fraction of the Ge-Ge bonding state for Ge coverage less than 1 ML. The shift in the Ge-Ge peak binding energy to the bulk Ge value at less than 1 ML suggests that the islands are three dimensional for deposition at 625°C.

I. Introduction

The heteroepitaxial growth of Ge on sapphire may offer the same technological benefits as silicon on sapphire (SOS). These include lower cross-talk, latch-up, and parasitic capacitance compared to a device built on a non-insulating substrate [1]. Single crystal Ge(110) films can be grown on the (1$\bar{1}$02) crystal face of sapphire, but the crystallinity of the epilayer depends upon the substrate temperature during deposition [2].

Both Ge [2] and Si [3] form islands on the (1$\bar{1}$02) sapphire surface at thicknesses as low as 50 Å. However, in order to determine the microscopic growth mode, one must probe the deposition of the first few monolayers. Typically, heteroepitaxy is described by the Frank-van der Merwe, Stranski-Krastanov, or the Volmer-Weber growth modes [4]. The Volmer-Weber mode proceeds by the nucleation of three dimensional islands directly on the substrate. Conversely, in the Frank-van der Merwe mode the epilayer grows in a layer-by-layer fashion. The Stranski-Krastanov model is a combination of the two previous models with the first few layers growing in a layer-by-layer fashion and subsequent layers growing via three dimensional islands.

This paper describes the interfacial growth of Ge on (1$\bar{1}$02) sapphire. X-ray photoelectron spectroscopy (XPS) is used to monitor the interfacial chemistry for submonolayer coverage of Ge. The Ge deposition is performed at room temperature (25 - 50°C) and at 625°C.

II. Experimental

The Ge films were deposited in a UHV molecular beam epitaxy (MBE) instrument with a base pressure of 5×10^{-11} Torr. The Ge was deposited at a rate of 2-4 Å/min with a monolayer of Ge defined to be 1.415 Å. After deposition, the sample was transferred under UHV to the VG ESCALAB analysis chamber which contained the XPS system. The photon energy used was 1486.6 eV (Al K_α) and the pass energy of the hemispherical analyzer was set to 10 eV. At this pass energy, the FWHM of the Ag $3d_{5/2}$ peak from Ag foil is 0.88 eV. The XPS scans were collected in 0.1 eV steps. The 3" diameter sapphire wafers were chemically cleaned [2] and then annealed in vacuum at 1400°C for 30 minutes. The cleaned wafers exhibited a sharp low energy electron diffraction (LEED) pattern. In addition, the substrate was free of chemical contamination to the sensitivity of the XPS analysis.

Charging of the sapphire causes a large and variable peak position shift from run to run. In order to compensate for this variability, the Ge binding energies are referenced to the sapphire Al $2p_{3/2}$ and O 1s peaks at 74.7 eV and 531.6 eV respectively [5]. Neither the separation between the sapphire peaks nor their peak shapes change with Ge coverage. This is a result of the long inelastic mean free paths of these photoelectrons (approximately 15Å) and the thin overlayers. Only at the thickest Ge coverage studied here are the sapphire peaks significantly attenuated.

III. Results and Discussion

Figure 1 displays the Ge $2p_{3/2}$ region of the XPS spectrum for the room temperature deposition of Ge on sapphire. There are two peaks at all Ge coverage indicating two distinct bonding states of Ge. The peak with a binding energy of 1220.7 eV is shifted to higher binding energy from the bulk Ge value of 1217.2 eV by 3.5 eV [5]. This peak shift is due to Ge bonded to the sapphire substrate and is consistent with Ge-oxygen bonding but it cannot be definitively assigned to a species such as GeO or GeO_2. The position of the second peak changes from 1218.1 eV at 0.1ML coverage to 1217.3 eV at 3ML coverage and is associated with

Ge-Ge bonding. It experiences a binding energy shift at low Ge coverage as a result of the electronic interaction with the substrate. This peak is assumed not to be from Ge-Al bonding since the binding energy shift for this species should be towards lower binding energy relative to bulk Ge [6].

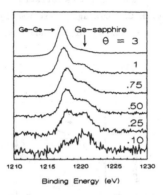

Figure 1 Ge $2p_{3/2}$ peaks for various surface coverages. The substrate temperature is 25°C. Each spectrum is normalized to the peak height above baseline.

At 0.1ML coverage, the majority of the Ge is in the Ge-sapphire bonding state. As the Ge coverage increases, the relative amount of Ge involved in the two bonding states changes. The fractional amount of Ge-Ge bonding increases rapidly for coverage less than 1 ML (fig. 2). In fact, the bonding states have equal intensities at 0.25 ML coverage. The sharp rise in the relative amount of the Ge-Ge peak indicates that Ge is agglomerating into islands. Therefore, the depositing Ge atoms have sufficient surface mobility at a substrate temperature of 25°C to diffuse across the sapphire surface until they attach to a nucleation site or a growing island. These nucleation sites may be surface steps, kink sites or other surface defects [4]. The filling of the nucleation sites is reflected in the saturation of the signal level of the Ge-sapphire bonding state (fig. 2). In contrast, the amount of Ge-Ge bonding increases linearly with total Ge coverage.

Elevating the substrate temperature during growth provides thermal energy for several activated processes, such as surface diffusion, desorption, and structural rearrangement of the growing film. In fact, the Ge films are not single crystalline for deposition temperatures below 700°C [2]. At a deposition temperature of 625°C, the higher surface mobility is evident by the greater fractional amount of the Ge-Ge bonding state (fig. 3). The majority of the Ge is agglomerated even at a coverage as low as 0.1 ML. Similar to the room temperature deposition, the Ge-sapphire bonding state saturates but at a lower concentration (fig. 3). Also, the concentration of the Ge-Ge bonding state continues to increase with Ge coverage. Presumably, the increased surface mobility results in a lower density of islands distributed on the surface but increases the island size.

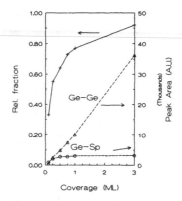

Figure 2 The relative concentrations of the two bonding states of Ge, Ge-Ge and Ge-sapphire (Ge-Sp). The substrate temperature is 25°C.

The Ge-Ge peak undergoes a shift to lower binding energy with increasing Ge coverage for both deposition temperatures (fig. 4). The interaction of the Ge island with the substrate perturbs its chemical environment causing a binding energy shift from the bulk Ge value. As the island size increases, this interaction diminishes and the binding energy shifts towards the bulk Ge value. For the room temperature deposition, the Ge-Ge peak shifts to the bulk value at 3 ML coverage. However, depositing the Ge at a substrate temperature of 625°C causes the peak to shift to the bulk value at less than 1 ML coverage. This suggests that Ge forms three dimensional islands at submonolayer coverage when the substrate temperature is 625°C and that the growth mode is of the Volmer-Weber type at this temperature.

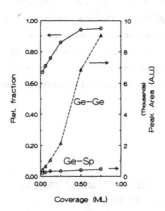

Figure 3 Same legend as figure 2 except that the substrate temperature is 625°C.

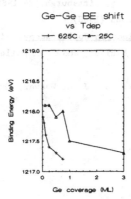

Figure 4 The change in the binding energy of the Ge-Ge peak versus Ge coverage for substrate temperatures of 25° and 625°C.

IV. Conclusions

Ge exists in two bonding states at the sapphire interface at substrate temperatures of 25° and 625°C. At both temperatures, Ge agglomerates into islands on the (1$\bar{1}$02) sapphire surface at submonolayer coverage. The formation of the islands results in a rapid increase in the relative fraction of the Ge-Ge bonding state at less than 1 ML. For a substrate temperature of 625°C, the shift in the Ge-Ge peak position to the bulk Ge value at submonolayer coverage suggests that the islands are three dimensional and that Ge grows by the Volmer-Weber mode.

V. References

1. P. K. Vasudev, in Epitaxial Silicon Technology, ed. by B. J. Baliga, 233 (Academic Press Inc., NY 1986)
2. D. J. Godbey, S. B. Qadri, M. E. Twigg, and E. D. Richmond, Appl Phys Lett, 54, 2449 (1989)
3. J. G. Pelligrino, E. D. Richmond, and M. E. Twigg in SOI and Buried Layers, (Mater. Res. Soc. Proc. 116, Pittsburgh, PA 1988), 389
4. K. Reichelt, Vacuum, 38, 1083 (1988)
5. Handbook of X-Ray Photoelectron Spectroscopy, ed. by C. D. Wagner, W. M. Riggs, L. E Davis, J. F. Maulder, and G. E. Muilenberg (Perkin-Elmer, Eden Prairie, MN 1978)
6. M. del Giudice, J. J. Joyce, and J. H. Weaver, Phys Rev B, 36, 4761 (1987)

ORDERING AND ENERGETICS OF Hg OVERLAYERS ON Cu(001)

C.W.HUTCHINGS[a], P.A.DOWBEN[a], Y.J.KIME[a], W.LI[a], M.KARIMI[b], C.MOSES[c] AND G.VIDALI[a]

a) Syracuse University, Physics Department, Laboratory of Surface and Low temperature Physics, Syracuse, N.Y. 13244-1130
b) Alabama A&M University, Physics Department, Normal, Ala.35762
c) Utica College, Physics Department, Utica, N.Y. 13502

ABSTRACT

The growth, ordering, and energetics of Hg overlayers on Cu(001) have been studied using atom beam scattering, LEED and angle-resolved photoemission. Two stable ordered phases have been identified: one phase is a c(2x2) and the other phase is a higher density square lattice which has a coincidence c(4x4) structure. A phase diagram has been determined using LEED and atom beam diffraction data for surface temperatures between 180 and 330 K. ABS data suggest out that there is a narrow coexistance region between these two phases. The isosteric heat of adsorption has been determined as a function of coverage.

I. INTRODUCTION

There have been many studies of metal overlayers on different substrates and several excellent reviews have been compiled recently [1,2]. In most of these studies, the structure of the overlayer is identified, but less is known about the energetics of adsorption or the electronic properties of the overlayer. In this report we show that the system Hg on Cu(001) can indeed be characterized in its structural, electronic and thermodynamic properties.

Our interest in Hg on Cu(001) is due to the fact that Hg adsorbed on metal surfaces can display insulating as well as conducting properties. A few studies have been carried out on Hg adsorbed on transition metal surfaces [3]. Hg is expected to only weakly interact with metal surfaces. Indeed, mercury adsorption energies on many transition metal surfaces are in the range of less than 1 eV. The weak interaction and the large atomic size of Hg, provide the rich phases which Hg layers display on many metal surfaces [3]. It is also worth noting that Hg is one of the few metal adsorbates which can be studied under equilibrium conditions with its own vapor.

II. EXPERIMENTAL

Structural investigations were carried out in our atom beam scattering apparatus [4]. Briefly stated, this apparatus consists of a beam line and a scattering chamber which is in communication with a preparation chamber. The beam line consists of three differentially pumped stages, in which high pressure ultra pure helium is made to expand through a 10 micron nozzle into the first chamber. Beam energies can be changed by controlling the temperature of the nozzle. In the present studies we used E_i=21 and 63 meV. The beam is then mechanically chopped and collimated before entering the UHV scattering chamber which has a reverse view LEED and a differentially pumped detector for the He beam. The detector can rotate by 190 degrees around the sample holder. The sample can be moved from the preparation chamber to the scattering chamber by a sample manipulator; the manipulator can change the polar angle (made by the He beam with the surface normal) and the azimuthal angle. The sample can be cooled down to 140 K or heated to 1200 K.

The sample was oriented, cut and polished. Prior to each experiment the surface was cleaned using Ar ion bombardment and annealing up to 700 to 800 K. He beam reflection was used to check the condition of the surface, as well as Auger electron spectroscopy. Temperatures were measured using a chromel-alumel thermocouple. Triply distilled mercury was admitted into the UHV chamber through a UHV leak valve. Reported mercury partial pressures are uncorrected for gauge calibration and position. No evidence from UPS spectra [3] or thermal desorption was found of alloying of Hg with copper for the pressure (up to 1×10^{-7} torr) and temperature (180 to 330 K) range used.

III. RESULTS

For a range of surface temperatures between 180 K and 330 K the first ordered structure is a c(2x2) net, see Fig. 1. Upon further exposure, we

Figure 1. Helium diffraction scans for the c(2x2) (top) and c(4x4) (bottom) Hg overlayers. θi=70° (top) and θi=71.2° (bottom); Ei=21 meV. Real space structures are shown in the insets, where Cu atoms are shown as crosses. Hg atoms are shown with circles if they are in registry with the substrate; otherwise they are at the intersection of lines (bottom inset).

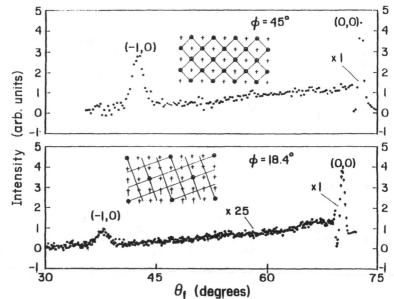

obtain a complex LEED pattern at about 10 Langmuir. By combining atom beam scattering, which probes only the topmost layer, with the obtained LEED pattern we identified this structure. We found that at 10 Langmuir exposure the mercury net is rotated by 18.4 degrees with respect to the <100> direction of Cu and the lattice spacing is 3.22 Å (see Fig. 1, bottom panel). This structure is a high density c(4x4) since, as can be seen from the inset of Fig.1, there is a coincidence c(4x4) superlattice. The fractional coverage of this phase is $\theta=0.62$ of a Cu(001) layer.

We used LEED to obtain the phase diagram shown in Fig.2. The symbols

Figure 2. Phase diagram obtained using LEED and ABS (inset); symbols indicate maximum intensity of c(2x2) (●) and c(4x4) (✚) phases.

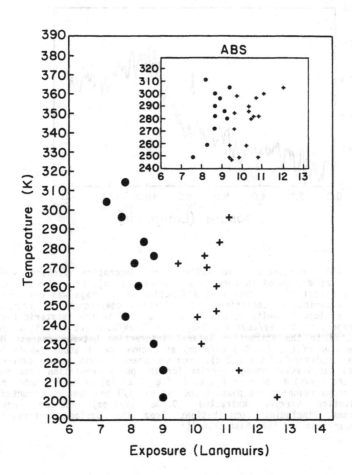

shown represent the <u>maximum</u> intensity of the LEED pattern at that given
temperature. For the range of temperatures presented, no annealing of the
layer was necessary to obtain the maximum intensity. In the inset of Fig.2
we present atom beam scattering diffraction data. The symbols represent the
maximum intensity of the diffraction peaks. In a typical run, the detector
is set at the angular position of a Bragg diffraction peak pertinent to the
structure of the overlayer to be observed. Then the exposure of Hg is
increased while the intensity is monitored. Fig. 2 shows all the data
taken; the scatter is due to errors in timing the exposure and gauge
reading reproducibiltiy. In Fig. 3 we show a run for the azimuthal
direction ϕ=18.4.° Except at the lowest temperatures and highest exposures,
see Fig.2, the (0,-1) diffraction intensity remains nearly constant upon
further adsorption, indicating that no other phases are formed.

Figure 3. Helium diffraction peak vs. exposure at 1×10^{-8} torr of Hg;
diffraction peak is the (0,1) for the c(4x4) phase. Arrow indicates maximum
intensity for the (0,-1) helium diffraction of the c(2x2) phase. Ei=21 meV,
θi=60°, Ts=250 K.

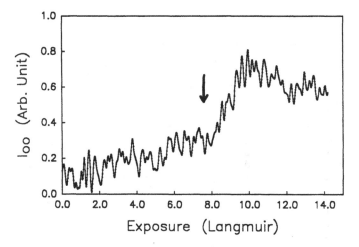

Using ABS intensities, adsorption and desorption isobars were
constructed, as discussed in more detail elsewhere [5]. From 16 isobars,
the isosteric heat of adsorption as a function of coverage was obtained.
Under our experimental conditions, the sticking coefficient is roughly
constant and close to unity [5,6]. There is a rise in the isosteric heat
with coverage from 0.5 eV/atom (at θ=0) to 0.71 eV/atom (at θ=0.45) which
is attributed to the attractive lateral interaction between nearest Hg
atoms. The drop of q_{st} (to 0.1 eV/atom) at a coverage of about θ=0.45 is
due to the completion of the c(2x2). From the phase diagram we can extract
the "global" activation energy barrier for the phase transition from the
c(2x2) to the c(4x4) as seen in Fig.4. In Fig. 4 a plot of ln R (number of
impinging atoms reaching the phase boundary) vs. 1/T has been constructed.
The activation barrier extracted (0.008 eV/atom) includes many
contributions, including contributions from the surface potential,
adsorption process and the phase transition.

Fig.4 Plot of impinging rate R (atoms/cm^2 sec) vs. 1/T from data of Fig.3

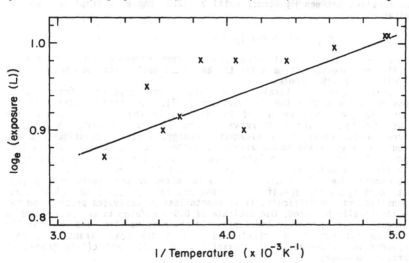

IV. DISCUSSION

Similarly to other Hg/metal systems [3], and because of the large size of the Hg atom compared to the typical size of metal atoms, simple 1x1 overlayers are not observed; rather, a complex sequence of ordered phases is seen instead. The nearest neighbor distance in solid mercury is 3.00 Å, which is substantially larger than the Cu-Cu distance of 2.55 . Therefore Hg atoms are initially accomodated in an expanded lattice, a c(2x2) phase, of lattice constant 3.61 Å. At θ=0.62 a high density c(4x4) phase has been identified. In this case the lattice spacing of the overlayer is still significantly greater than that observed for solid Hg. No higher density phases have been seen in the range of Hg exposure and surface temperature explored. Evidently, the influence of the substrate potential, and in particular of its modulation across the unit cell, is strong enough to constrain Hg atoms to occupy well defined adsorption sites and to impose a square symmetry on the overlayer. Photoemission results demonstrate that the overlayer is an insulator [3,7], in accord with other work on Hg on other metal substrates which show that a triangular symmetry and smaller nearest neighbor distances are required to obtain a conductive layer [3].

The arrow in Fig.3 indicates where the c(2x2) ABS diffraction peak reaches the maximum. This and other data to be presented elsewhere [6] suggest that the c(2x2) and high density c(4x4) phases coexist. LEED pictures show an incipient, faint c(4x4). We believe this to be evidence of a coexistence of the c(2x2) and the high density c(4x4) phase. Similar data have been obtained at other temperatures. Note that LEED patterns cannot be used to show the coexistence of phases, since the c(2x2) spots form a subset of the high density c(4x4) pattern. Complete I-V analysis of the LEED data cannot be presently carried out because of the complex unit cell that has to be considered in the calculations.

The rise in the isosteric heat with coverage is due to the attractive interactions between Hg atoms, until a c(2x2) phase is completed. We can write:

$$q_{st} = (d \ln P/dT)_\theta \, k_b T^2 = Eo - W\theta,$$

where Eo is the heat of adsorption at zero coverage and P is the Hg equilibrium pressure. From a fit to the experimental data, we obtain: Eo=48 KJ/mole and W=-39 KJ/mole.

These strong lateral interactions are expected from strain considerations within the Hg overlayer [5,7]. The sharp reduction in the isosteric heat is a result of two changes in the contributions to the isosteric heat. With a decrease in the lattice constant, there is a corresponding decrease in adsorption energy. Further investigations are required to ascertain which effect is larger but preliminary calculations suggest [7] that the Hg-Hg lateral interactions are a major contributor to the isosteric heat. Defect density increases and entropy terms related to the domain edges of the c(2x2) can be reflected in the isosteric heat [5], but identifying the specific contributions to the c(2x2) to c(4x4) phase transition may be difficult. It is nonetheless an activated process and can be kinetically hindered. Our estimate of 0.008 eV/atom is an upper bound on the rate limiting kinetic barrier to the phase transition, and while small, does suggest domains of c(2x2) will resist the phase transition. This supports our observation of coexistence of c(2x2) and c(4x4) phases at certain coverages.

ACKNOWLEDGMENTS

Support from the National Science Foundation, grants DMR 8802512 (to G.V.) and 8820779 (to P.A.D.), is gratefully acknowledged. We thank Ms.Lin for technical support.

REFERENCES

[1] See, for example: J.C.C.Fan and J.M.Poate, Eds., Heteroepitaxy on Silicon, MRS Symposia Proceedings, v. 67, 1986.
[2] K.Heinz, Progr.Surf.Science 27, 239 (1988); W.N.Unertl, Comments Cond.Mat.Phys. 12, 289 (1987); P.A.Dowben, M.Onellion, and Y.J.Kime, Scann.Microsc. 2, 177 (1988).
[3] N.K.Singh and R.G.Jones, Chem.Phys.Lett. 155, 463 (1989); R.G.Jones andA.W.L.-Tong, Surf.Sci. 188, 87 (1987); R.G.Jones and D.L.Perry, Vacuum, 31, 493 (1981); P.A.Dowben, S.Varma, D.R.Muller, and M.Onellion, Z.Physik B73, 247 (1988) and references therein.
[4] G.Vidali and C.W.Hutchings, Phys.Rev.B37, 10374 (1988).
[5] P.A.Dowben, Y.J.Kime, C.W.Hutchings, W.Li, and G.Vidali, Surf.Sci. in press.
[6] G.Vidali, C.W.Hutchings, P.A.Dowben, M.Karimi, C.Moses, and M.Foresti, in Proceedings of the 36th American Vacuum Society Meeting (Boston, 1989); C.W.Hutchings, W.Li, M.Karimi, C.Moses, P.A.Dowben and G.Vidali, in preparation.
[7] P.A.Dowben, Y.J.Kime, D.LaGraffe, and M.Onellion, to appear in Surf. and Int. Analy. 15 (1990); M.Onellion, Y.J.Kime, P.A.Dowben, and N.Tache, J.Phys.C: Solid State, 20, L633 (1987).

PART III

Silicide Interface
Reactions and Structure

AN INTERFACIAL PHASE TRANSFORMATION
IN CoSi$_2$/Si(111)

D.J. Eaglesham, R.T. Tung, R.L. Headrick, I.K. Robinson and F. Schrey

AT&T Bell Labs,
600 Mountain Avenue,
Murray Hill,
NJ 07974

ABSTRACT

A new type of phase transformation at the interface is described in CoSi$_2$/Si(111) B (i.e. twinned) epilayers. Thin (25Å) CoSi$_2$ films are codeposited at room temperature on Si (111) with a Si-rich surface layer, and subsequently annealed. Plan-view transmission electron microscopy (TEM) shows that these films have low symmetry, the interface being characterised by a shift "R" between (220) planes in CoSi$_2$ and Si. X-ray diffraction from "R" films differs from otherwise identical films either grown without Si-rich surface layer, or not annealed, which have the conventional cubic structure ("C"); R-CoSi$_2$ cannot be indexed on a single reciprocal lattice. Cross-section high-resolution TEM suggests the presence of a separate (non-cubic) layer ~9Å thick at the interface in these films. Annealing of R-CoSi$_2$ in-situ in the TEM shows a reversible transformation R⇔C occuring at temperatures varying from 180°C to 150K, depending on layer stoichiometry. R⇔C is thus a quasi-equilibrium, diffusionless transformation. We propose that R-CoSi$_2$ lowers interfacial free energy for certain stoichiometries, but that bulk constraints stop the entire layer from transforming.

CoSi$_2$ and NiSi$_2$ are the archetypal metal-semiconductor contacts, growing epitaxially on Si with low lattice mismatch to form a controllable Schottky barrier (see [1] for a recent review). Considerable effort has been devoted to the structural characterisation of CoSi$_2$/Si(111) interfaces in the twinned, or B orientation, with the interface structure being the subject of some debate. High Resolution Electron Microscopy (HREM) [2], and X-ray Standing Wave (XSW) [3,4] studies determined rigid shifts consistent with either the 5-fold or (with the same rigid shift) the 8-fold coordinated interface: the 8-fold is more probable from energetics calculations [5,6], is consistent with recent Surface Extended X-ray Absorption Fine Structure (SEXAFS) data [7], and matches the form of HREM images [8]. The system thus seemed to be structurally well-understood. Here we show that CoSi$_2$/Si(111) continues to provide surprises: under certain circumstances the epilayer can undergo a remarkable phase transformation in which a massive structural rearrangement appears, reversibly, at the interface.

We have used Transmission Electron Microscopy (TEM) of plan-view specimens to characterise thin B-CoSi$_2$/Si(111) films grown under a wide variety of conditions [1]. Films grown by UHV codeposition at room temperature and subsequently annealed *in situ* at temperatures >300°C can show a characteristic "domain" contrast, as shown in Fig. 1. Layers in which this domain contrast is visible shall be referred to here as "R-CoSi$_2$". Films which are Co-rich, (or unannealed, or grown above 300°C), do not show domain contrast, and we shall refer to these conventionally structured films as "vanilla" or C-CoSi$_2$.

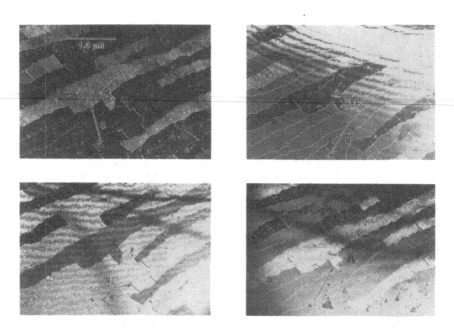

Fig. 1: Dark-field images in 3 symmetry-related <220>s at the [111] zone axis, with a -g image showing inversion

There are generally three levels of domain contrast in weak-beam <2$\bar{2}$0> dark-field images. A given domain cycles round the 3 contrast levels in images using the three <2$\bar{2}$0>s related by the 3-fold symmetry axis at [111]. This suggests a set of displacements **R** running along [$\bar{2}$11] [1$\bar{2}$1] and [11$\bar{2}$], which, imaged in **g** = [2$\bar{2}$0] will give 3 distinct values of **g.R**. The g-vector which produces neutral contrast in a given domain presumably has **g.R**=0 (i.e. g perpendicular to the displacement field) Weak beam images in <$\bar{4}$22> reflections show the domains, but with only 2 levels of contrast, suggesting that **g.R** takes only 2 values for this g. The observation of 3 levels of contrast in <2$\bar{2}$0> but only 2 in <$\bar{4}$22> images confirms that the contrast must arise from a displacement field lying along the <$\bar{2}$11>s. Reversing either the imaging g-vector or the sign of s_g (the deviation from the perfect Bragg position), is observed to invert the contrast of <2$\bar{2}$0> images. The domains are closely linked to the interface morphology, and in particular the boundaries between domains almost invariably occur at $1/6<\bar{2}11>$ interfacial dislocations, which mark the position of steps at the interface [10].

This domain structure is open to a variety of interpretations. At the simplest level, the domain contrast is an indication of lowered symmetry in the bicrystal: the 3 <220> reflections are no longer equivalent for a given domain. The two most straightforward ways to modify the bicrystal symmetry are either by modifying the epilayer structure, or by adding a rigid shift at the interface. Thus the silicide could be distorted, e.g. by shearing the lattice along a particular $<\bar{2}11>$ in a given domain (so that the epilayer becomes monoclinic). Alternatively, a cubic epilayer whose interface has one of three displacements parallel to a $<\bar{2}11>$ would also lower bicrystal symmetry. These two classes of structural changes can be distinguished by imaging in an inclined $<11\bar{1}>_B$ reflection: rigid shift contrast at the interface arises from interference between diffraction from the upper and lower crystals, and thus domains should not appear in reflections from the silicide alone unless structural distortions dominate. Fig. 2 shows a typical pair of $<11\bar{1}>_B$ images (at high and low tilt from the [111]), and the contrast suggests that the domains do not arise from rigid shifts. (Contrast is similarly seen in other inclined silicide reflections.)

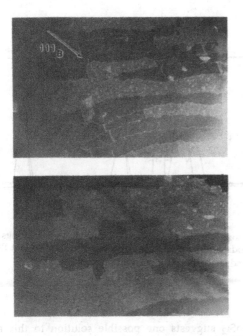

Fig. 2 Dark-field images in a <111>B reflection: since contrast comes from the epilayer alone, this at first sight suggests that "R" is not an interface phenomenon

X-ray diffraction provides a more quantitative picture of the changes in the reciprocal lattice, although it will average over all the domains. Fig. 3 shows typical rod scans for R and C structures along the 20*l* and T0*l* rods (i.e. 2/3<$\bar{4}$22>+l[111] and -1/3<$\bar{4}$22>+l[111] rods in reciprocal space, expressed in the hexagonal indices appropriate for the Si(111) surface [9]). The structural rearrangement is evident in the strong modification of the 20l rod, where the <31$\bar{1}$>$_B$ and <040>$_B$ R-CoSi$_2$ peaks are shifted and split about the C positions. The 10*l* and 11*l* rods show similar splittings and shifts, though to a rather lesser degree. The explanation for the splittings seems likely to be related to the presence of three inequivalent domains, which are averaged in X-ray diffraction. Thus 10*l*, 11*l*, and 20*l* rods are consistent with a distortion of the epilayer in which the lattice is either sheared or rotated (although it is difficult to find a distortion which is exactly consistent with all three rods). The problem with this picture lies in the T0*l* rod, where the <1$\bar{1}$1>$_B$ and <022>$_B$ silicide reflections are left invariant. In order to move the positions of the <31$\bar{1}$> and <040> without splitting any of the 3 <1$\bar{1}$1>$_B$ reflections (i.e. without moving the <1$\bar{1}$1>$_B$ in any of the 3 domains), the effective reciprocal lattice of R-CoSi$_2$ must be buckled, i.e. aperiodic. Thus X-ray diffraction appears only to be consistent with a picture in which the structural distortion of R-CoSi$_2$ is not uniform through the epilayer, but has a shape-transform that extends far from the silicide reciprocal lattice. Simultaneously, the strength of the modifications to the silicide diffraction pattern suggests that a single layer alone may not be sufficient (note that the C-CoSi$_2$ diffraction closely resembles the expected behaviour for the strained bulk).

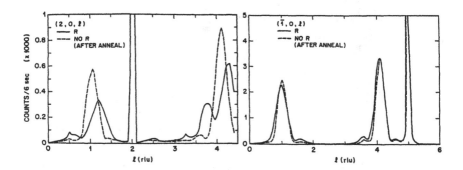

Fig. 3: X-Ray diffraction along the (20*l* and T0*l* rods, showing large splittings in all reflections along certain rods, while others are unaffected: this implies that R-CoSi$_2$ lacks a reciprocal lattice, and suggests that R is an interface phenomenon.

HREM of this R-CoSi$_2$ suggests one possible solution to this apparent dilemma. Fig. 4 shows HREM images from R-CoSi$_2$ cross section samples prepared by 4kV ion milling at 25°C (a) and 2kV ion milling at 100K (b). While (a) shows all the features we would expect of C-CoSi$_2$, (b) appears to show a separate layer, more than one monolayer thick. While interpretation of HREM is difficult, the interface region in 4(b) appears to be drastically different from 4(a): while 4(a) matches simulated images for the C structure [8], no match can be found for 4(b): thus it currently seems possible (in view of HREM, and the apparent conflict between X-ray diffraction and plan-view TEM) that R-CoSi$_2$ has a broad interface region whose structure differs greatly from that of either silicide or Si: this interface region lowers the symmetry of the bicrystal, and thus gives rise to a domain structure.

During the TEM experiments described above, it was noted that electron irradiation of R films over prolonged periods leads to a phase transformation into the high-symmetry structure which we observe in unannealed films. The rate of this damage transformation increased rapidly with increasing voltage, occurring in a few seconds at 200kV. A similar transformation can be produced by He ion irradiation, the characteristic interface peak in RBS disappearing during observation after doses as low as 5×10^{14} cm^{-2}: TEM of irradiated material confirms that the loss of the interface peak coincides with loss of R.

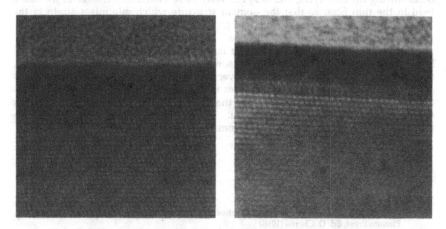

Fig. 4: HREM images of the silicide-Si interface for (a) C-CoSi$_2$ and (b) R-CoSi$_2$. Note that the {111} planes in (a) change direction abrubtly at the interface, while those in (b) appear vertical over a finite distance near the interface.

The transformation from R to C is irreversible when initiated by radiation damage effects. However, we have also observed a reversible transformation of the same kind occuring at ≈100-200°C on heating these films *in situ* in the electron microscope. Both the R⇒C (heating) and the C⇒R (cooling) phase transformations proceed slowly, but the transformation rate increasing rapidly with increasing overheating or undercooling. As for the irradiation-induced transformation, there is a marked tendency for nucleation of the C phase to occur at domain boundaries (i.e. at 1/6<$\bar{2}$11> dislocations). There is evidence for strong pinning of the transformation front, as might be expected for a highly-strained transformation. Not all domains in a given region transform at the same time, and we have observed large local variations in the transformation temperature. Indeed, many films are mixed in character at room temperature: on heating, the fraction of C increases, and on cooling to liquid nitrogen temperatures the interface transforms fully into the R phase. In addition, a thickness-dependence of the transformation was observed in the thin TEM samples, with areas where the substrate was ≈1μm thick undergoing the transformation (in both directions) at temperatures on average ≈10-20°C higher than regions where the sample thickness was only 500Å. This effect seems to imply a strong link between the transformation temperature and the epilayer strain; the most obvious effect of sample thickness is strain relaxation in the thin films, with an appreciable fraction of the epilayer strain being taken up in the substrate for the thinnest regions of sample. Extrapolation of this thickness dependence suggests that the transformation is likely to occur at ≈220°C(±30°C) in bulk material. The transformation was reversible, with domains of a given type almost invariably reverting to the same type after heating into the unsheared form and subsequent quenching back into the sheared state. On a few occasions a domain was observed to switch type on thermal cycling. This tended to occur during the first cycle, and is again likely to be linked to a change in the strain field in the thin film, so that the domain structure which minimises strain in the TEM sample may be different from that produced following the anneal in the bulk.

In conclusion, we have observed an interfacial phase transformation in annealed $CoSi_2/Si(111)$ epilayers. The low temperature phase has an interfacial layer several monolayers thick. At some strain- and composition-dependent temperature, the silicide transforms reversibly into the cubic phase with a high-symmetry interface structure which does not yield an interface peak in RBS. We deduce that the cubic interface is not the equilibrium structure for films at this stoichiometry at room temperature. The driving force for this phase transformation must arise principally from the interfacial energy.

REFERENCES

[1]   R.T. Tung in *Electron Microscopy Evaluation of Electronic Materials*, NATO Adv. Workshop, Plenum Press, Ed. D. Cherns (1988)

[2]   J.M. Gibson, J.C. Bean, J.M. Poate, and R.T. Tung, *Appl. Phys. Lett.* **41**, 818 (1982).

[3]   A.E.M.J. Fischer, E. Vlieg, J.F. van der Veen, M. Clausenitzer, and G. Materlik, *Phys. Rev*, **B36**, 4769, (1987).

[4]   J. Zegenhagen, K.-G. Huang, B.D. Hunt, and L.J. Schowalter, *Appl. Phys. Lett.* **51**, 1176 (1987).

[5]   D.R. Hamann, *Phys. Rev. Lett.* **60**, 313 (1988).

[6]   P.J. van der Hoek, W. Ravenek, and E.J. Baerends, *Phys. Rev. Lett.* **60**, 1743 (1988).

[7]   G. Rossi, X. Jin, A. Santaniello, P. DePadova, and D. Chandesris, *Phys. Rev. Lett.* **62**, 191 (1989).

[8]   C.W.T. Bulle-Lieuwma, A.F. de Jong, A.H. van Ommen, J.F. ven der Veen, and J. Vrimoth, *Appl. Phys. Lett.* **55**, 648 (1989).

[9]   I.K. Robinson, W.K.Waskiewicz, R.T. Tung, and J. Bohr, Phys. Rev. Lett. 57, 2714 (1986).

[10]   D.N. Jamieson, G. Bai, Y.C. Kao, C.W. Nieh, M.-A. Nicolet, and K.L. Wang, *MRS Symp. Proc.*, **91**, 479 (1987).

[11]   R.T. Tung, J.L. Batstone, and S.M. Yalisove, *J. Electrochem. Soc* (1989), in press.

[12]   R.F.C. Farrow, D.S. Robertson, G.M. Williams, A.G. Cullis, G.R. Jones, I.M. Young, and P.N.J. Davies, *J. Cryst. Growth* **54**, 507 (1981).

ATOMIC SCALE STUDY OF CoSi/Si (111) AND CoSi$_2$/Si (111) INTERFACES

A. CATANA, M. HEINTZE, P.E. SCHMID, AND P. STADELMANN
Institute of Applied Physics, Swiss Federal Institute of Technology
1015 Lausanne, Switzerland

ABSTRACT

High Resolution Electron Microscopy (HREM) was used to study microstructural changes related to the CoSi/Si-CoSi/CoSi$_2$/Si-CoSi$_2$/Si transformations. CoSi is found to grow epitaxially on Si with $[1\bar{1}\bar{1}]$Si // $[1\bar{1}\bar{1}]$CoSi and < 110 >Si // < 112 >CoSi. Two CoSi non-equivalent orientations (rotated by 180° around the substrate normal) can occur in this plane. They can be clearly distinguished by HRTEM on cross-sections (electron beam along [110]Si). At about 500°C CoSi transforms to CoSi$_2$. Experimental results show that the type B orientation relationship satisfying [110]Si // $[1\bar{1}2]$CoSi is preserved after the initial stage of CoSi$_2$ formation. At this stage an epitaxial CoSi/CoSi$_2$/Si(111) system is obtained. The atomic scale investigation of the CoSi$_2$/Si interface shows that a 7-fold coordination of the cobalt atoms is observed in both type A and type B epitaxies.

INTRODUCTION

Recently, an increasing number of studies have focused on the understanding of the formation of cobalt silicide and characterization of the CoSi$_2$/Si interfaces. The great interest generated by this system is due both to its applicability in the electron device technology and to the fundamental questions related to its formation path [1,2,3]. It has already been reported that low temperature annealing (< 500°C) of Co layers on Si(111) under UHV ambient results in the formation of metal rich silicides (Co$_2$Si, CoSi) which progressively transform to CoSi$_2$ [4,5]. Since CoSi transforms directly to CoSi$_2$, a better knowledge of the CoSi/Si interface should improve the understanding of the complex mechanisms of atomic interfacial intermixing and rearrangement. There are two epitaxial orientations of CoSi$_2$ on Si (111): type A is aligned with the Si substrate and type B is 180° rotated around the substrate normal. It is of interest to know whether the final CoSi$_2$ orientation with respect to the Si substrate depends on the CoSi/Si interface structure. In the second part of the paper, results are presented on the atomic configuration at the A- and B-type CoSi$_2$/Si(111) interfaces.

EXPERIMENTAL

Thin cobalt layers (2-10 nm) were deposited by a differentially pumped e-gun onto heat-cleaned, <111>-oriented Si substrates in a UHV chamber with a base pressure of 2×10^{-8} Pa. The chemical composition and contaminants were checked by Auger spectroscopy. The samples were then annealed by Joule heating of the substrate to 400, 500, 600 and 900°C for 2 minutes. From each substrate both flat-on and cross-sectional samples were prepared for high resolution electron microscopy (HREM) investigations using mechanical polishing and conventional ion-thinning techniques. The observations were performed on a Phillips 430ST microscope with a point resolution better than 0.2 nm. The crystallographic analysis and image calculations were carried out using a software package developed by one of us [6].

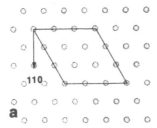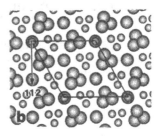

Figure 1: (a) Si and (b) CoSi crystal projections along the interface normal. On both projections, nearly coincident supercells are represented.

RESULTS AND DISCUSSION

1. CoSi/Si - CoSi$_2$/Si transformation

Annealing at 400°C results in the formation of epitaxial CoSi. Flat-on diffraction investigations show $[1\bar{1}\bar{1}]$CoSi // $[1\bar{1}\bar{1}]$Si. The relative rotation between both crystals around this axis is such that $<112>$CoSi is parallel to $<110>$Si. This orientation corresponds to the best two-dimensional lattice match in the common $(1\bar{1}\bar{1})$ interface planes. Projections of both crystals normal to the interface plane are displayed in Fig. 1. In order to understand whether the original CoSi/Si structure has an impact on the final CoSi$_2$/Si epitaxy, a first step is the identification of the orientation of CoSi on Si. Since the exact orientation relationship between CoSi and Si cannot be determined unambiguously from the flat-on diffraction pattern, cross-sectional samples were prepared. From a theoretical point of view, two orientations of CoSi satisfy the diffraction results obtained on flat-on samples, namely $[\bar{1}1\bar{2}]$CoSi // $[110]$Si (type A) or $[1\bar{1}2]$CoSi // $[110]$Si (type B). They are related by a 180° rotation around the interface normal. Note that an identical rotation characterizes the CoSi$_2$/Si (111) epitaxies type A and type B. Fig. 2 shows structural and atomic potential projections of CoSi along both $[\bar{1}1\bar{2}]$ and $[1\bar{1}2]$ axis. It is interesting to note that in this orientation, the silicide shows planes of either Co or Si atomic composition parallel to $(1\bar{1}\bar{1})$Si. HREM images were calculated using the atomic potentials reported in Fig. 2 for both CoSi orientations (Fig. 3). The calculations were performed for a 6.5 nm thick sample at 4 different defocus values, namely 57 nm, 72 nm, 87 nm and 102 nm. The microscope parameters are the following: spherical aberation coefficient = 1.1 mm, spread of focus = 10 nm and beam semiconvergence = 1 mrad. At 57 and 102 nm defocus atoms are imaged in black. The contrast is reversed at 87 nm. In that case, bright spots mark atomic positions and brightness maxima correspond to metal atoms. At 72 nm defocus Co atoms are imaged as bright spots whereas Si and Si-Co groups of atoms are imaged in black. It is therefore interesting to have high resolution observations recorded at this defocus and thickness, since this will help for the exact positioning of the interface and identification of the silicide atomic species at the interface. This study is presently in progress. In order to distinguish the CoSi twin orientations (A and B) the 72 and 102 nm defocus settings are the most appropriate. A HREM micrograph of a sample annealed at 400°C is shown in Fig. 4a. In this case $[1\bar{1}2]$ CoSi is parallel to $[110]$ Si (type B). Annealing at 450-500°C results in a partial transformation of CoSi to CoSi$_2$. Since

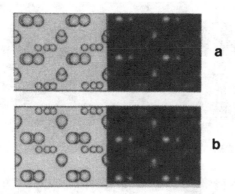

Figure 2: Projections of the CoSi structure (left) and corresponding calculated atomic potentials (right) for Type A (a) and type B (b) epitaxies. The larger dots correspond to Co atoms.

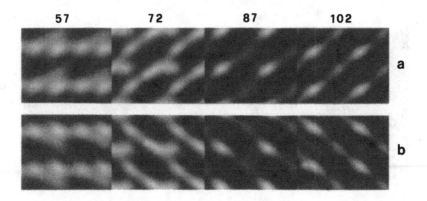

Figure 3: Image calculations along (a) $[\bar{1}1\bar{2}]$ and (b) $[1\bar{1}2]$ for a sample thickness 6.5 nm. The numbers indicate the defocus in nm.

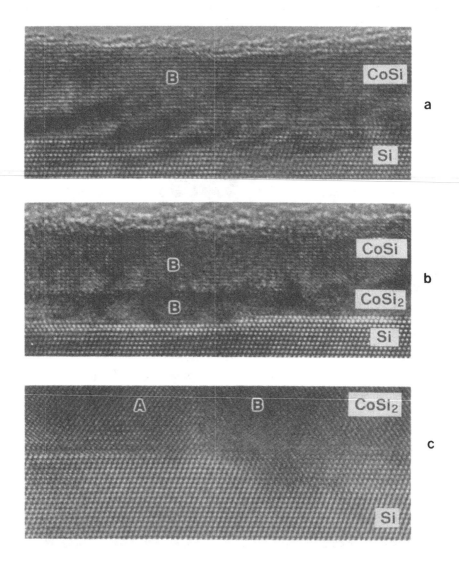

Figure 4: Cross-sectional micrographs of Co/Si samples annealed at 400°C (a), 500°C (b) and 600°C (c).

Figure 5: Simulated images (left) and experimental image (right) for a 4 nm thick sample at 48 nm (a,c) and 93 nm (b,d) defocus for A-type (a,b) and B-type (c,d) CoSi$_2$/Si epitaxies. At the left of the calculated images dots mark the positions of atomic columns.

no region with CoSi$_2$ on top of CoSi could be observed we conclude that the nucleation of CoSi$_2$ proceeds at the CoSi/Si interface. This model is supported by observations of epitaxial CoSi on top of the CoSi$_2$ layer (Fig. 4b). On the basis of these results, the formation of CoSi$_2$ at the upper CoSi interface by diffusion of Si through the intermediate CoSi layer can be ruled out. The epitaxial relationships of the type B CoSi/CoSi$_2$/Si system are described by $[1\bar{1}1]$CoSi$//[1\bar{1}1]$CoSi$_2//[1\bar{1}1]$Si and $[1\bar{1}2]$CoSi$//[\bar{1}10]$CoSi$_2//[110]$Si. Annealing at temperatures about 600°C results in a mixed A and B type CoSi$_2$ layer (Fig. 4c). Further investigations of incompletely reacted samples should provide observations of type A epitaxies. These are less frequently observed since the annealing process that transforms CoSi into CoSi$_2$ is also promoting the conversion of type A to the more stable type B epitaxy. After processing at higher temperatures (> 600°C) the silicide shows exclusively B-type oriented CoSi$_2$ regions.

2. HREM at the CoSi$_2$(A)/Si (111) interface

For the study of the interface bonding, we consider 3 models that differ in the coordination of the metal atoms at the interface: 5-fold, 7-fold and 8-fold. HREM micrographs were recorded under various focusing conditions and compared to calculated images of the 3 interface models. Previously, a careful examination of the specimen thickness and calibration of the focusing conditions were carried out. Experimental and calculated images for two focusing conditions (48 nm and 93 nm) and a sample thickness of 4 nm are displayed for both A and B type $CoSi_2$ epitaxies in Fig. 5. A more detailed study of the B-type $CoSi_2$/Si (111) epitaxy is reported elsewhere [7]. At the top of the calculated image, a projection of the atomic columns is superimposed. The results show that for both type A and type B epitaxies, the interface is supported by Si-Si bonds and a 7-fold coordination of the first metal layer. This configuration is the same as the one described previously for $NiSi_2$ [8]. From a theoretical point of view, the 7-fold model corresponds to the most stable A type interface and to the second most stable B type interface [9]. Other groups are in favor of 5 [10] and 8-fold [11,12,13] coordinations of the Co atoms at the interface.

CONCLUSION

In summary, structural changes related with the $CoSi$/Si-$CoSi_2$/Si transformation were investigated using cross-sectional HREM. CoSi is found to grow epitaxially on Si(111) and it remains epitaxial after the early stages of $CoSi_2$ formation, preserving its orientation on Si. Two different orientations of CoSi can occur on Si(111). They correspond to a 180° rotation around the interface normal and can be clearly identified on cross-sectional HRTEM micrographs under proper imaging conditions. Whether a given CoSi orientation conditions the growth of A or B type $CoSi_2$ still remains open. This is due to the lack of experimental data on $CoSi$/$CoSi_2$(A)/Si structures. However our results show that B-type $CoSi_2$ grows at B-type CoSi/Si interfaces. The investigation of the atomic structure at both A and B type $CoSi_2$/Si interfaces shows the presence of 7-fold coordinated Co and Co-Si interfacial bonds.

BIBLIOGRAPHY

1. R.T. Tung, A.F.J. Levi and J.M. Gibson Appl. Phys. Lett. **48**, 635 (1986)
2. J.M. Phillips, J.L. Batstone, J.C. Hensel, M. Cerullo and F.C. Unterwald
   J. Mater. Res. **4**, 145 (1989)
3. C. d'Anterroches Surf. Sci. **168**, 751 (1986)
4. J.M. Gibson, J.L.Batstone and R.T. Tung Appl. Phys. Lett. **51**, 45 (1987)
5. A. Appelbaum, R.V. Knoell and S.P. Murarka J. Appl. Phys. **57**, 1880 (1985)
6. P. Stadelmann, Ultramicroscopy **21**, 131 (1987)
7. A. Catana, P.E. Schmid, S. Rieubland, F. Lévy and P. Stadelmann
   J. Phys. Condens. Matter **1**, 3999 (1989)
8. D. Cherns, G.R. Anstis and J.L. Hutchison Phil. Mag. **46A**, 849 (1982)
9. D.R. Hamann Phys. Rev. Lett. **60**, 313 (1988)
10. J.M. Gibson, J.C. Bean, J.M. Poate and R.T. Tung
    Appl. Phys. Lett. **41**, 818 (1982)
11. G. Rossi, X. Jiu, A. Santaniello, P. DePadova and D. Chandesris
    Phys. Rev. Lett. **62**, 191 (1989)
12. A.E.M.J. Fisher, T. Gustafsson and J.F. Van der Veen
    Phys. Rev. **B37**, 6305 (1988)
13. C.W.T. Bulle-Lieuwma, A.F. de Jong, A.H. van Ommen, J.F. van der Veen
    and J. Vrijmoeth to be published in Appl. Phys. Lett.

MICROSTRUCTURAL ASPECTS OF NICKEL SILICIDE FORMATION IN EVAPORATED NICKEL-SILICON MULTILAYER THIN FILMS

KAREN HOLLOWAY AND LARRY CLEVENGER
IBM Thomas J. Watson Research Center, Yorktown Heights, NY 10598

ABSTRACT

The early stages of the nucleation and growth of nickel silicides in Ni-Si multilayers evaporated onto oxide-stripped < 100 > Si substrates and annealed at 150 °C have been studied by cross-section transmission electron microscopy (TEM). Observed differences in the interaction of evaporated Ni with amorphous silicon and single crystal < 100 > Si have been explained by thermodynamic modeling of the Ni-Si system. The as-deposited films show a 3 nm amorphous Ni-Si intermixed layer at all Ni-Si interfaces, including that with the single-crystal Si substrate. Crystalline Ni_2Si formed in all annealed films, consuming the elemental Ni layers. The amorphous alloy layer grows concurrently with Ni_2Si during the reaction with a-Si; however, there is no amorphous phase present at the Ni_2Si - < 100 > Si interface. Thermodynamic calculations show that at 150 °C metastable equilibrium might be expected between a-Si, the amorphous phase, and Ni_2Si; but not between Ni_2Si, the amorphous phase, and crystalline Si. The composition of the amorphous phase is very close to a-$Ni_{50}Si_{50}$. After a 6 hour anneal at 150 °C, crystalline NiSi forms between the a-Si and the Ni_2Si layers by crystallization of the amorphous phase. Further annealing is necessary to form NiSi at the < 100 > Si - Ni_2Si interfaces.

INTRODUCTION

The formation of an amorphous silicide phase at a metal-silicon interface during the process of deposition has been reported for such a large number of systems, that this may be regarded as a common phenomenon [1]. Further interdiffusion on annealing of some metals with amorphous silicon results in the growth of this amorphous alloy prior to the nucleation of the first crystalline silicide phase [3-6]. In a similar reaction, the amorphous interlayer formed on deposition of Ti or Rh onto a single-crystal Si substrate, grows to a thickness on the order of 10 nm [7,8].

Such an amorphization reaction has been most frequently discussed for reactions of metals with amorphous Si; however, the technological applications of silicides usually require metal reactions with crystalline Si. For this reasons, we have examined the early stages of interaction at the metal-single crystal Si interface and the a-Si interface for the Ni-Si system. Many aspects of the Ni amorphization reaction with a-Si have already been investigated. After the formation and growth upon anneal of the amorphous alloy between the Ni and a-Si layers, a layer of crystalline Ni_2Si nucleates at the interface between the amorphous alloy and polycrystalline Ni. Further heating causes the two intermixed layers to grow concurrently [9,10]. Earlier studies of the interaction between thin Ni layers and single-crystal Si have indicated that Ni_2Si is the first crystalline silicide to form and that NiSi forms once the elemental Ni is consumed [5,11]. In the present study, we have used cross-section TEM to study the evolving microstructure of a-Ni_xSi_y, Ni_2Si, and NiSi formation and growth in evaporated Ni-Si multilayer thin films. Our observations are explained by a thermodynamic model of the Ni-Si system approximating the amorphous alloy as an under-cooled liquid.

EXPERIMENT

Nickel-silicon multilayers were produced by evaporating alternating layers of the elements onto one inch <100> Si substrates in a multiple-hearth e-beam system. Its base pressure is 8 x 10⁻⁸ torr, and the vacuum is about 10⁻⁷ torr during deposition. Each wafer was given a standard Huang cleaning procedure, except that the final DI rinse was omitted. Instead, the wafers were pulled from a dilute HF solution, and immediately loaded and pumped down in the vacuum system. Nickel was evaporated first, then Si, and so on until the last layer, always Si, was deposited. Between layers, the samples were shuttered and the hearth was allowed to cool so that intermixing does not occur as an artefact of deposition. The sample used in the present study consisted of 8 Ni-Si bilayers; the nickel layers were nominally 10 nm thick, and the Si was 18 nm thick, for an overall composition ratio of 1:1 Ni to Si. The actual composition, determined by Rutherford Backscattering, was 53 at.% Ni, 47 at.% Si.

Cross-section samples for transmission electron microscopy (TEM) were prepared by a low-temperature modification of the Bravman-Sinclair process [12,13]. The temperature reached during any part of this procedure was not higher than 100 °C. An as-deposited 1 inch sample was subjected to the initial mechanical thinning and polishing steps of TEM sample preparation before annealing in air at 150 ±2 °C in a conventional convection furnace. The annealed samples then underwent final thinning and ion-milling in an LN₂-cooled stage. All TEM experiments were performed in a JEOL 200CX operating at 200 kV. Each cross-section sample was aligned edge-on with respect to the electron beam using the <110> zone axis of the <100> silicon substrate.

RESULTS AND DISCUSSION

Microstructural Observations

The evolution of the microstructure of the Ni-Si multilayers in the as-deposited state and after annealing for 2,6, and 24 hours at 150 °C is portrayed in Figure 1(a-d). The unannealed stack (Figure 1a) consists of 7 nm thick polycrystalline nickel layers and 10 nm amorphous silicon layers. The selected area diffraction (SAD) pattern taken from this area (not shown) shows that the fcc Ni has a strong (111) texture in the growth direction. At each Ni-amorphous silicon interface, there is a 3 nm uniform and planar amorphous intermixed layer [1,11]. There is also an amorphous layer at the interface between Ni and the <100> Si substrate. Its contrast level is similar to that of the other intermixed Ni-Si layers; it does not show the typical lighter appearance of amorphous SiO₂.

After annealing at 150 °C for 2 hours (Figure 1b), the Ni layers have been consumed by the formation of δ-Ni₂Si. Electron diffraction from this area shows that the silicide has a moderate (400) texture in the growth direction. The a-Si layers have also been partly consumed; they are now only about 3 nm thick. Between each Ni₂Si and a-Si layer is a 5 nm amorphous alloy interlayer. The concurrent growth of this phase along with the crystalline dimetal silicide during the reaction of Ni with a-Si has been described earlier [9,10]. Note, however,

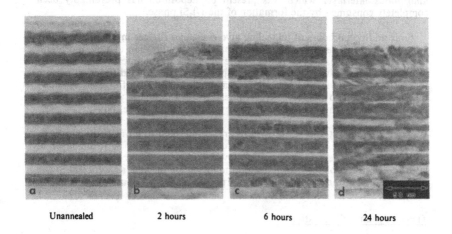

| Unannealed | 2 hours | 6 hours | 24 hours |

Figure 1(a-d) Cross-section TEM micrographs of the Ni-Si multilayer sample on deposition (a), and after annealing at 150 °C for 2, 6 and 24 hours (b-d, respectively).

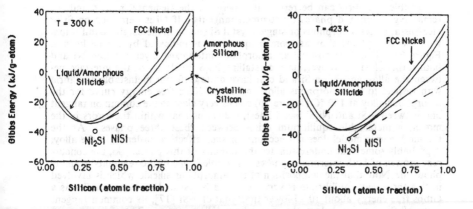

Figure 2(a) Calculated Gibbs free-energy version composition diagram of the Ni-Si system at room temperature and (b) at 150 °C.

that there is no amorphous phase present at the Ni_2Si - <100> Si interface. The amorphous interlayer which was present on deposition has presumably been completely consumed by the formation of the Ni_2Si phase.

Four additional hours of annealing give rise to no further intermixing (Figure 1c). However, there is evidence of the partial crystallization of the amorphous phase. In some areas, bend contours are apparent in the interlayers. High resolution imaging shows lattice fringes in places. An electron diffraction pattern taken from a plan-view TEM sample of a similarly annealed film, which allows the elecron beam to intersect several interfaces in the multilayer, shows several new rings which can be indexed for the crystalline monosilicide, NiSi.

After 24 hours at 150 °C, growth of crystalline NiSi to 6 nm is the most evident feature of the Ni-Si multilayer structure (Figure 1d). A 6 nm layer of Ni_2Si remains in the center of the top seven Ni layers. The bottom-most Ni_2Si layer has been completely consumed by the formation of NiSi, as sufficient Si can be supplied from the substrate. No amorphous intermixed phase remains in this sample.

Thermodynamic Calculations

The Gibbs free energies at 150 °C of elemental (fcc) Ni, covalently-bonded amorphous Si, crystalline Si, crystalline Ni silicides, and the amorphous intermixed phase, modeled as an undercooled liquid, were calculated as a function of composition based on data reported by Mey [14], Donovan, etal.[15], and DeAvillez, et al.[16]. The microstructural observations can be explained by consideration of this diagram, shown in Figure 2(a-b). If the formation of one or more of the crystalline silicides is suppressed by a nucleation or kinetic barrier, metastable equilibria can be represented between the elemental phases and the remaining intermixed phases by common tangents. If Ni is deposited on Si or vice-versa, either amorphous or single-crystal Si can be in metastable equilibrium with the amorphous alloy at room temperature as indicated by the tie lines in Figure 2a. Thus, we observe an amorphous intermixed layer between Ni and both forms of Si. However, once crystalline Ni_2Si nucleates at the interface between the alloy and Ni, a-Si and c-Si have quite different metastable equilibria with respect to the amorphous alloy and Ni_2Si. The free energy curve for the amorphous alloy at 150 °C (Figure 2b) comes very close to the commmon tangent line between a-Si and Ni_2Si (solid lines), indicating that within the errors in the model, a metastable equilibrium exists between these three phases. As the thermodynamic model does not account for such effects as ordering in the alloy, it probably somewhat underestimates the stability of this phase. So, the concurrent growth of both intermixed phases is possible if one of the reactants is amorphous Si. Note that the composition of the amorphous silicide which is involved in this three-way equilibrium is very close to a-$Ni_{50}Si_{50}$. However, as c-Si has a Gibbs free energy about 10 kJ lower than that of a-Si [17], its common tangent with Ni_2Si lies far below the amorphous alloy curve. Therefore, even a metastable equilibrium is not possible between c-Si and the amorphous phase once Ni_2Si is present. This is consistent with our observation that there is no amorphous phase at the interface between the bottom Ni_2Si layer and the <100> Si substrate. Once the Ni layers have been consumed by the formation of the dimetal silicide, a further decrease in free energy can only be obtained through the formation of a more Si-rich crystalline phase, NiSi. This occurs first in contact with a-Si by the crystallization of the amorphous silicide, which is equiatomic in composition. Further annealing is necessary to nucleate NiSi at the Ni_2Si - c-Si interface. The common tangents between NiSi and both a-Si and c-Si (dashed lines) lie far below the free energy of the amorphous phase; accordingly, we observe that no amorphous phase remains once NiSi is established.

CONCLUSIONS

1. A 3 nm amorphous Ni-Si alloy is formed on deposition at the interface of Ni with both amorphous Si and single-crystal < 100 > Si.
2. Crystalline Ni_2Si forms on annealing at 150 °C. It grows concurrently with the amorphous alloy at the interface with a-Si.
3. No amorphous alloy is present at the interface with < 100 > Si after Ni_2Si formation. Thermodynamic calculations show that once Ni_2Si is present, the alloy can be metastable with respect to a- Si, but not with crystalline Si. The amorphous silicide composition is close to a-$Ni_{50}Si_{50}$.
4. The amorphous alloy crystallizes to form NiSi after 6 hours at 150°C. The monosilicide subsequently grows to consume the a-Si and Ni_2Si.

ACKNOWLEDGEMENTS

We wish to thank Ron Petkie for deposition of the multilayer samples, and Grant Coleman for RBS measurements. Also, the contributions of R.R. deAvillez for the calculation of the Ni-Si free energy diagrams, and C.V. Thompson, K.N. Tu and E. Ma for helpful discussions are gratefully acknowledged.

REFERENCES

1. K. Holloway, R. Sinclair, M. Nathan, J. Vac. Sci. Technol. A 7, 1479 (1989).
2. S. Herd, K.N. Tu, and K.Y. Ahn, Appl. Phys. Lett. 42, 599 (1983).
3. K. Holloway and R. Sinclair, J. Appl. Phys. 61, 1359 (1987).
4. M. Natan, J. Vac. Sci. Technol. B4, 1404 (1986).
5. M.O. Aboelfotoh, H.M. Tawancy, and F.M. d'Heurle, Appl. Phys. Lett. 50, 1453 (1987).
6. M. Nathan, J. Appl. Phys. 63, 5534 (1988).
7. W. Lur and L.J. Chen, Appl. Phys. Lett. 54, 1217 (1989).
8. S.R. Herd and K.N. Tu, private communication.
9. E. Ma, W.J. Meng, W.L. Johnson, M.-A. Nicolet and M. Nathan, Appl. Phys. Lett. 53, 2033 (1988).
10. L.A. Clevenger and C.V. Thompson, to be published in J. Appl. Phys.
11. H. Foll, P.S Ho and K.N. Tu, Phil Mag. A 45, 31 (1982).
12. J.C. Bravman and R. Sinclair, J. Electron Microsc. Tech. 1, 53 (1984).
13. K. Holloway and R. Sinclair, Mater. Res. Symp. Proc. 77, 357, (1986).
14. Sabine an Mey, Z. Metallkde. 77, 805 (1986).
15. E.P. Donovan, F. Spaepen, D. Turnbull, J.M. Poate and D.C. Jocobson, J. Appl. Phys. 57, 4208 (1984).
16. R.R. DeAvillez, L.A. Clevenger, C.V. Thompson, and E. Ma, unpublished.
17. J.C. Fan and C.H. Anderson, J. Appl. Phys. 52, 4003 (1981).

STRANSKI-KRASTANOV GROWTH OF Ni ON Si(111) AT ROOM TEMPERATURE

J. R. BUTLER AND P. A. BENNETT
Physics Department, Arizona State University, Tempe, Arizona 85287

ABSTRACT

We introduce quantitative Auger lineshape analysis methods to study the room temperature reaction of nickel on Si(111). We show that coexisting phases may be separated by numerically fitting the composite lineshapes using a linear combination of single phase "fingerprint" spectra, obtained by scraping bulk compounds in situ. The reaction proceeds in three stages. For nickel coverage below 1 Å, the growth is layerwise, forming $NiSi_2$. For nickel coverage from 3 to 12 Å, islands of Ni_2Si are formed. For nickel coverage above 12 Å, islands of pure nickel are formed. The overlayer reactions appear to be a kinetically limited form of Stranski-Krastanov growth, with multiple compound formation.

BACKGROUND

Compound formation in contact reactions of ultrathin film silicides (h $<\sim$ 50 Å) is typically kinetically limited, with the result that the path of reaction can be altered by manipulation of the initial or growth conditions. Examples of this effect include the template selection of A-type or B-type $NiSi_2$ on Si(111) [1], and the epitaxial growth at room temperature of $NiSi_2$ or CoSi2 by codeposition of metal and silicon [2, 3]. In the latter case, it is found that the crystalline quality of the overlayer is markedly improved with a pre-deposition of 2.5 Å of nickel, before initiating codeposition [4]. In both cases, the structure of "usefully thick" layers (h $>\sim$ 100 Å) is determined by reactions occurring at low coverage and low temperature, prior to growth of the overlayer. Such effects are increasingly under study on an empirical basis, yet a fundamental and/or predictive understanding is lacking. For this it is first necessary to properly characterize the nature of reactions at low temperature and coverage.

Room temperature silicide formation for Ni/Si(111) has been studied often in the past, but with conflicting results. It is clear that some form of silicide reaction occurs for the first 10 Å of nickel deposited, but the basic nature of the reaction remains controversial. Thus, the material first formed at room temperature is variously claimed to be: $NiSi_2$, using SEXAFS [5], low energy ion backscattering (ICISS) [6] or UHV-TEM techniques [7]; NiSi using XPS [8]; or Ni_2Si using RBS [9], RHEED [10] or Auger techniques [11]. A fundamental difficulty shared by many such techniques is the poor ability to identify and separate coexisting compounds. At low temperatures, it can be expected that a large composition gradient exists, due to limited diffusion, and that multiple phases of silicide compounds will coexist through the depth of the film. In this paper we show that Auger lineshape analysis can quantitatively separate such coexisting phases, and we use this to monitor with submonolayer sensitivity the thickness dependence of reactions in the Ni/Si(111) system.

METHOD

The lineshapes of CVV Auger transitions clearly are rich in chemical information [12], and have often been used in a _qualitative_ fashion to monitor silicide reactions [13, 14, 15, 16]. We note that Auger spectroscopy is more useful than photoemission for the problem at hand since the latter tends to emphasize the nickel derived d-band states, while the Auger measurement is very sensitive to changes in s-p bonding and ignores the metal d-bands. On the other hand, calculations of the two-hole CVV spectra are quite complicated and are not readily compared with measured data [14, 17]. We bypass this problem by fitting the data with linear combinations of measured "fingerprint" spectra, taken from single phase reference compounds. In doing so, most of the difficulties of "quantitative" Auger analysis, such as unknown backscattering factors, attenuation lengths, inelastic scattering, detector resolution, etc. are avoided [18, 19]. To the extent that inter-compound scattering is small, these quantities are all included in the reference spectra. The reference spectra will also include vacuum "interface states" [20]. We note that, in conventional "quantitative Auger analysis", one is concerned with extracting concentrations of elements, while in our procedure, we are concerned with extracting concentrations of compounds.

In Fig. 1 we show the spectra obtained for 5 reference silicide compounds. Samples were prepared in an RF furnace and characterized with X-ray powder diffraction to verify purity of phase. Clean surfaces were prepared _in situ_ by scraping with a carbide file. We measure $E*N(E)$ signals with a single pass CMA with a floating electrometer, but we display and fit the computer differentiated curves, ie d/dE $(E*N(E))$.

We check to see that surface segregation does not occur by comparing the measured stoichiometries with the known bulk stoichiometies, as shown in Fig. 2. We have plotted the ratio of integrated line strengths, Ni/Si, for both the high energy lines (Si KLL @ 1614 eV and Ni LVV @ 844 eV, with mean free paths of ~ 15 Å), and the low energy lines (Si LVV @ 89 eV and Ni MVV @ 56 eV, with mean free paths of ~ 4 Å). A linear background is subtracted before integrating the peaks. The ratio is insensitive to the choice of background. In the absence of surface segregation, these plots will be linear and pass through the origin, at least in the formalism of Hall and Morabito [19], since the backscattering and escape depth variables for each element are identical

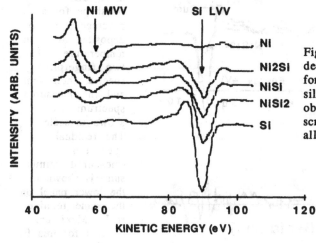

Figure 1. Auger derivative spectra for 5 reference silicide compounds, obtained by in-situ scraping of bulk alloys.

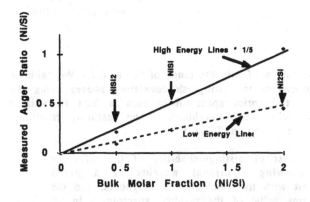

Figure 2. Measured vs. known (bulk) soichiometry, showing the ratio of integrated line strengths, Ni/Si, for both the high energy lines

and cancel in the ratio. The only appreciable deviation of stoichiometry seen in the data occurs for the low energy lines of $NiSi_2$, which implies a 20% enrichment of silicon within 4 Å of the surface for this sample.

An example of the lineshape fitting is shown in Fig. 3. This corresponds to a coverage of 0.4Å deposited at room temperature, which results in an incomplete layer of $NiSi_2$ on a silicon substrate. The lineshape is best fit with fractional weights of Si = 0.64 and $NiSi_2$ = 0.31. For a perfect fit, the sum of fractional weights would be exactly 1.0. For the uptake experiment, we obtain values of this sum that range from .95 to 1.05. The residual spectrum (composite curve minus measured curve) is shown in the lower panel. We characterize the quality of the fit with the parameter δ, the rms value of the residual spectrum. For the uptake

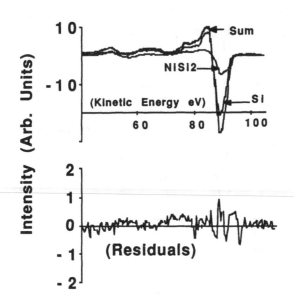

Figure 3. Auger dN(E)/dE lineshape for a coverage of 0.4 Å deposited at room temperature. The two fingerprint spectra, their sum and the measured spectrum are shown to scale. The residual spectrum (measured minus sum) is shown in the lower panel on the same relative scale. Maximum residual for this fit is 4%, and the rms value is δ = 0.49%

experiment, we obtain values of δ in the range of 0.4 to 0.7. We calibrate the ability of our procedure to distinguish coexisting phases using the following test: Each fingerprint spectrum in turn is best fit with a combination of its stoichiometric neighbors. The resulting fractional weights and residuals are shown in Table 1.

Table I. Test of distinguishability of the reference spectra, showing fractional weights of a given spectrum fit with its stoichiometric neighbors, and the resulting rms value of the residual spectrum. In all cases, the residuals are well above the statistical value of 0.3.

	Ni_2Si	NiSi	$NiSi_2$
δ	1.8	1.0	1.6
Ni	0.26		
Ni_2Si	x	0.43	
NiSi	0.81	x	0.85
$NiSi_2$		0.49	x
Si			0.25

The portion of δ due to statistical noise is $\delta \sim 0.35$. The latter is determined by fitting a repeated measurement of a given reference spectrum. We note that the most difficult compound to distinguish in the presence of others is NiSi. Since the values of δ obtained in the uptake experiment are well below the values shown in the table, we conclude that the resulting fractional amplitudes are physically meaningful. That is, we are able to quantitatively separate coexisting phases in the Ni-Si system using Auger lineshape analysis.

The uptake experiment is performed by recording the Auger spectra after each of 25 depositions of nickel, spaced initially by 0.1 Å, and later by 1.5 Å for a total coverage of 17 Å. Nickel thickness is determined from Auger peak heights, which are cross referenced to published RBS and Auger data from vanLoenen et al [9]. A 1 Å coverage corresponds to 9.1×10^{14} cm^{-2} of nickel. The numerical fitting is done in two steps. Initially, the amplitudes of all 5 reference spectra are unconstrained. In this iteration, we occasionally observe negative values of amplitude. In a second iteration, we constrain to zero those amplitudes that were less than 0.05 in the first fitting. In both cases, the energy scale of the set of reference spectra is unconstrained in order to accommodate spectrometer alignment and changes of band bending (in the substrate) during deposition and reaction. Since the silicide compounds are all metallic, no relative energy shifts between the silicides and pure nickel reference spectra are allowed.

RESULTS:

The results of the uptake experiment are shown in Fig. 4, where we plot the fraction of each silicide component in the Auger lineshape versus thickness of nickel deposited. We find that the reaction proceeds through definite stages as follows: For thickness below 1.0 Å, predominantly $NiSi_2$ is formed by reaction with the substrate. The 1/e thickness for attenuation of the substrate signal (pure silicon) is 1 Å. This may appear unphysically small, but it must be noted that the substrate is consumed in the reaction that produces the overlayer. If we assume that an epitaxial $NiSi_2$ layer is formed, the corresponding attenuation length (of overlayer material) is ~ 4 Å. This signifies that the growth occurs in a layer-wise fashion during this stage of the reaction. The symmetry of the RHEED pattern at this point is streaked 1x1, indicating that some degree of surface disorder is present. From 3 to 12 Å of deposited nickel, the surface is predominantly Ni_2Si. The 1/e length for growth of this material is ~ 5 Å of Ni, corresponding to 8 Å of Ni_2Si. This signifies that the growth mode during this stage corresponds to island formation. The RHEED pattern at this point is diffuse, implying either a very fine grain structure for the Ni_2Si islands, or a coexistence of materials with different structures. Above 12 Å of deposited nickel, the

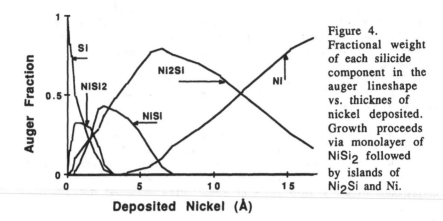

Figure 4. Fractional weight of each silicide component in the auger lineshape vs. thicknes of nickel deposited. Growth proceeds via monolayer of $NiSi_2$ followed by islands of Ni_2Si and Ni.

surface is predominantly Ni. The 1/e length for growth of pure nickel is ~ 8 Å, which signifies that the growth mode corresponds to island formation in this stage, also. This is consistent with the RHEED pattern, in which transmission diffraction spots from nickel clusters are evident. The nickel coverage range from 1 to 3 Å is difficult to interpret. It appears that NiSi is formed in a layer-wise fashion, according to the rapid growth of the signal. On the other hand, the fitting procedure is questionable in this region, since there are 4 coexisting phases, and NiSi, the most difficult to resolve, is dominant. An alternative interpretation is that the NiSi material corresponds to a "skin" of silicon covering islands of Ni_2Si, as has been described in RBS experiments [9].

DISCUSSION:

Our results help put into context a number of earlier experiments on this system. The claims of different stoichiometries for the room temperature silicide reaction are readily explained by the coverage dependence we have found: Experiments with limited surface sensitivity would find Ni_2Si (as has been seen with RBS [9] and RHEED [10]), while experiments with submonolayer sensitivity would find $NiSi_2$ (as has been seen with SEXAFS [5] and ICISS [6]).

From our results one can also gain insight into the template phenomenon observed in the room temperature growth of $NiSi_2$ by codeposition. At the coverage corresponding to the best template (~2.5 Å) , the overlayer is made up of finely dispersed, nickel rich silicides covering a monolayer of $NiSi_2$. Apparently, a nickel rich environment is preferred for the epitaxial growth of $NiSi_2$ by codeposition, although the reason for this is not apparent.

The coverage dependence of this reaction is similar to the familiar Stranski-Krastonov growth mode, in which the growth of an intermediate layer is followed by the growth of islands. However, this system is certainly far from thermal equilibrium at room temperature, and a comparison of surface and bulk free energies that might be inferred from the growth mode would probably not be appropriate. It is possible that the formation of progressively more nickel rich silicides with increasing coverage is simply a reflection of diffusion lengths that are only a few atomic spacings. At the same time, it is puzzling that sufficient mobility exists to form a dispersed and possibly ordered monolayer of $NiSi_2$, given the short thermal diffusion length. It is possible that the good epitaxial match between silicon and NiSi2 favors initial growth of the latter. Comparison experiments with crystalline vs. amorphous silicon substrates may illuminate this question. A second interesting question is whether the low coverage material is buried or consumed in the formation of nickel rich silicides at higher coverages. Further experiments and model calculations to explore this question are underway.

ACKNOWLEDGEMENTS:

This work is supported by AFOSR grant no. 87-0367. The assistance of the Materials Preparation and X-Ray facilities in the Center for Solid State Science at ASU is gratefully acknowledged.

REFERENCES:

1. R. T. Tung, J. M. Gibson, J. M. Poate, Phys. Rev. Lett.50, 429 (1983).
2. C. d'Anterroches, H. N. Yakupoglu, T. L. Lin, R. W. Fathauer, P. J. Grunthaner, Appl. Phys. Lett. 52, 434-6 (1988).
3. R. T. Tung, F. Schrey, Appl. Phys. Lett.55, 256-258 (1989).
4. R. T. Tung, F. Schrey, Appl. Phys. Lett.54, 852-854 (1989).
5. F. Comin, J. E. Rowe, P. H. Citrin, Phys. Rev. Lett.51, 2402 (1983).
6. T. L. Porter, C. S. Chang, U. Knipping, I. S. T. Tsong, J. Vac. Sci. Technol. A6, 2034-6 (1988).
7. J. M. Gibson, J. L. Batstone, Surf. Sci. 208, 317-50 (1989).
8. V. Hinkel, L. Sorba, H. Haak, K. Horn, W. Braun, Appl. Phys. Lett. 50, 1257-9 (1987).
9. E. J. vanLoenen, F. vanderVeen, F. K. LeGoues, Surf. Sci. 157, 1-16 (1985).
10. A. E. M. J. Fischer, P. M. J. Maree, F. vanderVeen, Appl. Surf. Sci. 27, 143-50 (1986).
11. P. A. Bennett, J. R. Butler, X. Tong, in Mat. Res. Soc. ((in press), 1989),
12. H. H. Madden, in Surf. Sci. 1983), pp. 80-100.
13. G. W. Rubloff, P. S. Ho, Thin Solid Films 93, 21-40 (1982).
14. C. Calandra, O. Bisi, G. Ottaviani, Surf. Sci. Rep.4, 271-364 (1985).

166

15. U. DelPennino, P. Sassaroli, S. Valeri, C. M. Bertoni, O. Bisi, C. Calandra, J. Phys. C 16, 6309-19 (1983).
16. J. A. Roth, C. R. Crowell, Jour. Vac. Sci. Technol.15, 1317-1324 (1978).
17. P. A. Bennett, J. C. Fuggle, F. U. Hillebrecht, A. Lenselink, G. A. Sawatzky, Phys. Rev. B 27, 2194-209 (1983).
18. M. P. Seah, in Practical Surface Analysis by Auger and X-ray Photoelectron Spectroscopy, D. Briggs, M. P. Seah, Eds. (J. Wiley, 1983), pp. 181-216.
19. P. M. Hall, J. M. Morabito, D. K. Conley, Surf. Sci.62, 1-20 (1977).
20. P. Ho, P. E. Schmid, H. Foll, Phys. Rev. Lett.46, 782785 (1981).

X-RAY ABSORPTION STUDIES OF TITANIUM SILICIDE FORMATION AT THE INTERFACE OF Ti DEPOSITED ON Si.

D.B. Aldrich*, R.W. Fiordalice*+, H. Jeon†, Q. Islam*, R.J. Nemanich*†, and D.E. Sayers*
* Department of Physics, North Carolina State University, Raleigh, NC 27695-8202
† Department of Materials Science and Engineering, North Carolina State University, Raleigh, NC 25695
+ Current Address; Motorolla APRDL, Austin, TX 78721

ABSTRACT

Near edge X-ray absorption spectra (XANES) have been obtained from the Ti K-edge for several series of titanium silicide samples produced by different techniques. Samples were fabricated by depositing Ti on silicon wafers and subsequently annealing them up to temperatures from 100°C to 900°C in UHV, vacuum furnace, or in a Rapid Thermal Annealing system. Measurements were done in the fluorescence and total electron yield modes. The XANES measurements were correlated with Raman scattering measurements. The XANES data of several reference compounds were obtained, and the data showed a high sensitivity to changes in the film structure. Ti metallic bonding and Ti-Si bonds can be distinguished and their evolution as a function of annealing is related to previous results. For the samples with increased impurities, Ti regions were stable at higher temperatures. The XANES spectra of samples annealed under N_2 indicate the formation of a surface nitride.

INTRODUCTION

The focus of this paper is on the application of X-ray absorption near edge spectroscopy (XANES) to study the stages of titanium silicide formation during the reaction of a Ti film with a Si substrate. Previous studies have indicated that at low temperature (~200°C), interdiffusion occurs at the Ti-Si interface, and a disordered intermixed phase forms at the interface. At ~450°C, the metastable C49 phase of $TiSi_2$ nucleates, while at temperatures greater than 650°C, the film transforms to the stable C54 phase of $TiSi_2$ [1-5].

In this study the XANES and Raman spectral signatures of different silicides and oxides are identified, and used to examine the structures which form during the titanium silicon reaction. The XANES measurements are sensitive to local atomic configurations while the Raman spectra displays the vibrational modes associated with different crystal structures. Raman spectroscopy has proved to be a very useful probe of the different crystalline structures which form from annealing of Ti/Si structures [6]. Spectra associated with both C49 and C54 $TiSi_2$ have been identified [1,4]. The two techniques are used to follow the Ti-Si reaction from initial Ti deposition through C54 $TiSi_2$ formation. UHV cleaning, deposition, and annealing were employed to form the silicides which were used to characterize the interaction process. Samples were also prepared by Rapid Thermal Annealing (RTA) and furnace annealing techniques to examine the effects of particular annealing processes on silicide formation. The results are described in terms of Ti-Si composition, crystal structures, and formation temperatures.

Mat. Res. Soc. Symp. Proc. Vol. 159. ©1990 Materials Research Society

XANES refers to the region of the x-ray absorption spectrum in the neighborhood of an inner shell absorption edge (10eV below to 100eV above). The structure in the absorption spectra at higher energy above the edge is called EXAFS. The shape of the absorption spectrum in the near edge region is due to longer range atomic structure than EXAFS. While direct analysis of the XANES spectrum is not easily accomplished, the spectra can be used as a fingerprint to identify structures in the sample. When compared with the data from known structures the XANES of an unknown sample can provide qualitative and semi-quantitative information about the composition of the sample [7].

Three different XANES data collection techniques were used (transmission, fluorescence, and total electron yield). The transmission mode uses gas ionization detectors to measure the attenuation of the incident x-ray beam as it passes thru a thin sample. In the fluorescence mode a gas ionization detector is used to measure the photons fluorescing from the sample. The total electron yield data collection technique consists of a grid, biased relative to the sample, placed in front of the sample to collect electrons emitted by the sample [8,9]. The transmission data is, by its very nature, sensitive to the crystal structure through the entire thickness of the sample. For the incident x-ray energies and sample compositions being used in this study, the fluorescence data is sensitive to a depth on the order of thousands of Angstroms and the total electron yield data is sensitive to a depth on the order of hundreds of Angstroms [8]. It is, therefore, possible to distinguish between surface and bulk formations by comparing total electron yield and fluorescence data.

EXPERIMENT

To investigate the titanium-silicon reaction samples were deposited in UHV and by standard evaporation techniques. In both cases the Ti films were 40 nm thick.

The UHV samples were prepared in a load-locked deposition chamber with a base pressure of ~1 x 10^{-10} Torr. The Si wafer cleaning process included precleaning by exposure to uv generated ozone to remove hydrocarbons bonded to the surface, HF etching to remove the native oxide, and then annealing to >800°C in the UHV system to remove the remaining oxide from the surface. The LEED pattern of the samples cleaned by this method consistently showed the 2x1 silicon surface reconstruction, and AES showed no detectable oxygen or carbon [5]. The Ti was deposited by filament evaporation from a Ti wire onto unheated substrates. The samples were then annealed, in situ, at their respective temperatures for 20 minutes. In another series, samples were cleaned by HF etch, deposited in the UHV system, and annealed in a turbo-pumped vacuum annealing system with a base pressure of ~1 x 10^{-6} Torr.

The samples used in the Rapid Thermal Annealing were prepared by evaporation from a resistive heated boat in a diffusion pumped system with a base pressure of <1 x 10^{-6} Torr. a one step HF etch was used to clean the subsrate prior to deposition. The cleaning process and relatively low vacuum used in the creation of these samples, and the furnace annealed samples mentioned above, will lead to an increase in the level of oxygen and carbon contamination. The samples were annealed in a flowing gas Rapid Thermal Annealing system with either Ar or nitrogen.

The XANES measurements were carried out on beamline X-11A of the National Synchrotron Light Source (NSLS) at Brookhaven National Laboratory (BNL). Data for the standards C54 $TiSi_2$, Ti_5Si_3, TiN, and Rutile TiO_2 was collected in the transmission mode from powder samples. The metallic titanium standard and the titanium-silicon samples were individually mounted in a detector which simultaneously collects data in both the fluorescence and total electron yield modes. Further description of the detector has appeared elsewhere [9]. All of the measurements were made at room temperature.

Raman spectra were excited with ~150 mW of 514.5nm Ar ion laser radiation. The scattered light was dispersed with a triple grating monochromator to reduce stray light at the laser frequency.

RESULTS AND DISCUSSION

Characterization of Titanium Silicide Formation

The XANES spectra of several standard titanium compounds are shown in Fig. 1. Each of the standards has a unique spectrum with unique spectral features. The XANES spectrum of metallic titanium has a sharp peak at the base of the absorption edge. For the samples being examined in this study the presence of this pre-edge peak is indicative of the presence of metallic titanium. The spectra of Ti_5Si_3 and TiN though very similar have distinguishable absorption edge features. The relative amplitudes of the first and second peaks are different for TiN and Ti_5Si_3 and the shoulder is more prominent in Ti_5Si_3. The XANES spectrum of Rutile TiO_2 is easily identifiable by the three pre-edge peaks at relative energies of ~1.5 eV, ~3.7 eV, and ~6.5 eV. The spectrum of C54 $TiSi_2$ has a gradually sloping absorption edge with a prominent shoulder. The presence of these spectral features in the XANES spectrum of a sample is indicative of the presence in the sample of the associated structures.

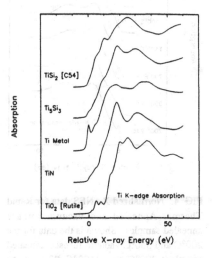

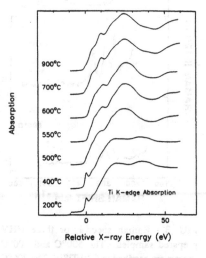

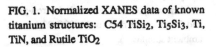

FIG. 1. Normalized XANES data of known titanium structures: C54 $TiSi_2$, Ti_5Si_3, Ti, TiN, and Rutile TiO_2

FIG. 2. Normalized XANES data of UHV prepares samples. Samples were annealed for 20 minutes at the specified temperature.

The XANES spectra of the UHV annealed samples are shown in Fig. 2. The predominant titanium structure present in the sample progresses from metallic titanium for annealing temperatures below 400°C to the C54 phase of TiSi$_2$ for annealing temperatures above 700°C. Samples annealed at temperatures below 400°C did not show evidence of any silicide structures. The presence of the pre-edge peak in the spectra of these samples indicates the presence of metallic titanium regions. The smoothing of the first and second peak in the 400°C sample could be caused by the interdiffusion of Ti and Si. Annealing to temperatures between 500°C and 600°C produced a XANES spectrum very different from that seen at the lower temperature anneals. The spectra of these samples have some of the same features as those of the C54 TiSi$_2$, but the features are not quite as distinct. The Raman analysis of these samples, shown in Fig. 3, indicated that this structure is the C49 phase of TiSi$_2$. The non-distinct features of the XANES data and the reduced Raman peak amplitude are a clear indication that the 500°C sample is an incomplete TiSi$_2$ reaction.From the XANES, it might be suggested that disordered Ti-Si, TiSi or Ti$_5$Si$_3$ may be present in the sample. Comparison of the XANES spectra of the C54 TiSi$_2$ standard and the 700°C and 900°C annealed samples shows that in those samples the titanium film has completely reacted with the silicon substrate to form the C54 phase of TiSi$_2$. The Raman analysis for both the 700°C and the 900°C annealed samples also shows spectral features characteristic of the C54 phase.

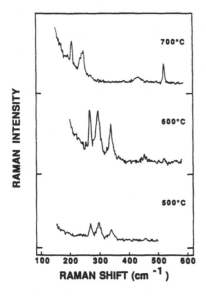

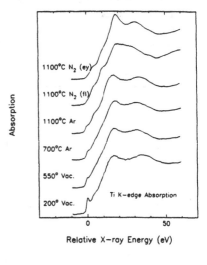

FIG. 3. Raman spectra for three UHV prepared samples. The 500°C and 600°C spectra are ascribed to C49 TiSi$_2$. The 700°C spectrum is ascribed to C54 TiSi$_2$.

FIG. 4. Normalized XANES data for Rapid Thermal Annealed and vacuum furnace annealed samples. Shown is the data for the 200°C and 550°C vacuum furnace annealed samples, 700°C and 1100°C RTA in Ar ambient samples, and the total-electron-yield and fluorescence data for 1100°C RTA in N$_2$ ambient sample.

Effects of Furnace Annealing

The XANES data from two of the samples prepared by vacuum furnace annealing is shown in figure 4. The 200°C furnace annealed sample shows the same titanium structure (metallic titanium) as the 200°C UHV prepared sample. After 550°C annealing the results are different. In the analysis of the UHV prepared samples it was found that the C49 phase of TiSi$_2$ began to form at, or near, 500°C, and that at 550°C the C49 TiSi$_2$ was the predominant titanium structure present. The XANES data of the 550°C vacuum furnace annealed sample does not contain any of the spectral features characteristic of TiSi$_2$ indicating that the silicide has not begun to form or is present in only minute amounts. The lack of sharp spectral features suggests that this sample contains several different titanium structures (i.e. the spectral features of several different structures can average together resulting in a spectrum with no distinct features).

The differences in sample preparation between the UHV and the vacuum furnace annealed samples leads to a higher level of contamination in the vacuum furnace annealed samples. As noted previously, oxygen contamination can inhibit the formation of TiSi$_2$, resulting in an increased formation temperature [2,6]. The XANES results show that dispite higher annealing temperatures the Ti and Si have not fully reacted. From the results we can suggest that the increased level of contamination in the sample inhibits the nucleation of Ti-Si

Effects of Rapid Thermal Annealing

The effects of Rapid Thermal Annealing were studied using samples annealed at high temperatures for 10 seconds in either nitrogen ambient or argon ambient. The XANES data are shown in Fig. 4. The XANES spectra of the samples annealed in argon ambient do not have any distinct features which match the features of the standards presented previously. The absence of the Ti peak is an indication that significant Ti-Si interdiffusion has occurred. The 10 second annealing time may, however, have been too short for the Ti-Si reaction to reach an equilibrium. The reacted sample region could be a mixture of silicide phases or a disordered intermixed region in which there has been no crystalline silicide nucleation..

For the samples annealed in nitrogen, reaction can occur with both the Si and the nitrogen. The differences in depth sensitivity of the different data collection techniques was used to characterize the reactions. The bulk sensitive fluorescence data shown in Fig. 4 displays features ascribed to a mixture of TiN and silicide structures. The absorption edge shoulders are indicative of C54 TiSi$_2$ and the first peak shape can be represented as a combination of the absorption peaks of TiN and C54 TiSi$_2$. The surface sensitive total electron yield data matches the data of the TiN standard. The results thus indicate that a surface nitride is forming.

CONCLUSIONS

Using UHV prepared, clean samples the formation temperatures of the phases of TiSi$_2$ were determined. The initial formation of the C49 phase of TiSi$_2$ is detected at ~500°C and the transition to C54 TiSi$_2$ occurs between 600-700°C. The formation of TiSi$_2$ is preceded by intermixed silicide formation. The formation temperatures were seen to increase as the level of impurities in the sample increased, and interdiffusion/mixed silicide formation was detected at higher temperatures. For short annealing times the Ti-Si reaction is incomplete resulting in a non-equilibrium mixed silicide state. Annealing at high temperatures in the presence of nitrogen results in the formation of a surface nitride. The results of this study show that XANES and Raman spectroscopy can be used to characterize the structures that form during the titanium silicon reaction.

ACKNOWLEDGEMENTS

We gratefully acknowledge Cathy Sukow and Dr. Gerraldine Lamble for their assistance in conducting the Raman and X-ray experiments respectively. This study was supported in part by the Division of Materials Science of the Department of Energy under contract DE-AS05-80ER10742, and the National Science Foundation through grant DMR-8717816.

REFERENCES

[1] R. Beyers, and R. Sinclair, J. Appl. Phys., 57, 5240 (1985).

[2] M. Berti, A.V. Drigo, C. Cohen, J.Siejka, G.G. Bertini, R. Nipoti, and S. Guerri, J. Appl. Phys, 55, 3558 (1989).

[3] S.P. Murarka, and D.B. Fraser, J. Appl. Phys., 51, 342 (1980).

[4] R.J. Nemanich, R.T. Fulks, B.L. Stafford, and H.A. VanderPlas, J. Vac. Sci. Technol., A3, 938 (1985).

[5] H. Jeon, and R.J. Nemanich, Thin Film Solids, 182 (1989).

[6] R.J. Nemanich, R.W. Fiordalice, and H. Jeon, IEEE J. Quant. Elect., 25, 997 (1989).

[7] D.C. Koningsberger, and R. Prins, *X-ray Absorption: Principles, Applications, Techniques of EXAFS, SEXAFS, and XANES.* (John Wiley & Sons, Inc., New York, 1988).

[8] A. Erbil, G.S. Cargill, III, R.Frahm, and R.F. Boehme, Phys. Rev., B37, 2450 (1988).

[9] F.W. Lytle, R.B. Greegor, D.R. Sandstrom, E.C. Marques, J. Wong, C.L. Spiro, G.P. Huffman, and F.E. Huggins, Nucl. Instr. and Meth., 226, 542 (1984).

ELECTRONIC AND STRUCTURAL STUDY OF Ni(Co) SILICIDE/Si(111) CONTACT SYSTEM STUDIED BY SOFT X-RAY EMISSION SPECTROSCOPY

H. WATABE[*], H. NAKAMURA[**], M. IWAMI[+], M. HIRAI[+] AND M. KUSAKA[+]
* Matsushita Electric Industrial Co., Ltd., Twin 21 National Tower, 1-61 Shiromi 2-Chome, Chuo-ku, Osaka 540, Japan
** Osaka Electro-Communication University, Hatsu, Neyagawa, Osaka 572, Japan
+ Research Laboratory for Surface Science, Faculty of Science, Okayama University, Okayama 700, Japan

ABSTRACT

An electron excited Si $L_{2,3}$ valence band soft x-ray emission spectrum (SXES) for Ni(or Co)Si$_2$ showed a clear modification from that for Si. From the SXES study, a fair amount of the Si(3s) valence band density of state (VB-DOS) is concluded to be included in the upper part of the VB-DOS for the transition metal(TM) disilicides due to the TM-Si bond formation, which is a clear contrast to proposals given so far. Non-destructive structural analysis of a NiSi$_2$(tens of nm)/Si(111) contact is also carried out successfully using the SXES.

1. INTRODUCTION

Transition metal(TM) disilicides are expected to be widely used in semiconductor devices for their interesting characteristics, e.g., the possible epitaxial growth on a Si single crystal substrate at rather low temperatures without losing the high conductivity. In order to have highly reliable and reproducible device, it is necessary to have knowledge on their physical properties. There have been, therefore, quite a lot of fundamental works on transition metal silicides[1]. However, there are still left problems to be solved.

In order to clarify the valence band density of state (VB-DOS) of transition metal silicides(TMSi's), a soft x-ray emission spectroscopy (SXES) has been carried out. Characteristics of the SXES method are as follows: (1) the method is especially suitable for the determination of VB-DOS, because a core level involved in the soft x-ray emission has little wave number dependence. (2) It can provide information corresponding to different electronic states, e.g., s-, p- and/or d-electronic DOS, separately due to the selection rule in an photon-emitting electron transition.

In the present study, we have also intended to clarify the interface structure of a TMSi/Si contact using an interesting application of the SXES method for a non-destructive analysis of a thin-film/substrate contact system as has been reported[2] and the VB-DOS of a TM(Ni, Co)Si$_2$.

2. EXPERIMENTAL

Ni(or Co)Si$_2$(thin film)/Si(substrate) contacts were prepared by a solid phase epitaxy, where TM films with thicknesses of 10 ~ 100 nm were deposited on Si substrates followed by a heat treatment in an electric furnace under N$_2$ + H$_2$ gas flow.

Growth of a Ni(or Co)Si$_2$ crystal for a thick film was ascertained by an x-ray diffraction and an x-ray microanalysis. A combined system of a vacuum chamber with an x-ray tube and an x-ray spectrometer was used for the SXES study. A grazing incidence spectrometer was used for the spectral analysis of emitted soft x rays. An energy resolution of the spectrometer is better than 0.6 eV. Details of the SXES experiments are given elsewhere[2].

3. RESULTS AND DISCUSSION

Si L$_{2,3}$ soft x-ray emission band spectra are shown in Fig. 1 for NiSi$_2$(a), CoSi$_2$ (b) and single crystalline Si (c)[3, 4].

3.1 A non-destructive Analysis of a NiSi$_2$/Si(111) Contact

Characteristics shown in Fig. 1 can be used for a structural analysis of a TMSi/Si contact system[2, 4]. An example is shown in Fig. 2(a) for a specimen of a NiSi$_2$/Si(111) heterocontact system, where the incident electron energy, E_p, is 2 keV, together with a synthesized one [Fig. 2(b)]. In the latter, the composition of the spectrum is carried out by simply adding spectra for NiSi$_2$ and Si at E_p = 4 keV, i.e., (1-x)Si + xNiSi$_2$, where x is chosen to be 0.35. A proper data on the x ray production depth of an energetic electron in a material will tell either the surface layer thickness or the interface structure with a resolution of ~ 1 nm[2]. A tentative estimation of the NiSi$_2$ layer is ~ 25 nm in the present specimen studied.

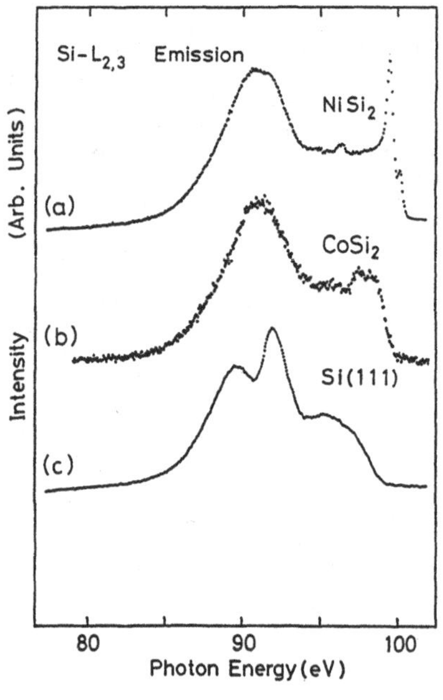

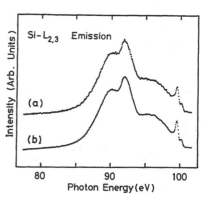

Fig. 1 Si L$_{2,3}$ soft x-ray emission band spectra: (a) NiSi$_2$, (b) CoSi$_2$ and (c) Si crystals.
Fig. 2 (a) An experimental spectrum for NiSi$_2$/Si(111) at E_p = 2 keV. (b) A composed one as 0.65-Si + 0.35-NiSi$_2$.

3.2 VB-DOS for Ni(or Co)Si$_2$

For the sake of discussion, several experimental and theoretical spectra are shown in Fig. 3[3, 5] for NiSi$_2$: (a) the Si L$_{2,3}$ SXES spectrum, (b) theoretical Si(3s) partial DOS, (c) theoretical total VB-DOS and (d) theoretical Ni(3d) partial DOS. The spectrum (b) in Fig. 3 is the one for CoSi$_2$, but, according to the authors, it is almost the same for NiSi$_2$[5]. The E$_F$ position for the SXES spectrum is set at hν = 99.7 eV considering the binding energy (E$_b$) of Si(2p) core line for NiSi$_2$ to be 99.7 eV from an XPS study[3].

As for the Si L$_{2,3}$ soft x-ray emission band spectrum, it should be noted that the Si L$_{2,3}$ spectrum reflects a valence band density of state with s- and d-symmetry. The spectrum of Fig. 3(a), or Fig. 1(a), for NiSi$_2$ has following characteristics: (1-1) A broad peak at ~ 91 eV. (1-2) A terrace at 94-97 eV. (1-3) Another sharp peak at ~ 100 eV. Similar is the case for CoSi$_2$, although the peak near the top of valence band is less clear than the one for NiSi$_2$; see Fig. 1 (b) in comparison with Fig. 1(a). For a Si single crystal [Fig. 1(c)], features to be noted are as follows: (2-1) A double peak at ~ 89 eV and ~ 92 eV. (2-2) A terrace extending up to ~ 100 eV. (2-3) No special peak at the top of VB. There is a clear difference between the Si L$_{2,3}$ spectrum for Ni(or Co)Si$_2$ and that for Si. Differences to be pointed out are those in spectral shapes at ~ 90 eV, features (1-1) and (2-1), and those near the top of valence band, features (1-3) and (2-3).

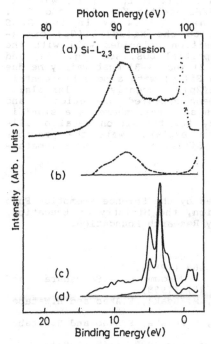

Now we will discuss the valence band electronic structure of TMSi's. So far, it seems that, using coincidences between a theory and a PES study, it has been concluded for the VB-DOS of a transition metal silicide (TM$_x$Si$_y$) as follows. Namely, it looks like to be widely believed in a TM$_x$Si$_y$ compound that TM(nd), n is the principal quantum number for highest occupied d-electron of TM, and Si(3p) electronic states give rise to a bonding and anti-bonding state and a nonbonding TM(nd) state sits in between them, where the Si(3s) electrons contribute only to the lower isolated part of the

Fig. 3 Comparison between an experimental spectrum and theoretical VB-DOS's: (a) Si L$_{2,3}$ emission band spectrum for a NiSi$_2$ crystal. (b)-(d) Theoretical VB-DOS's for Ni(or Co)Si$_2$[5]; (b) Si(3s) partial VB-DOS (dotted line), (c) total VB-DOS (solid line, upper) and (d) Ni(3d) partial DOS (solid line, lower).

VB-DOS, but not to the upper part of the VB-DOS[1, 5]. In the case of a $NiSi_2$ and a $CoSi_2$ crystal, a TM(3d) state has been believed to be more isolated than those in other silicides. However, the Si $L_{2,3}$ SXES spectrum has a meaningful intensity in the energy region corresponding to the upper part of the VB-DOS for Ni(or Co)Si_2, i.e., the features (1-2) and (1-3) above. Therefore, we propose the followings for an explanation of the VB-DOS of a Ni(or Co)Si_2 crystal.

Since the Si $L_{2,3}$ SXES spectrum should reflect s- and/or d-electronic VB-DOS, the spectrum may be composed of Si(3s) and/or TM(3d) electronic DOS's. If we assume that Si(3s) DOS has little contribution to the upper half of the VB-DOS as has been believed, the characteristics (1-2) and (1-3) must be due to TM(3d) DOS, i.e., due to the cross-transition from TM(3d) to Si(2p) core hole, which might be the case because of the consideration that there must be TM-Si bonds in Ni(or Co)Si_2. However-er, as can be seen in the theoretical DOS of Fig. 3(d), TM(3d) DOS is concentrated mainly at a so called non-bonding d-state and the electronic state due to TM(3d)-Si(3p) bond formation spreads below and above it with contributing only a little to the VB-DOS. Consequently, the features (1-2) and (1-3) cannot be explained by considering the transition from TM(3d) to Si(2p) only, because the TM(3d) DOS has a peak at $E_b \sim 3$ eV, which is hard to be seen in the present SXES spectra[Fig. 1(a) and (b)]. The small signal peaked at $h\nu \sim 96.4$ eV in the SXES spectra[Fig. 3(a)] may be due to the Ni(3d)-Si(2p) cross-transition, because the peak position coincides well with the peak of the theoretical Ni(3d) partial DOS, see Fig. 3(a) and (d). Therefore, the features (1-2) and (1-3) must mainly be due to the transition from Si(3s) to Si(2p) with a possible contri-bution of that from TM(3d) to Si(2p). Recently, similar claims like the present study have been appeared theoretically and experimentally[6, 7], where it is said that there is a signifi-cant Si(3s) derived VB-DOS at E_F for transition metal disili-cides. The present consideration explains well the previous experimental finding by NMR that Si(3s) electrons have a metal-lic DOS at E_F[8].

ACKNOWLEDGMENTS

This work is partly supported by the Science Promotion Fund of Japan Private School Foundation, the Ministry of Education, Science and Culture and the Toray Research Foundation.

REFERENCES

1. G.W.Rubloff, Surf. Sci. 132, 268 (1983).
2. M.Iwami, M.Kusaka, M.Hirai, H.Nakamura, K.Shibahara and H.Matsunami, Surf. Sci. 199, 467 (1988).
3. H.Nakamura, M.Iwami, M.Hirai, M.Kusaka, F.Akao and H.Watabe, submitted.
4. M.Iwami, H.Nakamura, M.Hirai, M.Kusaka, F.Akao and H.Watabe, Vacuum, in press.
5. J.Tersoff and D.R.Hamann, Phys. Rev. B28, 1168 (1983).
6. W.Speier, private communication, to be published.
7. L.Martinage, N.Cherief, A.Pasturel, J.Y.Veullen, D.Papacon-stantopoulos and F.Cryot-Lackmann, presented at the 11th IVC/7th ICSS, Koln, 1989.
8. K.Okuno, M.Iwami, A.Hiraki, M.Matsumura and K.Asayama, Solid State Commun., 33, 899 (1980).

PART IV

Oxide Interfaces

IN-SITU TRANSMISSION ELECTRON MICROSCOPY STUDIES OF THE OXIDATION OF SI (111) 7X7

J.M. GIBSON
AT&T Bell Laboratories, 600 Mountain Avenue, Murray Hill, NJ 07964

ABSTRACT

The kinematical approximation is valid for High-Energy Transmission Electron Diffraction from monolayers in plan-view. We use this fact to study quantitatively the attack of Si (111) 7x7 by O_2. Oxygen is found to bind in the bridging position of the adatom backbonds and render the structure very stable during subsequent O_2 exposure. Electron-beam exposure during dosing additionally creates rapid disordering which is presumed to represent SiO_x formation.

INTRODUCTION

The initial stages of silicon oxidation are of interest in controlling the properties of very thin oxides and their interfaces, which are needed for VLSI devices. The use of ultra-high vacuum and very clean surfaces, although not directly relevant to commercial oxidation, provides atomic-level insight into oxidation mechanisms. The Si (111) 7x7 is a good template for such study, as it provides a variety of sites for attack and the structure is well-known. The surface structure was solved by transmission electron diffraction[1], which is the technique we apply here to the study of the oxygen/Si (111) 7x7 interaction. The advantage of this technique comes from the kinematical nature of high-energy electron diffraction from monolayers in plan-view, combined with relatively high signal intensity. Our data shows that there is a relatively stable Si(111) 7x7:O structure, formed by O_2 exposure, whose principal difference from the clean surface is the bonding of oxygen atoms at bridging sites in the backbonds of adatoms. The ordered 7x7:O structure is stable in excess of 10^5L O_2 exposure, yet forms at considerably less than 1L O_2 dose. If the 100kV electron beam is present during the oxygen dosing, then simultaneous rapid disordering of the surface occurs. The latter process is believed to represent the formation of SiO_x, and has been observed with lower-energy (1.5kV) electron beam exposure[2].

EXPERIMENTAL

The experiments were accomplished with a modified JEOL 200CX transmission electron microscope (TEM), which has a base vacuum of 10^{-9} torr in the specimen region[3]. Thin specimens were prepared from p-doped 10 ohm cm Si wafers by mechanical and chemical thinning. In-situ cleaning is performed by resistive heating to approx 1200°C. Molecular oxygen is introduced throughout the chamber through a controlled leak-valve, and its pressure is measured with a quadrapole mass analyzer. There are no hot filaments within line-of-sight of the specimen. Diffraction patterns were recorded from thin (< 2000 Å) regions by direct exposure of photographic emulsion. An optical microdensitometer, employing background subtraction and peak integration, was then used to obtain quantitative data on scattered intensity. This method permits relative intensity measurements which are accurate to better than 5%. Absolute measurements with about 10% accuracy could be made by examining a series of photographic exposures with constant intensity but different exposure times.

The validity of the kinematical diffraction approximation for the specific case of 100kV electron diffraction from the Si (111) 7x7 has been examined by numerical simulation[4]. We find very similar results. Diffraction patterns are typically taken with the incident beam several degrees from the exact (111) zone axis. Reflections are averaged over the P6mm symmetry of the clean surface. The resulting error in the derived structure factors is on average 13%, and can be attributed primarily to dynamical diffraction from the substrate. Absolute measurements are susceptible to considerably larger errors due to dynamical diffraction in the substrate. Data shows that these errors are typically less than 50% in suitably thin specimens. Absolute measurements from the same specimen, during dosing for example, are accurate to better than 5% on a relative scale.

The original structural analysis performed on the 7x7 by Takayanagi et. al.[1] relied principally on Patterson maps. The Patterson function is the fourier transform of the measured intensities

$$P(\mathbf{r}) = \sum_{g} |F_g|^2 e^{2\pi i g.\mathbf{r}} \tag{1}$$

where F_g is the structure factor of the reciprocal lattice vector $\mathbf{g}$.

The fourier difference map is very useful in studying small differences between two structures. This is most often applied in comparing an experimental and model structure, however, we have found it useful to observe the changes introduced into the 7x7 by oxygen exposure. In this application, one assumes that the phases of the structure factors (Φ_g), obtained from a model calculation, remain unchanged in the perturbed structure, so that the fourier difference function

$$F(\mathbf{r}) = \sum_{g} (F_g{}^1 - F_g{}^2) e^{2\pi i (g.\mathbf{r}+\Phi_g)} \tag{2}$$

provides a real-space map of the difference in scattering power between the two structures.

As the structure is solved, the structural refinement method allows the identification of detailed atomic positions and occupancies by minimization of the least squares residual

$$R = \frac{\sum_{g} w_g |F_g{}^o - F_g{}^c|}{\sum_{g} |F_g{}^o|} \tag{3}$$

where w_g is a weighting factor, and the superscripts o and c refer to observed and calculated structure factors, respectively. This method also permits an absolute estimation of the statistical reliability of a model. In this regard, the parameter X (chi), defined as the ratio of the mean-square residual to the experimental error (and weighted by the number of fitting parameters), is most useful. A chi value < 2 indicates a good fit.

RESULTS

The structure of the clean Si (111) 7x7 surface was
examined carefully as a prerequisite to the study of the
oxygen exposed surface. Four independent samples were examined
in detail. We found generally good agreement with the Dimer-
Adatom-Stacking fault (DAS) model[1] (whose minimum asymmetric
unit is shown in fig. 2). Typically 75-100 symmetry
inequivalent reflections were obtained after data reduction,
with an average experimental error of 13%. Figure 1 shows a
set of such reflections plotted as semi-circles whose radius
is proportional to the structure factor. A second set of data
from the same specimen area after oxygen dosing is also shown
in the figure.

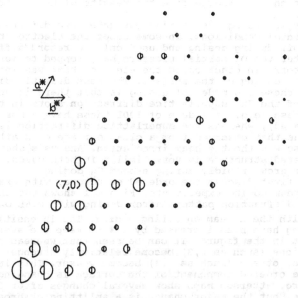

Figure 1: Measured intensities of some TED reflections from a
clean Si (111) 7x7 surface (left semicircles) and the same
area exposed to 0.6L O_2 (right semicircles). The electron-beam
was on during dosing. The area of each semi-circle is
proportional to the intensity.

Our Patterson maps look very similar to those reported by
Takayanagi[1] and Robinson[5]. The strongest feature of the
Patterson map from the 7x7 is a peak corresponding to the 2x2
adatom net. Other features can be identified with the
structure of the 7x7, including the stacking fault and dimers.
Structural refinement has been carried out on the clean
7x7 surface data. The original DAS model has only one
structural positional parameter, the dimer bond length, and
refinement gives an average length 17(±2)% greater than the
equilibrium Si-Si distance of 2.35 Å. This is significantly
larger than the value of 9% reported by Takayanagi et. al.[1].

However, if we allow a full refinement of the structure along the lines of Robinson[5], with 15 independent positional variables, then the dimer bond strain is reduced to 11(±4)%. In other details, the structure after refinement is very similar to that reported by Robinson[5]. An average improvement of the R value from 0.33 (±0.02) to 0.28 (±.02) accompanied the refinement. This corresponds to an average improvement in the chi value from 2.4 to 2.0 (±0.4). In general, these results indicate that the DAS model is a good fit to our 7x7 structures. One notable discrepancy is that the clean surface refinement leads to an increased occupancy of the corner adatom Si1 (see fig. 2), to an effective value of 1.6 (±0.1) Si atoms, whereas the center adatom (Si2) occupancy is 1.1 (±0.1) Si atoms. The excess 0.6 Si atoms at the Si1, center adatom, position, is non-physical. However, a likely explanation will emerge from the results of deliberate oxygen exposure.

Oxygen dosing was carried out under two different experimental conditions. In some cases the electron beam was turned off during dosing and used only to record diffraction data after the O_2 partial pressure had dropped to less than 2×10^{-9} torr. In other cases the electron beam was left on during O_2 dosing of the surface. The most dramatic difference between these two modes of dosing is that there is very rapid fading of the 7x7 superlattice diffraction spots in the e-beam dosing case (e.g. at a dose of 100L these have almost disappeared), whereas the superlattice diffraction is as strong as that observed from a clean surface even after 10^5 L O_2 exposure without e-beam irradiation. Analysis shows that the ordered structure is very similar in both cases, but that disorder grows rapidly during e-beam+O_2 dosing.

We first discuss the ordered structure, with examples from e-beam dosing experiments. Figure 1 shows the changes in the 7x7 diffraction pattern after dosing with 0.6L O_2 (at 10^{-7} torr) with the e-beam on during dosing. The intensities after O_2 dosing have been increased by 22% to keep the average value constant in the figure. It can be seen that whereas some reflections (such as 0,3) become relatively stronger after O_2 exposure, others (such as 7,3) become weaker. This implies that the ordered component of the surface is changing its structure.Patterson maps show several changes after O_2 exposure, but the major change is a splitting or broadening of the 2x2 adatom peak. At the same time integrated peak intensity increases. These observations imply that additional scattering power is bonded near, but not exactly at, the position of the adatoms.

Further corroboration of this idea is provided by fourier difference maps, such as shown in figure 2. In this map we show the difference between O_2 exposed data and the calculated DAS model structure factors, using calculated DAS phases. The largest peak on the difference map lies in the backbond Si1-Si3 of the corner adatom. The most likely explanation of this peak is that an oxygen atom bonds in the bridging position of the backbond. This is also consistent with the Patterson maps. We invariably find that the preferred bonding position for the bridging oxygen is at the corner adatom, specifically the Si1-Si3 backbond. Other peaks occur, particularly at higher exposure, around the center adatom position. There is also evidence for some opening of the dimer bonds. However, the dominant effect at low oxygen exposures is in the corner

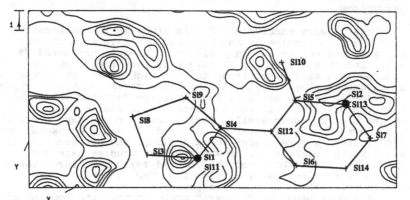

Figure 2: A Fourier difference map from a 0.6L O_2 exposed Si (111) 7×7 surface, compared with the clean surface and using phases calculated from the DAS model. Superimposed is the asymmetric unit of the DAS model. Adatom Si1 lies directly above Si11, and Si2 lies above Si13 (adatoms are shown with solid circles). The highest peak in the difference map lies in the backbond of the corner adatom, Si1-Si3, and is believed to correspond to oxygen in a bridging position.

adatom backbond Si1-Si3. Structural refinement has been carried out for a variety of O_2 exposures. For example, the 0.6L e-beam dosed surface described above, gives an R-value of 0.33 (chi = 1.8) when refined from the DAS model. However, the occupancy of the center adatom (Si1) rises to 1.8. The fit can be greatly improved by putting an oxygen atom at the position suggested by the Fourier difference map and permitting refinement. This gives an R-value of 0.28 (chi = 1.63). The occupancy of the oxygen atom is 0.9, and the adatom shifts away from the oxygen atom so that the Si-O bond-lengths are close to the normal values of 1.6 Å. It is this oxygen atom which gives rise to the increased occupancy of adatom Si1 in the fit to the DAS model. Clearly, the various methods of structural analysis employed here all give a consistent picture of the effect of oxygen attack on the 7x7 surface: the bonding of oxygen atoms in bridging positions of adatom backbonds.

Data as a function of exposure shows that the corner adatom site is saturated at a coverage of < 0.3L, for the case of e-beam dosing. Preliminary experiments suggest that the uptake of oxygen is equally efficient without e-beam exposure during dosing. Our data for the "clean" surface also shows that there is significant oxygen present in the corner adatom backbond, even without deliberate introduction. In this case we estimate that a background partial pressure of 2 x 10^{-9} torr is exposed to the surface for typically 100s after flash cleaning. This implies a typical oxygen exposure of 0.1L for a "clean" surface. Fourier difference maps and refinements suggest a typical oxygen presence in the corner adatom backbond of 0.5 atoms. Since saturation represents an oxygen coverage of 0.12L, a sticking coefficient of order unity is implied. Values determined from thermal desorption spectroscopy for this coverage range are 0.1-0.2[6], which is reasonably consistent within the experimental errors.

DISCUSSION

The primary conclusion of this study is the observation of the oxygen bonding in the backbond of adatoms in the 7x7 surface. The corner adatom position is clearly preferred.We also find that the uptake of oxygen is very rapid, but that at ordered structure persists up to very high doses. The presence of an electron beam during dosing leads to an additional rapid disordering of the surface. Photoemission data has indicated that the bridging position in adatom backbonds is favoured for oxygen bonding under similar conditions[7]. In the same study it was shown that an O_2 layer is adsorbed at higher exposure over the surface (a "precursor" to oxidation). The presence of this overlayer probably explains why previous studies have not recognized the stabilty of the ordered 7x7:O structure. These studies were almost exclusively with LEED[7] which has very little penetration through overlayers.

A recent paper by Ohdomari[8] proposes that the bonding of oxygen in the bridging position of adatom backbonds can drastically reduce the strain in the 7x7 surface. This would be very consistent with our observations. However, Ohdomari proposes that the 7x7 is only stable when oxygen is present, and our data does not support this contention. Nevertheless, it is very difficult to obtain truly clean surfaces because of the very low oxygen exposures (<0.1 L) necessary to saturate the corner adatom position. Experiments in which the total exposure to oxygen exceeds 0.05L, should take into account the possibility of an oxygen contaminated surface. Although our base pressure is over an order of magnitude poorer than most commonly employed, our experiments are executed with an unusually short time delay after cleaning. Experiment that should be more closely scrutinized in this regard might be STM studies of 7x7 surfaces, where long periods of time are employed before imaging, for reasons of thermal stability.

These experiments lead to interesting observations of the oxygen/silicon interaction and demonstrate the usefulness of quantitative TEM and diffraction from monolayers in the plan-view geometry.

The author is grateful for discussions with Ian Robinson, Stuart Wolff and David Loretto and the technical assistance of Mike McDonald and Don Bahnck.

REFERENCES

1. K. Takayanagi, Y. Tanashiro, S. Takahashi and M. Takahashi, Surf. Sci. **164**, 367 (1985).
2. B. Carriere, J.P. Deville and A. El Maachi, Phil. Mag. B **55**, 721 (1987).
3. M.L. McDonald, J.M. Gibson and F.C. Unterwald, Rev. Sci. Inst. **60**, 700 (1989).
4. Y. Tanashiro and K.Takayanagi, Ultramic. **27**, 1, 1989.
5. I.K. Robinson, J. Vac. Sci. Technol. **A6**, 1966 (1988).
6. P. Gupta, C.H. Mak, P.A. Coon and S.M. George, Phys. Rev. B **40**, 7739 (1989).
7. P. Morgen, U. Hofer, W. Wurth and E. Umbach, Phys. Rev. B **39**, 3720 (1989) and ibid. **40**, 1130 (1989).
8. I. Ohdomari, Surf. Sci., to be published.

THE USE OF FRESNEL CONTRAST TO STUDY THE INITIAL STAGES OF THE in situ OXIDATION OF SILICON

FRANCES M. ROSS, J. MURRAY GIBSON* AND W. M. STOBBS
University of Cambridge, Department of Materials Science and Metallurgy, Pembroke Street, Cambridge CB2 3QZ, U.K.
*AT&T Bell Laboratories, Murray Hill, NJ 07974

ABSTRACT

We describe the analysis of Fresnel contrast seen at a free silicon surface in order to characterise the initial clean surface and the formation in situ of the first atomic layers of oxide.

INTRODUCTION

The oxidation of silicon, particularly in the initial stages, is still a subject of great interest in spite of much study of both oxide structure and oxidation kinetics in this régime. In this paper we describe observations made in the electron microscope of the oxidation of a free silicon surface aiming to determine the structure of the initial oxide layer under controlled conditions of oxide growth. This was achieved by the use of a UHV microscope in which it was possible to heat up the silicon specimen so that crystallographic facets formed, creating an unusual specimen geometry with clean flat areas of silicon surface at a range of orientations. The behaviour of a facet which was *parallel* to the electron beam was then observed as the bare surface was oxidised by the introduction of controlled amounts of oxygen into the microscope. As well as observing the oxide growth at high resolution, we characterised at lower resolution the Fresnel fringes formed at the specimen edge, which we will show to be highly sensitive to the form of the surface. The in situ observations and preliminary data analysis, which we describe in the next section, demonstrate that interesting changes occur in the nature of the silicon surface which are related, in the final section, to its progressive oxidation.

THE in situ EXPERIMENT

The observations were made at 200kV in a JEOL 200CX microscope modified to achieve an ultra high vacuum specimen environment by the use of a helium cooled cryoshield, and to enable heating of the specimen in situ [1]. The specimen itself was a 3mm square piece of silicon of normal (110), dimpled and then chemically thinned to perforation, which had been kept in distilled water until insertion into the microscope to avoid contamination. It was found to be tilted by $4.5 \pm 0.3°$ from the 110 normal with one set of (111) planes remaining vertical (note that it was not possible to tilt the specimen in the microscope). Once in the microscope it was heated to about 930°C using 7.6W of direct resistive power. At this temperature the edges of the specimen faceted onto low-index crystallographic planes with clean areas visible between silicon carbide crystallites which also form at this time (figures 1 and 2). The microscope was aligned and the specimen was then further heated to 890°C for three minutes with the electron beam off to allow silica to desorb from the surface. (SiC can be removed only by higher temperature treatment, which would also remove any thin areas of specimen.) The procedure described here is known from plan view diffraction studies [2] to form clean reconstructed silicon surfaces. During the cleaning, the oxygen partial pressure was $4.1 \times 10^{-10} \tau$ with the major contaminant He ($8 \times 10^{-9} \tau$).

After the cleaning procedure, a low resolution through-focal series of micrographs was taken of a region of the faceted edge. A magnification of 160,000x was used with an objective aperture radius of 5.5mrad, which excluded all diffracted beams, and with parallel illumination to improve the visibility of the Fresnel fringes. The change in defocus between each image was approximately 88nm. Oxygen was then allowed into the microscope column to a pressure of $2.0 \times 10^{-6} \tau$ and an area of about $1 \mu m^2$ of the specimen was simultaneously irradiated by the electron beam at a current of 120mA. In the absence of electron beam stimulation oxidation does not occur at these pressures of oxygen [3] (as was verified by subsequent observations of areas not exposed to the beam). Two further through-focal series were taken of the same area after oxidation times of 19 and 50 minutes.

The three sets of micrographs were analysed after being digitised at a resolution of 1 pixel per 0.157nm (i.e. 25μm on the plate). The image processing to be described was done using the

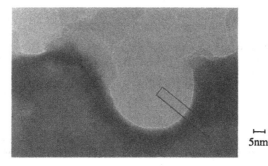

⊢——⊣
5nm

Figure 1 The silicon specimen after cleaning. This image is from the first through focal series at a defocus of -264nm (underfocus). Fresnel fringes are clearly visible at the edge. The area analysed is indicated. The circular features (some with moirés) are epitaxial SiC particles formed during specimen heating.

Figure 2 Schematic diagram of the likely form of the specimen edge after heating, at the position indicated in figure 1.

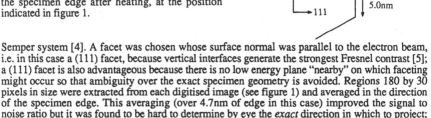

Semper system [4]. A facet was chosen whose surface normal was parallel to the electron beam, i.e. in this case a (111) facet, because vertical interfaces generate the strongest Fresnel contrast [5]; a (111) facet is also advantageous because there is no low energy plane "nearby" on which faceting might occur so that ambiguity over the exact specimen geometry is avoided. Regions 180 by 30 pixels in size were extracted from each digitised image (see figure 1) and averaged in the direction of the specimen edge. This averaging (over 4.7nm of edge in this case) improved the signal to noise ratio but it was found to be hard to determine by eye the *exact* direction in which to project; the direction was therefore calculated by cross-correlating rows of pixels from the extracted regions. It is important to project in the correct direction to prevent any artificial blurring of the Fresnel profiles. The profiles produced are shown in figure 3 after correction for the response of the photographic emulsion and division by the intensity level outside the specimen.

Distinct differences can be seen between the three series of profiles, particularly in the strength of the features marked P and Q. To interpret these differences in terms of changes in the silicon surface, the experimental profiles were compared with profiles from computer simulations. Atomistic multislice calculations were used with atomic positions (for the clean surface) shown in figure 4. This model, neglecting any reconstruction of the surface, is a suitable starting point.

It was necessary to measure certain imaging parameters for the simulation. The magnification was calibrated using lattice fringes and the beam convergence (semi-angle 0.45mrad) from the size of spots on a well focussed diffraction pattern taken with the condenser settings used for imaging. The effects of chromatic aberration and beam divergence were included in the calculation by incoherent summations over a range of energy and tilt angle and the correct crystal orientation was used. The defocus step size was calibrated using the method of Heidenreich et al. [6]. The most critical parameter, however, is the thickness of the specimen at its edge. This could not be measured by tilting (or even by matching high resolution images with simulations) and instead the intensity of the silicon relative to the background was measured from several near focus plates, from the first thickness fringe to within about 2nm from the specimen edge (where Fresnel effects alter the contrast) and was compared with simulated silicon thickness fringes at the correct crystal orientation. Extrapolation to the specimen edge suggested a thickness there of 5±1nm (figure 2). We feel that this is a reasonable measure of the specimen thickness, partly because there is no amorphous or contaminated layer on the surface, which if present could cause a large drop in intensity on entering the specimen by scattering to large angles, and partly because the effects of inelastic scattering, which are not included in the simulations, are likely to be very weak at this thickness. (The calculated thickness fringes in fact provided a good match to the experimental contrast, with the intensity at the first minimum agreeing to within 4%.)

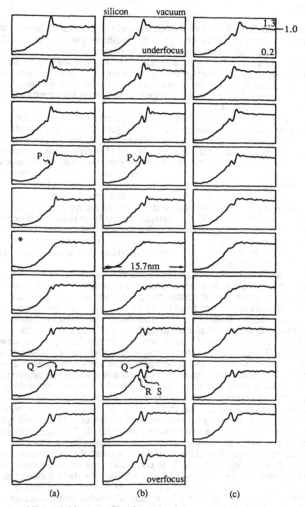

Figure 3 Experimental Fresnel fringe profiles from the three series: (a) the clean surface; (b) after 19 minutes oxidation; (c) after 50 minutes oxidation. The data has been corrected for the non-linear response of the photographic emulsion and then divided by the intensity outside the specimen. (Note that the plate labelled * was blurred during exposure so its lack of detail is not significant.)

The Fresnel fringes calculated from the simulation are shown in figure 5a. It can be seen that they reproduce at least qualitatively the features seen in the three experimental series. However it was found necessary, when trying to achieve this match, to include an imaginary (absorptive) potential in the simulations. This is because although the specimen thickness varies in the manner shown in figure 2, the multislice simulation models a rectangular specimen cross section. The contrast of the Fresnel fringes is strongly affected if they are superimposed on a changing background (for example the relative heights of features R and S in figure 3a) and the imaginary potential was therefore used to simulate the effect of the change in specimen thickness (at low thickness, $\ll \xi_{eff}$) by removing intensity from the central beam. Without the imaginary potential the match between simulation and data was much poorer. We now discuss the detailed analysis of the three data sets to obtain information about the changes in the specimen during oxidation.

DISCUSSION

It has been recognised for a long time that the Fresnel fringes seen at the edge of a specimen are highly sensitive to the form of the change in the projected potential at the edge. For example Fukushima et al. [7] demonstrated the difference in fringes simulated from square, rounded and tapered specimen cross sections. In the past this sensitivity has led to difficulties in interpreting Fresnel contrast due to the unknown geometry of the edge, but in the present experiment we can make use of the sensitivity, given the (fairly) well known specimen geometry, to characterise the behaviour of the projected potential near the specimen edge (as has already been done for vertical interfaces within a specimen [4,8]).

We consider firstly the clean silicon surface. In comparing the simulation (figure 5a) with the data (figure 3a) it is clear that although the match is generally good, particularly with respect to overall contrast levels and the spacing between fringes, certain discrepancies are present: features P underfocus on the silicon side of the main fringe and Q overfocus on the vacuum side are both weaker experimentally than theoretically. Neglecting the possibility of tilt of the interface away from vertical (the symmetry of the diffraction pattern showed this tilt to be well within $0.5\theta_{B(111)}$, so that an abrupt interface of height 5nm would be projected by less than 0.01nm), this could be due either to roughness of the surface (steps) or to a surface reconstruction. Both can be modelled simplistically by using fractional occupancy at the sites on the top planes in the model of figure 4: in the silicon 7x7 reconstruction [9], for example, the top layer is only 25% occupied (the heights of the adatoms above the surface are also modified slightly) and a similar model will apply if one or more steps are present on the facet. If the surface here were reconstructed, with adatoms only on certain sites, the averaging necessary in both the lateral and beam directions to obtain the Fresnel profiles would conceal the periodicity.

Figures 5b and c show the effect of fractional occupancy on the top two and four planes of atoms. A comparison of the differences between the three sets of fringes demonstrates the above mentioned sensitivity of the Fresnel contrast to details of the projected potential. It can be seen that these smoother models do in fact fit the data set of figure 3a much better than the abruptly terminated surface, because features P and Q are reduced with respect to figure 5a. The best fitting simulation is that of figure 5b because although feature P is then slightly too strong, feature Q is too weak compared with the experimental data. In fact the discrepancies between the *relative* strengths of P and Q suggest that the best fitting model might be one in which, instead of the top planes grading uniformly from 0 to full occupancy, the top silicon layer is already quite highly occupied and the occupancy then increases to 1 over a few additional atomic planes; this model would therefore have a sharp change in occupancy on the vacuum side of the partially occupied layer (stronger feature Q) but a more gradual change on the silicon side (weaker feature P). The behaviour of the Fresnel fringes from this interface is in fact consistent with it having a 7x7 reconstruction such as is seen in plan view on the surfaces of (111) normal specimens after the same heat treatment [2]. We have not attempted to model the 7x7 reconstruction because of the large size of its unit cell; the displacements of the adatoms and the fact that this top plane is analogous to plane B rather than plane A in figure 4 mean that differences would be expected between this and the models of figure 5. In conclusion, the Fresnel contrast from this surface suggests that it is not abruptly terminated but is diffuse over about 2 atomic planes. It is highly likely to be reconstructed: we do not think it probable that the diffuseness is caused entirely by surface steps as the number of steps on a $5nm^2$ facet is probably small, if not zero.

We now consider the behaviour after oxidation. Surprisingly, the 19 minute oxide appears to have a much sharper surface than the clean specimen. Note particularly the increase in the height of feature P between figures 3a and 3b. This interesting result holds irrespective of the detailed interpretation of the contrast of figure 3a. However, after 50 minutes oxidation the sharpness has

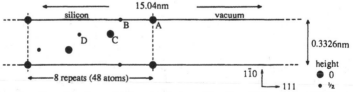

Figure 4 The unit cell used in the multislice simulations. The slice thickness was 0.384nm. This cell was sampled on a grid of 1024x64 points.

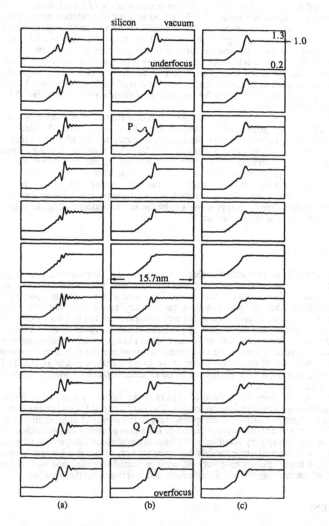

Figure 5 Simulated Fresnel fringe profiles: (a) from the model shown in figure 4; (b) from the same model, but with the top two planes of atoms (A and B in figure 4) having fractional occupancy $1/3$ and $2/3$ respectively; (c) from the same model but with the top four planes (A, B, C and D) having occupancies 0.2, 0.4, 0.6 and 0.8. An imaginary potential ramped from 0V at the Si surface to 5V 10nm into the silicon was used in these simulations to model the effects of the increase in specimen thickness. The images were also averaged over 0.16nm blocks to simulate the resolution of the densitometer.

diminished and this set of profiles appears to come from a more diffuse interface than that of figure 3b (though, interestingly, not figure 3a).

We interpret the sharpening seen in the change in projected potential after 19 minutes oxidation to be due to an increase in the fractional occupancy of the top atomic layer. Such an increase would indeed be expected if the first oxygen atoms attacked the bonds between the adatoms and the layer below, ending up near or on the plane of the adatoms, as has been suggested from plan view diffraction evidence [10]. By 50 minutes of oxidation we suggest that the oxygen atoms have penetrated more deeply into the bulk of the silicon (though still by only 2-3 atomic planes) so that the change in projected potential occurs over a more extended region. Because of the averaging process, any non-uniformity in the progress of oxidation would give the same appearence. Note that, although the fringe spacing is increased, even after 50 minutes of oxidation we suggest that the oxide film is still too thin to produce two clearly separate sets of Fresnel fringes, one for each interface present: in simulations of oxides thicker than 1-2nm, double fringes can be seen and have been used for an accurate determination of interfacial layer thickness at interfaces within materials [8].

We are currently carrying out high resolution studies of the same specimen geometry to observe directly the growth of the oxide film at the silicon surface. Studies of oxidation under similar conditions confirm the low thicknesses of oxide we see after these times. Electron beam stimulated oxidation at an oxygen pressure of $2 \times 10^{-7}\tau$ has been found using Auger spectroscopy to produce oxides of thickness about 0.5nm after 20 minutes and only 0.6nm after 50 minutes [11]. Further work at a pressure of $10^{-4}\tau$ has confirmed these slow growth rates, which are logarithmic in form, suggesting field-driven transport in the inital stages [12]. Growth rates are reduced if a higher electron voltage is used (these studies used voltages much lower than 200kV). The ultra-slow growth rates seen here appear to us to be the key to understanding the way in which oxygen first interacts with the silicon surface.

CONCLUSIONS

The analysis of Fresnel contrast is clearly a sensitive way to probe the structure of surfaces. The method has the potential of probing into the specimen by several atomic planes, as we have demonstrated with the simulations shown here, and can be used to investigate other surface reconstructions. Our analysis has been preliminary as we have not accurately modelled the reconstructions, but we have shown that when characterised in this way a clean silicon surface is not abruptly terminated, the scattering potential instead decreasing over about 2 atomic planes. When oxidation occurs, the nature of the surface changes subtly, becoming more abrupt. From these very interesting results we can speculate on where the initial oxygen attacks the silicon surface. The study of oxidation under these experimental conditions should be useful when considering oxidation in a plasma or in other circumstances where electron assisted surface processes are important.

We hope for even more detailed results in the future by comparing Fresnel analysis of a similar surface with high resolution images. The sensitivity of Fresnel contrast to overall projected potential changes at a surface is complementary to the power of high resolution imaging to detect crystallinity and to determine the *positions* of the atoms present - though without determining their *natures* or occupancy very accurately. We have not addressed the question of whether the oxide is ordered or not: ordered oxide models can generate very distinctive Fresnel fringes, particularly if the different atomic planes have very different projected potential, and we feel encouraged by the prospect of determining the initial oxide form and kinetics using these studies.

REFERENCES

1. M. L. McDonald, J. M. Gibson and F. C. Unterwald, Rev. Sci. Instrum. 60, 700 (1989)
2. J. M. Gibson, M. L. McDonald and F. C. Unterwald, Phys. Rev. Lett. 55, 1765 (1985)
3. B. Carriere, A Chouiyakh and B. Land, Surf. Sci. 126, 495 (1983)
4. W. O. Saxton, T. J. Pitt and M. Horner, Ultramic. 4, 343 (1979)
5. J. N. Ness, W. M. Stobbs and T. F. Page, Phil. Mag. A 54, 679 (1986)
6. R. D. Heidenreich, W. M. Hess and L. L. Ban, J. Appl. Cryst. 1, 1 (1968)
7. K. Fukushima, H. Kawakatsu and A. Fukami, J. Phys. D 7, 257 (1974)
8. F. M. Ross and W. M. Stobbs, submitted to Phil. Mag. A (1989)
9. K. Takayanagi, Y, Tanishiro, M. Takahashi and S. Takahashi, J. Vac. Sci. and Tech. A 3, 1502 (1985)
10. J. M. Gibson, this conference
11. B. Carriere, J. P. Deville and A. El Maachi, Phil. Mag. B 55, 721 (1987)
12. P. Collot, G. Gautherin. B. Agius, S. Rigo and F. Rochet, Phil. Mag. B 52, 1051 (1985)

ELECTRON SPIN RESONANCE STUDIES OF SILICON DIOXIDE FILMS ON SILICON IN INTEGRATED CIRCUITS USING SPIN DEPENDENT RECOMBINATION

M. A. JUPINA and P. M. LENAHAN

Pennsylvania State University, University Park, PA 16802

ABSTRACT

The technique of spin dependent recombination (SDR) allows the electron spin resonance (ESR) observation of electrically-active point defects in a single metal-oxide-semiconductor field-effect transistor (MOSFET) with surface areas of only 10^{-4} cm^2 and Si/SiO$_2$ interface point defect densities of ~10^{11}/cm^2. With SDR's enhanced sensitivity, devices with different processing details are explored. Differences in the E' spectra for variations in the oxidation processing are discussed.

INTRODUCTION

Over the last two decades, electron spin resonance (ESR) techniques have been quite successful in exploring the atomic scale structure of defects in large area metal-oxide-semiconductor (MOS) structures. However, ordinary ESR detection techniques are many orders of magnitude too insensitive to permit studies of individual metal-oxide-semiconductor field-effect transistors (MOSFETs). A technique known as spin dependent recombination (SDR) allows rapid (a few minutes) detection of low densities (~10^{11}/cm^2) of point defects in single MOSFETs in integrated circuits.

Using SDR, we observe radiation-induced P_b and E' centers in individual MOSFETs (10^{-4}/cm^2). Earlier ESR studies of extremely large (~1cm^2) capacitor structures have identified the P_b and E' centers as the dominant radiation-induced defects in MOS structures [1-3]. The P_b center is a trivalent silicon defect located at the Si/SiO$_2$ interface; the E' center is an oxygen deficient silicon defect in the oxide. Our SDR observations reconfirm and extend earlier ESR work on MOS structures. This present study on integrated circuits answers objections to the relevance of earlier ESR work on large area MOS structures in their relation to MOSFETs in integrated circuits.

It has been established for several years that SDR detection techniques have quite high sensitivity to P_b-like interface defects in several systems [4,5]. Chen and Lang [6] as well as Henderson [7] have shown that SDR may be applied to the Si/SiO$_2$ system. Recently, Vranch et al. [8] demonstrated that P_b centers could be detected via SDR in a gate-control diode with a large gate area and a high interface state density. Unlike our study, the Vranch study was unable to detect the presence of E' centers in their SDR measurements. Our measurements involve MOSFETs with different oxidation processing histories. Differences in the SDR spectra of the E' centers are discussed.

OUR EXPERIMENT

Just as in standard ESR, SDR detects the presence of paramagnetic point defects, that is, defects with an unpaired electron. The technique exploits that the capture cross-section of a paramagnetic trapping center is affected by its spin state. Several somewhat contradictory models [9-11] have been proposed to explain the spin dependent recombination process; however, a detailed description of these models is not appropriate for this discussion.

In our study, a MOSFET is used as a gate-controlled diode [12,13] (source and drain shorted together) to study radiation-induced paramagnetic trapping centers. The gate-controlled diode, or gated diode, is illustrated in figure 1. The p-n junction of the gated diode is slightly forward biased with the junction voltage $|V_J| <= 0.3V$, so that changes in the recombination current (ΔI) associated with deep trap levels can be monitored while their spin resonance condition is satisfied. For low forward biases, with the silicon surface under the gate depleted of carriers, the recombination current (I) due to interface states is the dominant current in the device. This study is primarily concerned with the radiation-induced centers at or near the Si/SiO_2 boundary of MOSFETs with different oxidation processing histories.

The MOSFETs used in this study were of three varieties, denoted as A, B, and C in the following. Device A and B had oxides grown in dry O_2 at 1000 C to a thickness of 37 nm, while device C's oxide was grown by wet oxidation at 900 C to a thickness of 62.5 nm. Device A's oxide was annealed in situ in N_2 at 900 C for 25 minutes while device B's oxide was annealed in situ in N_2 at 1000 C for 25 minutes. Device C's oxide received no post-oxidation anneal. The gate areas of all the devices were $\sim 10^{-4}$ cm^2. Device A and B had N-type substrates of (100) orientation and $1.5 \times 10^{16}/cm^3$ phosphorus doping. Device C had a P-type substrate of (111) orientation and $2 \times 10^{15}/cm^3$ boron doping

The SDR spectrometer employed in this study is schematically illustrated in figure 2. In our experiments, the integrated circuit was centered in a TE_{102} cavity of a model 8300 Micro-Now X-band ESR spectrometer. Care was taken to minimize the microwave electric field at the device to limit loading. The loaded cavity had a Q of about 5000. A 100 mW X-band solid state oscillator (9.5 GHz) applied a continuous microwave field to the resonant cavity while a large, slowly varying magnetic field (~3500 G) and a small ac magnetic field (~5 G_{pp} and $v <= 10$ KHz) were simultaneously also applied. The ac magnetic field permitted phase sensitive detection of the SDR signals. The gated diode was biased at a fixed gate and junction voltage while spin dependent variations in the recombination current (ΔI) were monitored using a current-to-voltage pre-amplifier and lock-in amplifier (Ithaco Dynatrac Model 393). The system easily detected SDR currents of order 10 fA. The observed spectra are approximately the first derivative of an "absorption-like" curve. For g-value determination of the paramagnetic recombination centers, an NMR gaussmeter (Micro-Now Model 515) was used in conjunction with a frequency meter.

Table I.   Summary of electrical characteristics of devices A, B, and C.

	A	B	C
Midgap D_{it} (10^{11}/cm²-eV)	2.0	3.5	5.0
ΔV_{mg} (Volts)	-1.6	-4.8	-5.7

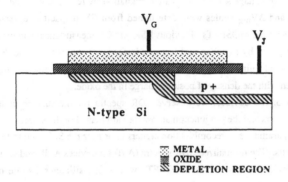

⬚	METAL
▨	OXIDE
◩	DEPLETION REGION

Figure 1. Schematic of the gate-controlled diode for low forward bias with the silicon surface depleted of carriers, where V_G and V_J denote gate and junction voltages, respectively.

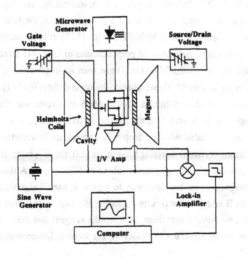

Figure 2. Block diagram of the SDR spectrometer.

DISCUSSION AND RESULTS

All devices were irradiated with ^{60}Co gamma rays. Device A and B were irradiated to a total dose of 5 Mrads while a gate bias of +5 volts was applied. Device C received a total dose of 3 Mrads while a gate bias of +6.25 volts was applied. The radiation damage process results in the creation of interface states at the Si/SiO$_2$ interface and the capture of holes in deep traps near the Si/SiO$_2$ interface [14-18]. Mid-gap interface state densities (D_{it}) and mid-gap voltage shifts (ΔV_{mg}) after irradiation are recorded in Table I for all three devices. D_{it} and ΔV_{mg} values were determined from CV (capacitance versus voltage) measurements on the MOSFETs. Previous ESR and CV measurements on very large area capacitor structures by Lenahan and Dressendorfer [1-3] showed that the density of P_b centers correlated quite well with the interface state density and that the density of E' centers tracked with the density of positive charge in the oxide.

In each of the devices, we observe SDR spectra for the surface under the gate depleted of carriers and the p-n junction at low forward bias. For the surface under the gate in depletion, paramagnetic recombination centers at or near the Si/SiO$_2$ interface generate the SDR signals. The normalized SDR spectra ($\Delta I/I$) for devices A, B, and C are shown in figure 3. Devices A and B are MOSFETs with (100) silicon substrate orientation. Therefore, with the magnetic field applied in the (100) direction, both the P_{bo} (g=2.006±0.0003) and E' (g=2.007±0.0003) centers are visible as in previous ESR studies done by Kim and Lenahan [19] on radiation-induced defects in (100) MOS structures. Just as they reported, we find that the radiation-induced interface state buildup consisted mostly of P_{bo} centers. On (100) surfaces with processed-induced interface states, Poindexter et al. [20] observed two P_b centers, termed P_{bo} and P_{b1}. The P_{bo} defect is a silicon bonded to three other silicons at the Si/SiO$_2$ interface; the structure of P_{b1} is not clearly established. Device C is a MOSFET with (111) silicon substrate orientation. Therefore, with the magnetic field applied perpendicular to the (111) direction, both the P_b (g=2.008±0.0003) and E' (g=2.007±0.0003) are seen as in previous ESR studies done by Lenahan and Dressendorfer [1-3] on radiation-induced defects in (111) MOS structures.

The E' signal observed by standard ESR in MOS structures was shown to be a hole trapped in an oxygen vacancy very near the Si/SiO$_2$ interface [2,3]. Whether the E' centers observed in SDR are associated with the deep hole trap cannot be ascertained with absolute certainty at this time. What is certain, though, is that the E' center must reside close enough to the Si/SiO$_2$ interface to play some role in an SDR event. Although the P_b and E' amplitudes are roughly equal for device A in figure 3, the E' signals are considerably smaller for devices B and C. Our results regarding E' must be regarded as preliminary at this time; however, we believe that these SDR results suggest that the E' centers reside, on the average, closer to the interface in device A, the most radiation-tolerant device of the three devices.

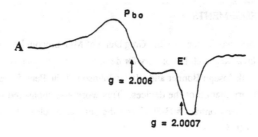

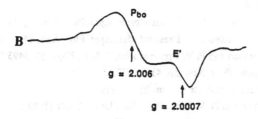

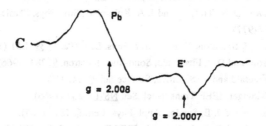

├───┤ 5 G (Magnetic Field)

⊥ 1.5 x 10⁻⁴ (ΔI/I)

Figure 3. SDR spectra of P_b and E' centers for devices A, B, and C.

CONCLUSIONS

In this study we used SDR to observe the radiation-induced buildup of P_b and E' centers in devices of various oxidation processing. For individual MOSFETs in integrated circuits where conventional ESR is impossible, the SDR technique allowed rapid detection of low densities of point defects. Differences in the magnitude of the E' spectra in oxides of different processing conditions is tentatively attributed to E' centers residing, on the average, closer to the Si/SiO_2 interface in a more radiation-tolerant oxide.

ACKNOWLEDGEMENTS

The authors would like to thank Greg Dunn of MIT Lincoln Laboratories and Nelson Saks of the Naval Research Laboratory for devices used in this study. The authors would also like to thank Joseph Bonner and Walter Johnson of the Penn State Breazeale nuclear reactor for the irradiation of the devices. This work was sponsored by Sandia National Laboratories under Contract #03-3999 and the defense Nuclear Agency under Contract #DNA002-86-0055.

REFERENCES

1. P. M. Lenahan and P. V. Dressendorfer, Appl. Phys. Lett. 41, 542 (1982).
2. P. M. Lenahan and P. V. Dressendorfer, Appl. Phys. Lett. 44, 96 (1984).
3. P. M. Lenahan and P. V. Dressendorfer, J. Appl. Phys. 55, 3495 (1984).
4. D. J. Lepine, Phys. Rev. B6, 436 (1972).
5. I. Solomon, Solid-State Comm. 20, 215 (1976).
6. M. C. Chen and D. V. Lang, Phys. Rev. Lett. 51, 427 (1983).
7. B. Henderson, Appl. Phys. Lett. 44, 228 (1984).
8. R. L. Vranch, B. Henderson, M. Pepper, Appl. Phys. Lett. 52, 1161 (1988).
9. R. M. White and J. F.. Gouyet, Phys. Rev. B16, 3596 (1977).
10. V. S. Livov, O. V. Tretyak, and J. A. Kolomiets, Sov. Phys. Semicond. 11, 661 (1977).
11. D. Kaplan, I. Solomon, N. F. Mott, J. Phys. Lett. (Paris) 39, L51 (1978).
12. A. S. Grove and D. J. Fitzgerald, Solid-State Electron. 9, 783 (1966).
13. D. J. Fitzgerald and A. S. Grove, Surface Sci. 9, 347 (1968).
14. K. H. Zaininger, IEEE Trans. Nucl. Sci. NS-13, 237 (1966).
15. J. R. Szedon and J. E. Sandor, Appl. Phys. Lett. 6, 181 (1965).
16. R. J. Powell and G. F. Derbenwick, IEEE Trans. Nucl. Sci. NS-18, 99 (1971).
17. P. S. Winokur and M. M. Sokoloski, Appl. Phys. Lett 28, 627 (1976).
18. F. B. McLean, IEEE Trans. Nucl. Sci. NS-27, 1651 (1980).
19. Y. Y. Kim and P. M. Lenahan, J. Appl. Phys. 64, 3551 (1988).
20 E. H. Poindexter, P. J. Caplan, B. E. Deal, and R. R. Razouk, J. Appl. Phys. 52, 879 (1981).

AN NMR STUDY OF HYDROGEN IN SiO$_2$ FILMS ON SILICON

DAVID H. LEVY and K. K. GLEASON, Department of Chemical Engineering, Massachusetts Institute of Technology, Cambridge, MA 02139

ABSTRACT

We have used solid state nuclear magnetic resonance (NMR) spectroscopy to study both "wet" and "dry" thermally grown films of SiO$_2$ on silicon substrates. For the 5000 Å wet film, grown at 1050 °C we observed a single Lorentzian line of 6 kHz HWHM (half width at half maximum). For the 500 Å dry film, we observed a convolution of two lines: a) a Lorentzian of 4 kHz HWHM and b) a Gaussian of 20 kHz HWHM. The hydrogen distributions in both oxides are interpreted as a function of these lines.

INTRODUCTION

Thin (100 Å to 1 μm) silicon dioxide films are used as dielectric layers in semiconductor devices. As device dimensions decrease, the silicon dioxide thickness must also decrease, leading to a situation where the Si/SiO$_2$ interface region (generally estimated to be approximately 30 Å thick[1]) plays a significant role in the structure and performance of a given device. It is clear that hydrogen plays a major role in the chemistry of this interface[2], whether through reactions with dangling bond defects at the interface[3-5], or through its presence in non-stoichiometric silicon bonding configurations near the interface[1].

Various techniques have thus far been used to investigate hydrogen at the Si/SiO$_2$ interface: internal reflection IR spectroscopy[6], secondary ion mass spectroscopy[7], and nuclear reaction techniques[8]. The latter two techniques both involve subjecting the sample to high levels of particle beam radiation, thus potentially changing bonding configurations in the sample, as well as causing hydrogen to migrate[8]. IR spectroscopy, on the other hand, provides insight into bonding configurations at the expense of quantitative data on the hydrogen since the oscillator strength of a particular bond is dependant upon its environment[9].

We have chosen to adapt solid state nuclear magnetic resonance (NMR) to the study of thermally grown SiO$_2$ films since it is a technique that uses very low energy radiation (radio frequencies), and is sensitive to hydrogen bonding configurations, as well as hydrogen internuclear spacings. In particular, we feel that since hydrogen is a quick, small diffuser, it may segregate near defects in the oxide, thus providing a microscopic probe to the structure of the Si/SiO$_2$ interface.

Mat. Res. Soc. Symp. Proc. Vol. 159. ©1990 Materials Research Society

EXPERIMENTAL

A) Evaluation of NMR Data

The NMR resonance line can reveal information on bonding, spacing, and the quantity of nuclei under investigation. The bonding configuration is determined from a shift in the resonance frequency of a nucleus (the chemical shift). Based on comparison to the $(C_6H_5)_n$-$SiH_{(4-n)}$ - $(C_6H_5)_n$-$SiOH_{(4-n)}$ system, we expect a resonance shift of approximately 3 ppm between SiH and SiOH bonding configurations[10]. This shift would be detectable on our spectrometer with the use of a line narrowing pulse sequence[11], and would permit us to establish whether SiOH or SiH is the dominant hydrogen species.

From the width of a hydrogen resonance, we can infer hydrogen internuclear distance or local hydrogen concentration. A resonance line of Gaussian shape indicates a regular spacing or high concentration of hydrogen atoms. According to Van Vleck[12], the second moment of the line, M_2, is given by:

$$M_2 = (HWHM/1.177)^2 = (3/5)\gamma^4 h^2 I(I+1) \Sigma_i (1/r_i^6) \qquad (1)$$

where γ is the hydrogen gyromagnetic ratio, h is the reduced Planck's constant, I is the nuclear spin of hydrogen (1/2), r_i is the internuclear distance to neighbor i, and HWHM is the half-width at half-maximum of the NMR resonance line. Assuming for simplicity that only one neighbor contributes to the line width, one can estimate the internuclear distance from the HWHM.

A resonance that is Lorentzian in shape, on the other hand, usually indicates a random spacing or dilute concentration of hydrogen nuclei. The effective concentration, n, of hydrogen can be calculated from the HWHM of the Lorenztian line by[13]:

$$HWHM = 3.8\gamma^2 hn \qquad (2)$$

B) NMR Probe

A large portion of work thus far has centered on constructing an NMR probe sensitive enough to study silicon dioxide films which, because of their growth conditions and small volumes contain little hydrogen. The probe that we have constructed is shown in schematic in Figure 1. In order to eliminate background noise due to a sample container, silicon sample slivers are placed directly in the coil. In addition, dry nitrogen from a liquid source is used to purge the inside of the probe during runs. Experimental data is usually averaged from approximately 40,000 scans, giving us a resolution on the order of 10^{16} protons per sample.

As mentioned earlier, our samples are generated from silicon slivers. In

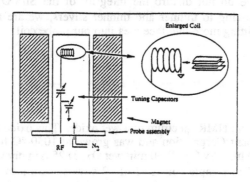

Figure 1: Schematic of the NMR probe

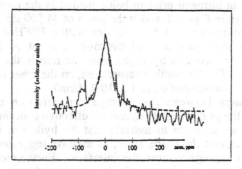

Figure 2: Wet oxide spectrum after 34,500
signal averages. Solid line: Actual data;
Dashed line: Fitted Curve

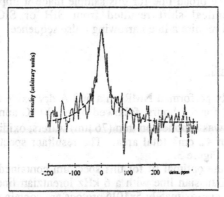

Figure 3: Dry oxide spectrum after 34,000
signal averages. Solid line: Actual Data;
Dashed line: Fitted Curve

this manner, we do not disturb the integrity of the Si/SiO_2 interface. In addition, by moving to thinner and thinner slivers, we are able to increase sensitivity by fitting more interface area into our probe coil.

RESULTS

A) Wet Oxide

We first took an NMR spectrum of a wet oxide. This oxide was supplied by Digital Equipment Corporation and was grown at 1050 °C for the following sequence: 1) 10 min dry O_2, 2) 30 min wet O_2, 3) 25 min dry $O_2/3\%$ HCl, 4) 4 min dry O_2. Our sample consisted of 12 slivers, with a total oxide area of approximately 7 cm^2. The sample was ultrasonically cleaned in methanol then dried at 170 °C in nitrogen prior to being loaded in the probe. The resulting spectrum is shown in Figure 2, and is the result of 34,500 signal averages.

We fit the spectrum to a lorentzian line with a HWHM of 6 kHz. The integrated peak area gave a total hydrogen content of 1×10^{17} protons, corresponding to an average hydrogen concentration in the oxide of 3×10^{20} H/cm^3 (0.6 at%). The linewidth measurement, on the other hand, indicates an effective hydrogen concentration of 1.3×10^{22} H/cm^3.

The discrepancy between the average concentration of hydrogen, as calculated from the peak integration, and the effective concentration calculated from linewidth would seem to indicate that the hydrogen is not uniformly distributed throughout the oxide layer, as may be suggested by a model in which hydrogen segregates near the interface. Furthermore, if the entire concentration of protons were to exist in a region of average concentration 1.3×10^{22} H/cm^3, this region would be approximately 110 Å thick (out of a total of 5000 Å of oxide).

Unfortunately the broad line for this sample made it impossible to resolve whether the chemical shift resulted from SiH or SiOH groups. This determination will require a line narrowing pulse sequence.

B) Dry Oxide Results

We subsequently performed NMR on a 500 Å dry oxide that was grown in dry O_2 at 850 °C for 5.5 hours, followed by a 300 °C anneal in H_2 for 2.5 hours. The oxide was supplied on thin (76 μm) wafers, oxidized on both sides, giving a sample of 42 cm^2 total area. The resultant spectrum, after 34,000 scans, is shown in Figure 3.

We found that the optimal fit for this spectrum contained a convolution of 20 kHz HWHM Gaussian line with a 6 kHz lorentzian line. The integrated areas indicate that approximately 5×10^{16} protons are contained in each "phase". This type of two phase behavior has been seen in the a-SiH system, in which the broad component represents hydrogen clusters, whereas the narrow component belongs the dilute hydrogen spaced throughout the layer[14,15]. In our particular case, the broad component would represent, based on the Van Vleck model, hydrogen spacing on the order of 1.7 Å.

The linewidth of the dilute phase yields an effective concentration of 9×10^{21} H/cm^3, which, as with the wet oxide, is substantially higher than the average concentration found by spreading the 5×10^{16} protons over the entire 500 Å thick oxide (2.5×10^{20} H/cm^3). If we assume that all dilute phase hydrogens are contained in a layer containing 9×10^{21} H/cm^3, we arrive at a layer thickness of approximately 14 Å. This thickness is on the same order of magnitude as that of the 30 Å interface region.

CONCLUSION

We have obtained to our knowledge the first NMR spectra of hydrogen in thermally grown SiO$_2$ films on silicon. We have been able to eliminate background noise in our spectra by placing slivers of SiO$_2$ on silicon directly into our probe coil, as well as purging the probe with dry nitrogen to avoid effects from ambient water.

Both wet and dry oxide spectra showed Lorenztian lineshapes, indicating a random distribution of hydrogen. In both cases, effective hydrogen concentrations were substantially larger that average concentrations, indicating that hydrogen is probably not uniformly distributed in the oxide layers. Furthermore, the dry oxide spectrum showed a broad Gaussian component, perhaps indicating a clustering of hydrogen atoms.

In both cases, the spectral lines were too broad for a chemical shift determination. However, sufficient hydrogen is present to perform a line narrowing experiment and thus address the question of H bonding configuration in these layers.

ACKNOWLEDGEMENTS

We would like to thank Dr. Paul Riley from Digital Equipment Corporation for supplying our wet oxide sample, and Dr. Keith Brower from Sandia National Laboratories for supplying the dry oxide.

REFERENCES

1. Grunthaner, P.J. and M.H. Hecht, F.J. Grunthaner, J. Appl.Phys., **61**(2), 629(1986)
2. Revesz, A.G., J. Electrochem. Soc., **126**, 122 (1979)
3. Poindexter, E.H. and P.J. Caplan, J. Vac. Sci. Technol. A, **6**(3), 1352 (1988)
4. Poindexter, E.H. and P.J. Caplan, Progress in Surf. Sci., **14**, 201 (1983)

5. Pantelides, S.T., Phys. Rev. Lett., **57**, 2979 (1986)
6. Beckman, K.H. and N.J. Harrick, J. Electrochem. Soc., **118**, 614, (1971)
7. Tsong, I.S. and M.D. Monkowski, J.R. Monkowski, P.D. Miller, C.D. Mosk, B.R. Appleton, A.L. Wintenburg, "Investigation of Hydrogen and Chlorine at the Si/SiO$_2$ Interface" from The Preparation of MOS Insulators, Lucovsky ed. (1980)
8. Marwick, A.D. and D.R. Young, J. Appl. Phys. **63**(7), 2291 (1988)
9. Kaxiras, E. and J.D. Joannopoulos, Phys. Rev. B, **37**(15), 8842 (1988)
10. "Aldrich Library of NMR Spectra" Vol. 2, C.J. Pouchert ed., (Aldrich, Milwaukee 1983) p. 1004,1008
11. Van Vleck, J.H., Phys. Rev. 74, 1168 (1948)
12. Rhim, W.-K. and D.D. Elleman, R.W. Vaughn, J. Chem. Phys., **59**(7), 3740 (1973)
13. Abragam, A., Principles of Nuclear Magnetism, (Oxford University Press, 1961)
14. Reimer, J.A. and R.W. Vaughn, J.C. Knights, Phys. Rev. B, **24**(6), 3360 (1981)
15. Jeffrey, F.R. and M.E. Lowry, J. Appl. Phys., **52**(9), 5529 (1981)

ELECTRONIC STRUCTURE OF EPITAXIAL SiO₂/Si(100) INTERFACES

T. Motooka

Institute of Applied Physics, University of Tsukuba, Ibaraki 305

ABSTRACT

The local densities of states (LDOS) of epitaxial SiO_2
layers on Si(100) surfaces have been calculated using the
recursion method combined with the Harrison's universal tight-
binding model. The interface states associated with strained
epitaxial layers of β-cristobalite $(\sqrt{2}\times\sqrt{2})R45°$ and tridymite
$(10\bar{1}0)\langle0001\rangle \parallel$ Si(100)$\langle011\rangle$ were examined. In the β-cristobalite
layer, gap states due to the surface Si dangling bonds appeared
while they were eliminated by H termination. In the tridymite layer,
the interface states primarily composed of the surface Si back bonds
appeared near the Si conduction band minimum. Comparing the
calculated DOS with photoelectron spectra for initial oxidation
processes of clean Si(100), it was found that the valence band
spectrum from the initial oxide formed at $\sim300°C$ resembled that of
the β-cristobalite layer.

INTRODUCTION

The SiO_2/Si interface has been extensively investigated for its
importance in Si MOS technology.[1] Various kinds of measurements
such as photoelectron spectroscopy and transmission electron
microscopy (TEM) have been utilized to characterize the atomic and
electronic structure at the SiO_2/Si interface. It is generally known
that there exists a thin suboxide region, SiO_x (x<2) and typically
$\simeq10^{11}cm^{-2}$ gap states at the interface. Although several atomic
structure models of the SiO_2/Si interfaces have been given, almost
all of these investigations were based on the experiments for the
samples obtained in the conventional oxidation technique and the
initial Si surfaces possibly included contaminations, defects, and
steps. Therefore, the growth and structure of initial oxides are not
well understood on the atomic scale. However, it has recently been
suggested, based on TEM lattice image analysis, that there exists
a $\simeq5Å$ crystalline SiO_2 layer at the interface on atomically flat
Si(100) surfaces obtained by MBE.[2]

In this paper, we have investigated the electronic structure
of epitaxial SiO_2/Si(100) interfaces employing the recursion
method[3] based on Harrison's universal tight-binding method.[4]
Clusters comprising $\simeq10^4$ atoms were generated to describe possible
crystalline SiO_2 layers on Si(100) substrates. In an epitaxial
$(\sqrt{2}\times\sqrt{2})R45°$ structure of β-cristobalite, only half of the surface
Si atoms are connected to oxygen in SiO_2. The calculated LDOS

associated with the remaining surface Si atoms showed that localized states due to the dangling bonds appeared in the middle of the Si band gap. These gap states were eliminated by terminating the dangling bonds with hydrogen consistent with the results of photoelectron spectroscopy measurements for H_2 annealed $SiO_2/Si(100)$ interfaces.[1] On the other hand, all the surface Si atoms are connected to the O atoms in a strained tridymite structure with an epitaxial relationship, tridymite $(10\bar{1}0)\langle0001\rangle \parallel Si(100)\langle011\rangle$. The interface states appeared near the Si conduction band minimum and were primarily composed of the surface Si back bonds. The calculated DOS from 5-layer epitaxial β-cristobalite and tridymite were compared with measured valence spectra for initial oxidation of Si(100) substrates in a UHV chamber.[5] The results suggest that the initial oxidation layer is likely to form a β-cristobalite structure at ~ 300 ℃.

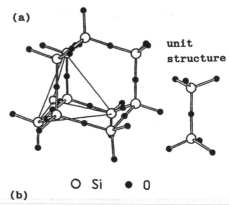

(a)

unit structure

O Si ● O

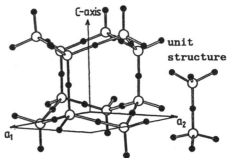

(b)

C-axis

unit structure

a_1 a_2

Fig.1 Crystalline SiO_2 structures: (a) β-cristobalite and (b)tridymite.

CRYSTALLINE SiO_2 AND STRAINED-LAYER-EPITAXIAL SiO_2 ON Si(100) SURFACES

Figure 1 shows bulk β-cristobalite and tridymite[6] which possibly form epitaxial layers on Si(100) surfaces as described below. Both of these crystalline structures are constructed by similar units as shown in Figure 1. The Si-O-Si angle is 180° in both cases, but the oxygen configuration around the Si-O-Si axis is staggered in β-cristobalite while it is eclipsed in tridymite. This results in a diamond (hexagonal) structure for β-cristobalite (tridymite). Bulk β-cristobalite is formed at ≈300℃ while tridymite at ≈200℃.[6]

Strained-epitaxial layers of β-cristobalite and tridymite on Si(100) are illustrated in Figure 2. The unit length of the f.c.c. in β-cristobalite is 7.16Å, approximately $\sqrt{2}$ times longer than that of the Si f.c.c. lattice, 5.43Å. Thus, strained-layer epitaxy is possible by rotating the ⟨100⟩ axis of β-cristobalite by 45° which is known as a ($\sqrt{2}\times\sqrt{2}$)R45° structure.[7] The lattice mismatch associated with this epitaxial interface is ≈7% and half of the surface Si atoms connect to the O atoms while the other half are free.

Tridymite has a hexagonal structure with a=5.03 and c=8.22Å.[6] Since the Si surface lattice vector ⟨011⟩ is 7.68Å long, Ourmazd et al.[2] proposed an epitaxial relationship in which the c-axis is parallel to the ⟨011⟩ direction of the Si(100) surface or tridymite (10$\overline{1}$0)⟨0001⟩ ∥ Si(100)⟨011⟩. The lattice mismatch is ≈7% in the c-axis while it is ≈24% in the a-axis. There is still an ambiguity in the interface oxygen position. In this paper, the oxygen position is assumed to be in a bridge site with a height of 1.54Å from the Si(100) surface.[8]

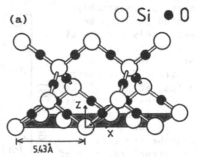

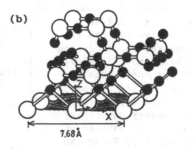

Fig.2 Epitaxial SiO₂/Si(100) interfaces: (a) β-cristobalite ($\sqrt{2}\times\sqrt{2}$)R45° and (b)tridymite (10$\overline{1}$0)⟨0001⟩ ∥ Si(100)⟨011⟩.

The local densities of states (LDOS) for the Si and O atoms in the strained-epitaxial layers were calculated using the recursion method. We used clusters comprising 48 layers of 12×12 Si f.c.c. (100) unit bases and 5 layers(≈5Å) of epitaxial SiO₂. Recursion coefficients were calculated up to 20 levels employing the Harrison's universal nearest-neighbor tight-binding matrix elements, $Vss\sigma$ =-10.06/d², $Vsp\sigma$ =10.82/d², $Vpp\sigma$ =16.92/d², and $Vpp\pi$ =-4.80/d² eV (d:distance between a pair of atoms in the unit of Å), and Hartree-Fock atomic term values, -14.79, -7.58, -34.02, and -16.72eV for Si 3s, Si 3p, O 2s, and O 2p, respectively.[4] The quadrature method[9] was used for calculating LDOS from the recursion coefficients.

CALCULATED RESULTS AND DISCUSSION

Figure 3 shows typical examples of the calculated LDOS for the Si and O atoms. The DOS of bulk SiO₂ (β-cristobalite) and Si are shown in Fig.3(a) and (b), respectively. The Si valence band maximum is taken to be the energy origin. The obtained band gaps were ≈7eV for β-cristobalite and ≈1eV for crystal Si. Since the present calculation scheme is not self-consistent, charge transfer effects cannot be reasonably treated and thus we tentatively put a -5eV shift on the atomic term values of the substrate Si in order to take account of the valence-band offset at SiO₂/Si interfaces.

In the β-cristobalite ($\sqrt{2}\times\sqrt{2}$) R45° structure, half of the surface Si atoms have dangling bond which give rise to two localized states in the Si band gap (Fig.3(c)). The low-energy states are primarily composed of 3s and $3p_z$ while the high-energy unoccupied states are composed of $3p_x$ and $3p_y$. These gap states are eliminated (Fig.3(d)) by terminat-

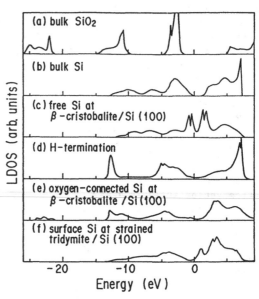

Fig.3 Calculated LDOS of epitaxial SiO₂/Si(100) interfaces.

ing the dangling bonds by H atoms since bonding orbitals, H 1s-Si 3s located at ≃-13eV and H 1s-Si 3p at ≃-5eV are formed. This gap states elimination is consistent with the results of photoelectron spectroscopy measurements for H₂ annealing effects on SiO₂/Si(100) interfaces.[1] The other surface Si atoms are connected to the O atoms in the SiO₂ layer and no localized states appeared in the gap (Fig.3(e)).

On the other hand, in the examined tridymite (10$\bar{1}$0)<0001> ∥ Si(100)<011> structure, the interface O atoms are located at a bridge site on Si(100) surface and the interface states primarily composed of the surface Si back bond[10] appeared near the conduction band minimum (Fig.3(f)). The difference in LDOS of the oxygen connected surface Si atoms in the β-cristobalite and tridymite interface may be attributed to the difference in the interface O position; i. e., in the former the O atom is in a non-tetrahedral configuration around the surface Si atom while it is in a tetrahedaral configuration in the latter.

Recently, initial oxidation processes of clean Si(100) have been investigated by high-resolution photoelectron spectroscopy using synchrotron radiation.[5] Typical examples of the valence band spectra from clean Si(100) and O₂ exposured surfaces are reproduced in Figure 4. The O₂ exposure conditions were 300L (L:Langmuir=10⁻⁶ Torr·sec) at room temperature and 20L at ~300℃. These spectra were compared with the calculated DOS corresponding to those of the ideal Si(100) surface, Si(100) with monolayer oxygen at a bridge site [8], and 5 layers of the epitaxial β-cristobalite and tridymite described above. In order to make this comparison, the calculated LDOS of the Si and/or O atoms from each layer were accumulated taking account of weighting factors for the electron escape depth and atomic densities in each layer. The calculated result for the Si(100) surface is in good agreement with the spectrum from clean Si(100). The arrow indicates the surface states due to the dangling bonds. The broad peak at ≃20eV is presently not assigned. The DOS of the Si(100) with oxygen monolayer is in fairly good agreement

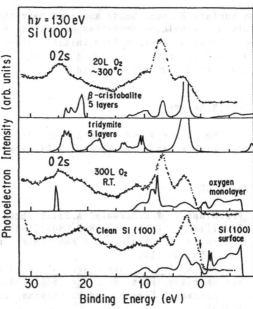

Fig.4 Comparison between the calculated DOS with photoelectron spectra for initial oxidation processes of clean Si(100) surfaces.

with the spectrum from the 300L O_2 exposed Si(100) surface at room temperature. The peak at ≃25eV corresponds to electrons from the O 2s orbital. The photoelectron spectrum from the 20L O_2 exposed Si(100) surface at ~300℃ looks more likely to be that of 5-layer β-cristobalite. Although the band offset is not well treated in the present calculations, the result might indicate that the initial thin oxide formed at ~300℃ on clean Si(100) is β-cristobalite like. For more decisive conclusions, self-consistent calculations as well as photoelectron spectroscopy measurements for initial oxides on atomically flat Si(100) should be necessary.

CONCLUSION

The electronic structure of epitaxial SiO_2/Si(100) interfaces, β-cristobalite (√2x√2)45° and tridymite (10⁻10)<0001> ∥ Si(100)<011>, has been investigated using the recursion method based on the Harrison's universal tight-binding method. A qualitative difference in the gap states was found between the β-cristobalite and tridymite interfaces. In the β-cristobalite case, gap states due to surface Si dangling bonds appeared in the Si band gap while they were eliminated after the hydrogen termination of the dangling bonds. In the tridymite case, interface states primarily composed of the surface Si back bonds appeared near the Si conduction band minimum. Calculated densities of states were compared with photoelectron spectrum from initial oxides formed at $\sim 300°C$ on clean Si(100) and it was found that the spectrum resembled that from 5 layer ($\simeq 5 Å$) β-cristobalite.

ACKNOWLEDGMENTS

The author would like to thank Dr. C. M. M. Nex for providing a copy of the Cambridge Recursion Library.

REFERENCES

1. F.J.Himpsel, F.R.McFeely, A.Taleb-Ibrahimi, J.A.Yarmoff, and G.Hollinger, Phys. Rev. B38, 6084(1988) and references therein.
2. A.Ourmazd, D.W.Taylor, J.A.Rentschler, and J.Bevk, Phys. Rev. Lett. 59, 213(1987)
3. R.Haydock, Solid State Phys., 35 (Academic, New York, 1980) p.215
4. W.A.Harrison, Phys. Rev. B24, 5835(1981)
5. M.Nakazawa, S.Kawase, and H.Sekiyama, J. Appl. Phys. 65, 4014 (1989)
6. R.W.G.Wyckoff, Crystal Structures (Interscience, New York, 1965)
7. F.Herman and R.V.Kasowski, J. Vac. Sci. Technol. 19, 395(1981)
8. V.Barone, Surface Sci. 189/190, 106(1987)
9. C.M.M.Nex, J. Phys. A: Math. Gen. 11, 653(1978)
10. S.Ciraci, S.Ellialtioglu, and S.Erkoc, Phys. Rev. B26, 5716(1982)

THE STRUCTURE OF INTERFACES IN OXIDE HETEROJUNCTIONS FORMED BY CVD

LISA A. TIETZ, SCOTT. R. SUMMERFELT, AND C. BARRY CARTER
Cornell University, Department of Materials Science and Engineering, Ithaca, NY 14853

ABSTRACT

Weak-beam imaging is used to characterize the interface structure of hematite (α-Fe$_2$O$_3$) / sapphire (α-Al$_2$O$_3$) heterojunctions parallel to (0001). The heterojunctions were prepared by low-pressure chemical vapor deposition of hematite directly onto electron-transparent sapphire substrates. Bright-field imaging and selected-area diffraction show that the growth is epitactic. The 5.5% lattice misfit at the interface is found to be accommodated by one of two different hexagonal dislocation networks: (1) $\mathbf{b} = 1/3<\bar{1}2\bar{1}0>$ or (2) $\mathbf{b} = 1/3<10\bar{1}0>$. The latter are associated with a "basal twin"-type orientation relationship of the hematite and the alumina.

INTRODUCTION

Interface structures at heterojunctions in metals and semiconductors have been extensively studied. Networks of lattice misfit-accommodating dislocations are commonly observed at such interfaces. In the case of interfaces parallel to two close-packed planes, hexagonal arrays of edge dislocations are typically observed. In addition to dislocation networks, planar defects such as stacking faults and twins are also common at such interfaces [1]. Other materials with one or more close-packed planes in their structures are expected to exhibit similar behavior.

Hematite (α-Fe$_2$O$_3$) and sapphire (α-Al$_2$O$_3$) were chosen as a model system for the study of the formation and structure of oxide heterojunctions. Both materials have the corundum-type crystal structure (Space Group = R$\bar{3}$c); hematite having a 5.5% larger lattice spacing. The (0001) plane is a close-packed plane in the oxygen sublattice. Defects in the individual materials, particularly in α-Al$_2$O$_3$, have been the subject of many investigations [e.g., 2-5]. In addition, these materials exhibit little interdiffusion and no reaction at the temperatures used in this study. The present work forms part of a larger investigation into the relationship between substrate surface structure and the formation of defects in thin films [6,7]. Here, the structure of the (0001) hematite / sapphire interface is characterized using transmission electron microscopy (TEM).

HETEROJUNCTION PREPARATION

Hematite / alumina heterojunctions were prepared using low-pressure chemical vapor deposition (LPCVD) to deposit a thin film of hematite directly onto electron-transparent, (0001)-oriented, sapphire substrates. The method has been described in detail elsewhere and is summarized here [6,7]. The thin-foil alumina substrates were prepared by conventional dimpling and ion-thinning to perforation of (0001)-oriented, 3-mm. sapphire discs. The samples were then acid-cleaned and annealed at 1400°C according to the method described by Susnitzky, et. al., to produce a stepped surface structure that could be imaged in the transmission electron microscope [8]. Hematite was produced by reacting FeCl$_3$ vapor, from a solid FeCl$_3$ source, with water vapor at 1150°C. After deposition, no further preparation was required before transferring the sample to the microscope, thus minimizing any possible damage to the interface. The interface structure was characterized by conventional TEM using a JEM1200EX operated at 120 kV.

OBSERVATIONS AND DISCUSSION

The LPCVD process resulted in a thin growth of hematite on both the top and bottom surfaces of the (0001) sapphire substrates as shown in Fig.1. Steps on the sapphire surfaces appear as lines separating regions of constant contrast (e.g., arrow in Fig.1) [8]. Two hematite growth morphologies are visible in Fig.1: islands and bands. The islands have grown along steps on the bottom surface of the substrate. Notice how they appear to run under the hematite bands. The island growth morphology has been described in detail elsewhere [6]. The hematite bands in Fig.1 lie parallel to steps on the top surface of the substrate. It is believed that they were formed by the growth and coalescence of islands at the steps during deposition.

The hematite growth was epitactic as shown by the selected-area diffraction (SAD) pattern (Fig.2) recorded from a hematite band. The misfit measured from the pattern is $\delta = 0.055$, indicating that nearly all of the lattice misfit at the interface is accommodated by dislocations. This observation is further supported by the measurements of the $\{11\bar{2}0\}$ moiré fringe spacing in Fig.1. Distortions in the hexagonal moiré fringe pattern indicate some bending of the hematite band. This bending could have been introduced during cooling due the difference between the hematite and alumina thermal expansion coefficients [9].

The misfit dislocations at the interface were imaged using both hematite and alumina reflections under weak-beam conditions. Figs 3a and b show the same area imaged under two different diffraction conditions using $g = (30\bar{3}0)$ and $g = (\bar{3}030)$, respectively. The images reveal two different sets of dislocations at the interface: one widely spaced, the other narrowly-spaced. The white dots in the widely-spaced dislocations correspond to the same position in Figs 3a and b. The spacing d_1 is approximately 14 nm. Fig. 4a illustrates a hexagonal network of edge dislocations

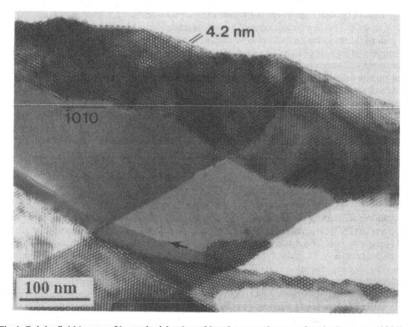

Fig.1. Bright-field image of hematite islands and bands grown heteroepitactically on an (0001) sapphire substrate. The spacing of the $\{11\bar{2}0\}$ parallel moiré fringes is indicated.

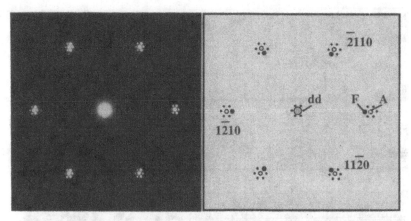

Fig.2. SAD pattern recorded from a hematite band with the electron beam parallel to [0001] and corresponding schematic. (F = α-Fe$_2$O$_3$; A = α-Al$_2$O$_3$; dd = double diffraction reflections)

with Burgers vectors **b** = 1/3<$\bar{1}$2$\bar{1}$0> which fits the experimental image. Dislocations with **b** = 1/3<11$\bar{2}$0> are perfect dislocations in the basal plane of the corundum-type structure [2]. The spacing d$_1$ calculated from the lattice misfits is 15.8 nm which corresponds well to the measured d$_1$ spacing. In Fig.3a, node (1) is in strong contrast while node (2) is in weak contrast. The situation is reversed in Fig.3b. The AC type dislocations are invisible in both images since **g.b** = 0.

The contrast form the narrowly-spaced dislocations can be ascribed to a misfit dislocation network of the type illustrated in Fig.4b. The hexagonal network consists of dislocations with Burgers vectors of the type **b** = 1/3<10$\bar{1}$0>. Since this Burgers vector is smaller than **b** = 1/3<11$\bar{2}$0>, the dislocations must be more closely spaced in order to accommodate the misfit at the interface. The calculated dislocation spacing d$_3$ = 5.1 nm corresponds well to the measured spacing of 4.5 nm (Fig.3a). Although, 1/3<10$\bar{1}$0> are not lattice translation vectors, they are perfect DSC vectors for the Σ = 3 boundary in the corundum-type structure [3]. The Σ = 3 boundary is the basal twin which can be described as a 180° rotation about [0001] which involves a stacking change only on the aluminum sublattice [4]. Dislocations with **b** = 1/3<10$\bar{1}$0> have been observed to accommodate small deviations from exact Σ = 3 coincidence in basal twin boundaries in α-alumina [3]. It is, therefore, proposed that in areas with the narrowly-spaced dislocations, the hematite is in a "basal twin"-type orientation with respect to the alumina substrate.

Further evidence for this proposal is given in Fig.5. Fig.5a shows a weak-beam image of a hematite band displaying both types of dislocation networks. Fig.5b shows a dark-field image of the same area recorded with only the **g** = (4$\bar{4}$0$\bar{4}$) hematite reflection. Bright and dark bands corresponding to the wide and narrow dislocation networks appear in the image. This contrast is expected since **g** = (4$\bar{4}$0$\bar{4}$) is an allowed reflection in the normally-oriented hematite, but the corresponding reflection in the twin-oriented material, **g** = ($\bar{4}$40$\bar{4}$), is forbidden.

CONCLUSION

Hematite thin films were grown directly on electron-transparent, (0001)-oriented sapphire substrates by LPCVD and the interface structure characterized by TEM. Selected-area diffraction and moiré fringe patterns showed that the hematite grew epitactically with little or no residual elastic strain at the interface. Weak-beam imaging revealed two types of misfit dislocation networks. The first type was a hexagonal network of **b** =1/3<11$\bar{2}$0> -type edge dislocations which were associated with a normal orientation of the hematite. The second type was a hexagonal

212

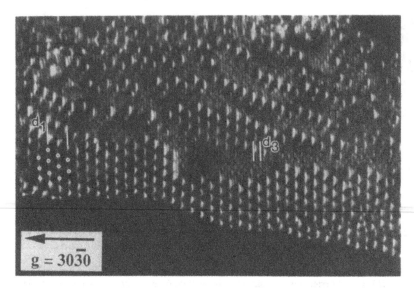

Fig.3a) Weak-beam image of a (0001) hematite / alumina interface recorded under the g-~4g condition where $g = (30\bar{3}0)$. Two types of interface misfit dislocation networks are visible. d_1 and d_3 correspond to the dimensions indicated in Figs 4a and b, respectively.

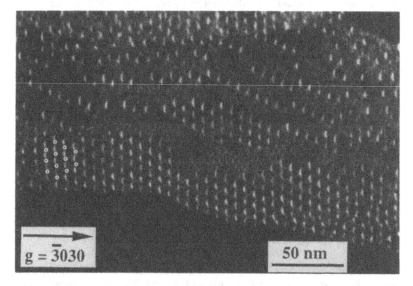

Fig.3b) Weak-beam image of the same area shown in Fig.3a also recorded with the g-~4g diffraction condition but with $g = (\bar{3}030)$. White dots indicate the same positions in Figs (a) and (b). Notice the reversal in the orientation of bright nodes in the two images.

network of $\mathbf{b} = 1/3<10\bar{1}0>$-type edge dislocations which were associated with a basal twin-type orientation relationship between the hematite film and the alumina substrate.

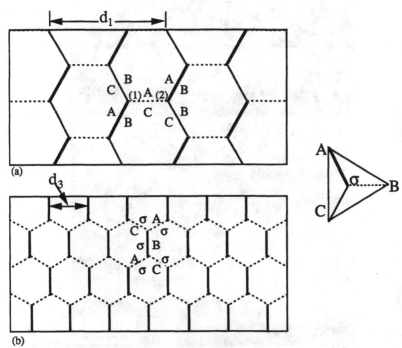

(a)

(b)

Fig.4. Models of misfit dislocation networks shown in Fig.3. (a) $\mathbf{b} = 1/3<\bar{1}2\bar{1}0>$ edge dislocations, d_1 (calculated) = 15.8 nm. (b) $\mathbf{b} = 1/3<10\bar{1}0>$ edge dislocations, d_3 (calculated) = 5.1 nm. AC, AB, BC = $1/3<\bar{1}2\bar{1}0>$; Aσ, Bσ, Cσ = $1/3<10\bar{1}0>$.

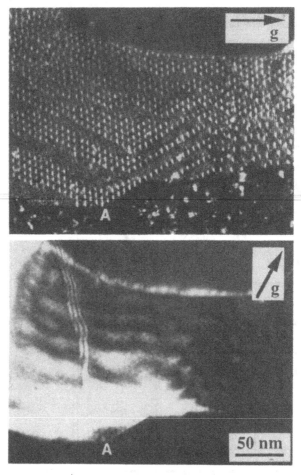

Fig.5. (a) Weak-beam image showing two types of dislocation networks ($g = (03\bar{3}0)$). (b) Strong-beam, dark-field image of same area recorded with the $g = (4\bar{4}0\bar{4})$ hematite reflection showing normal (bright) and twin (dark) -oriented regions of the hematite.

ACKNOWLEDGEMENTS

The authors thank Ms Margaret Fabrizio for photographic work and Mr Ray Coles for maintaining the electron microscopes. The electron microscopes are part of a Materials Science Center Facility supported in part by NSF. This work is supported by NSF through grant No. DMR-8521834.

REFERENCES

1. M.J. Stowell in Epitaxial Growth, Part B, edited by J.W. Matthews (Academic Press, New York, 1975) pp. 437-489.
2. A.H. Heuer and J. Castaing in "Structure and Properties of MgO and Al_2O_3 Ceramics," Advances in Ceramics, Vol. 10, edited by W.D. Kingery (The American Ceramic Society, Columbus, OH, 1984) pp. 238-257.
3. K.M. Morrissey and C.B Carter in "Character of Grain Boundaries," Advances in Ceramics, Vol. 6, edited by M.F. Yan and A.H. Heuer (The American Ceramic Society, Columbus, OH, 1983) pp. 85-95.
4. M.L. Kronberg, Acta Met. 5, 507 (1957).
5. L.A. Bursill and R.L. Withers, Phil. Mag. 40, 213 (1979).
6. L.A. Tietz, S.R. Summerfelt, G.R. English, and C.B. Carter, Appl. Phys. Lett. 55, 1202 (1989).
7. L.A. Tietz, S.R. Summerfelt, G.R. English, and C.B. Carter, these proceedings.
8. D.W. Susnitzky and C.B. Carter, J. Am. Ceram. Soc. 69, C-217 (1986).
9. The CRC Handbook of Physical Properties of Rocks, Vol. 3, edited by R.S. Carmichael (CRC Press, Boca Raton, FL, 1984) pp. 88-90.

TEMPERATURE DEPENDENT CURRENT–VOLTAGE CHARACTERISTICS IN THIN SiO$_2$ FILMS

JIN ZHAO AND N. M. RAVINDRA
Microelectronics Research Center, New Jersey Institute of Technology, Newark, NJ 07102

ABSTRACT

An analysis of the Fowler-Nordheim tunneling (FNT) theory and its application to temperature dependent current-voltage characteristics, of very thin films of SiO$_2$ on silicon, is presented. The final results are believed to provide the most complete examination of FN emission theory and predict the breakdown electric field in thin SiO$_2$ films. The role of the roughness, at the Si-SiO$_2$ interface, in determining the FNT current in these structures is also discussed.

Introduction

Oxidized silicon surfaces are used on virtually all of today's integrated circuits and silicon devices. It is very important to understand the electrical properties of SiO$_2$ for making high quality devices. From past studies, it is well known that the FNT contributes to current conduction in SiO$_2$ films. In the present work, an investigation of the current-voltage characteristics, in the temperature range of $100 - 350°$K, of thermally grown SiO$_2$ films on silicon, is presented.

At room temperature, our experimental data is in accordance with the FNT equation [1]:

$$J = CF^2 e^{-\beta/F} \qquad (1)$$

$$C = \frac{q^3 m_o}{16\pi^2 \hbar m_{ox} \phi_b}; \qquad \beta = \frac{4(2m_{ox})^{1/2}}{3q\hbar} \phi_b^{3/2} \qquad (2)$$

Where, F is the uniform electric field, q is the electronic charge, m_o, m_{ox}, are the electron mass in free space and in the oxide respectively, $2\pi\hbar$ is planck's constant, and ϕ_b is the barrier height. The FN plot parameters based on static current-voltage characteristics in this study are comparable to the other independent experimental results available in the literature. This is presented in table I.

TABLE I. Tunneling results from present and past work at $300°$K. Negative electrodes are n-type Si$<100>$.

Parameters	Present Work	Ravindra[2]	Weinberg[3]	Krieger[4]	Osburn[5]
Slope(MV/cm)	270	260	238.5	237	246
ϕ_b' (eV)	3.05	3.05	2.9	2.89	2.96
m_{ox}/m_0	0.46	0.45	0.5	0.36	0.5

However, at temperatures other than $300°$K, our data cannot fit the simplified FNT equation (1). This is because the barrier height and effective mass decrease with increasing temperature [5-7]. Such a variation leads to increase in the value of C given by equation (2).

Using the earlier reported results of high resolution transmission electron microscopy (HRTEM) [8, 9], a discussion of breakdowns caused by interface protrusions is presented.

Mat. Res. Soc. Symp. Proc. Vol. 159. ©1990 Materials Research Society

Theory

The high field imposed across the oxide, which is necessary for tunneling (approximately in the range of 7.5 MV/cm-10 MV/cm), results in a large density of electrons which are confined to a narrow potential well at the interface. This leads to quantization of the energy to the interface [10]. See Fig. 1(a).

E_o remains discrete while the higher energy levels are merged into a continuum beginning with E_1. The continuum is assumed to obey bulk statistics and tunneling from it will be small compared to tunneling from E_o. This is illustrated in Fig. 1(b).

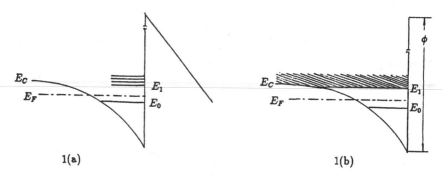

$$1(a) \qquad\qquad\qquad\qquad 1(b)$$

FIG. 1(a). The strong band bending in silicon leads to the formation of a narrow potential well with quantized energy levels or subbands ($E_o, E_1...$); 1(b). The energy levels are separated into two sources of emission. A continuum beginning with E_1 and a single subband beginning at E_o.

All the electrons within the same subband have the same energy associated with motion perpendicular to the interface. They will also have the same tunneling probability. The transmission probability current per electron (Q), for level n, is found by the following equations [3]:

$$Q = \frac{4\pi}{Z_n^{3/2}}(1 + \frac{m_{ox}E_n}{m_{si}\phi_b'})^{-1}(\frac{m_{ox}E_n}{m_{si}\phi_b'})^{1/2}\frac{E_n}{\hbar}e^{-\beta(E_n)/F_{ox}} \tag{3}$$

$$\beta(E_n) = \frac{4\sqrt{2m_{ox}}(\phi_b' - E_n)^{3/2}}{3q\hbar}; \qquad E_n = \frac{Z_n}{2^{1/3}}(\frac{q\hbar}{m_{si}^{1/2}}F_{si})^{2/3} \tag{4}$$

Where, E_n represents the energies of the quantized levels measured from the bottom of the conduction band, and Z_n are zeros of the Airy function, $Z_0 = 2.338$. To complete the solution of the tunneling current, Q has to be multiplied by carriers in the lowest subband. It is given by θN, where, $N = \epsilon_{ox}F_{ox}/q$, (Gauss law). N is the number of electrons per unit area at the interface and θ is its fraction residing in the lowest subband, $\epsilon_{ox}, \epsilon_{si}$ are the dielectric constants of the oxide and the silicon, respectively,

$$J(E_0) = q\theta N Q(E_0) = C F_{ox}^2 e^{-\beta(E_0)/F_{ox}} \tag{5}$$

$$Where \quad C = \theta(1 + \frac{m_{ox}E_0}{m_{si}\phi_b'})^{-1}(\frac{2q^2m_{ox}}{m_{si}^2\phi_b'})^{1/2}\frac{\epsilon_{ox}^2}{\epsilon_{si}} \tag{6}$$

$$\beta(E_0) = \frac{4(2m_{ox})^{1/2}}{3q\hbar}(\phi_b')^{3/2}; \qquad \phi_b' = \phi_b - E_0 \tag{7}$$

From earlier studies [6], it is known that just as the energy gap decreases with increasing temperature, the barrier height also decreases. Approximately, $d\phi_b/dT \cong dE_g/dT$,

$$\phi_{bt} = \phi_b' + \beta \triangle T \qquad (8)$$

$$Where \quad \beta = (\frac{\phi_{b0}}{E_g})(\frac{dE_g}{dT}); \quad dE_g/dT_{si} = -2.3 \times 10^{-4} \, eVK^{-1} \qquad (9)$$

Experimental details

The oxides were grown at 800°C on single crystal silicon wafers (n-type, Czochralski) of < 100 > orientation and 2 ohm.cm, resistivity. The silicon wafers were cleaned using conventional cleaning procedures including the RCA technique followed by a HF dip and thorough rinsing with deionized water. MOS capacitors, intended for breakdown measurements, were fabricated by evaporating 700 nm thick Al electrodes on the oxide (front) through a metal mask to form Al dots of 0.08 cm (32 mil) in diameter. Al was evaporated for a back contact after removing the oxide on the back of the wafers. Post-metallization anneal was done in forming gas (10% H_2 in N_2) at 450°C for 30 minutes.

Oxide thickness measurements were carried out, using a Gaertner automatic ellipsometer, at nine points on the wafer. The mean deviation in the thickness was found to be ±2%. For determination of the oxide thickness, a refractive index of 1.465 (at 6328 Å corresponding to the He-Ne laser) for SiO_2 was used. Breakdown voltage measurements were made using a Rucker and Kolls automatic prober stepper model 682. Steady state I-V measurements were made on these capacitors in order to determine the FNT current. Cross section samples for HRTEM were prepared using combination of mechanical and ion-beam polishing methods. High resolution phase contrast images of the interface were taken at Scherzer defocus value of 65 nm and < 110 > orientation using JEOL 200CX TEM at 200 KV with a 0.27 nm point to point resolution [8].

Results and discussion

Static current-voltage measurements were made in the temperature range of 100 − 350°K. The results of these measurements are fitted to the FNT equation (5). In Fig. 2(a), the FNT plots at temperatures of 100, 200 and 300°K are presented. As can be seen in the figures, within experimental errors, $ln(J/F^2)$ versus $1/F$ is indeed linear. This not only gives out temperature dependent FN slopes, but also implies that FNT current is temperature dependent. Although tunneling, in itself, is temperature independent, the number of electrons incident on the barrier is dependent on temperature.

Considering equations (6, 7), the parameters $\theta, \phi_b', E_0, m_{ox}, m_{si}$, and $\epsilon_{ox}, \epsilon_{si}$ are temperature dependent. However, E_0 is very small compared to ϕ_b, and $\epsilon_{ox}, \epsilon_{si}, m_{si}$ effect only the pre-exponent C. Therefore, we assume that the influence of temperature on $E_0, \epsilon_{ox}, \epsilon_{si}$ and m_{si} is negligible. The barrier height ϕ_b, from silicon conduction band edge to the oxide conduction band, has been determined to be 3.25 eV [11]. Temperature dependent ϕ_{bt} can be derived using equation (8). The slopes of FN plot , ϕ_{bt}, and equation (7) leads to evaluation of the effective mass m_{ox} given in table II.

TABLE II. Temperature dependent Slope, $\phi_{bt}, m_{ox}/m_0$, current ratio, and F_{max}.

Temperature (°K)	100	150	200	250	300	350
Slope (MV/cm)	281	280.5	277	274	270	266
ϕ_{bt} (eV)	3.17	3.141	3.111	3.081	3.050	3.018
m_{ox}/m_0	0.50	0.499	0.486	0.476	0.462	0.448
F_{max} (MV/cm)	13.1	13.09	12.99	12.91	12.97	12.67
$I_{180\lambda}/I_{200\lambda}$	20.51	20.40	19.7	19.12	18.37	17.65

Using the slope obtained from FNT plot, Fig. 2(a), barrier height and effective mass from table II and equation (5), current-voltage characteristics are calculated and presented in Fig. 2(2). It can be seen that the calculations are in excellent accord with experimental data at different temperatures. In Fig. 2(b), results of the comparison of experimental and calculated I-V characteristics are presented for temperatures of 100, 200 and 300°K.

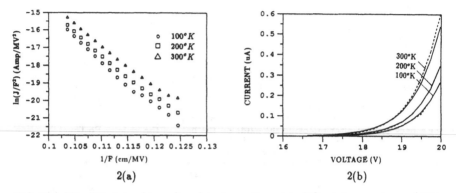

2(a) 2(b)

FIG. 2(a). The FN plots of tunneling data from silicon into SiO_2 at temperatures of 100, 200 and 300°K; 2(b). The comparison of the experimental data (solid line) with the calculated data (dotted line) of I-V characteristics at temperatures of 100, 200 and 300°K.

As an illustration of the influence of non-uniformity in oxide thickness on FNT current, we have considered a ~200 Å thick SiO_2 film. Assuming a 10% non-uniformity (decreasing) in the oxide thickness, and 20 V supply across the oxide, using equation (5), the current ratio $I_{180Å}/I_{200Å}$ is calculated. These results are presented in table II for different temperatures. As can be seen in the table, a decrease in T_{ox} by 10% has led to an increase in the FNT current by ~20%.

Furthermore, taking 1 mA current as the breakdown criteria [12], in equation (5), the temperature dependent breakdown fields are predicted in table II.

So far, we have only considered the ideal case of a flat Si-SiO_2 interface. In fact, no surface is entirely free of height irregularities. As shown in Fig. 3, the protrusions at the Si-SiO_2 interface is ~20 Å in ~200 Å SiO_2 films.

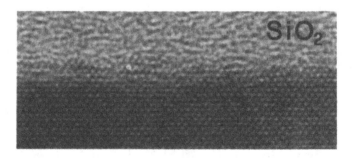

FIG. 3. High-resolution transmission electron micrograph of the Si-SiO_2 interface for an oxide thickness of 220 Å. The point to point distance is 3.14 Å[8].

Such protrusions enhance the field and produce locally intense tunneling currents. To simplify the problem, we assume a uniform distribution of electric field in SiO_2 films. Such a distribution is shown in Fig. 4.

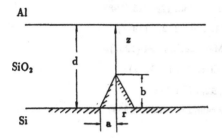

FIG. 4. A Silicon tip at the interface of Si-SiO_2.

$$I = 2\pi \int_0^a r J \, dr = 2\pi C \int_0^a r F_{ox}^2(r) e^{-\beta(E_0)/F_{ox}(r)} dr \tag{10}$$

$$where \quad F_{ox}(r) = \frac{V}{d-z} = \frac{V}{d-b+rb/a}; \quad z = -r\frac{b}{a} + b \tag{11}$$

Comparing I with I_0, a current gain caused by the protrusion can be derived as follows:

$$G = \frac{I}{I_0} = \frac{2 \int_0^a r F_{ox}^2(r) e^{-\beta(E_0)/F_{ox}(r)} dr}{a^2 F_{ox}^2 e^{-\beta(E_0)/F_{ox}}} \tag{12}$$

Where, I and I_o are currents with and without the protrusions respectively. From TEM micrographs in Fig. 3, for $d = 200$ Å, $b = 20$ Å, and $a = 10$ Å. Using equation (12), we obtain $G = 3.45$. Thus, a 3.45 times enhancement in current gain is expected to result in a local breakdown.

Conclusions

A detailed investigation of the application of FNT, to interpret the temperature dependent static current-voltage characteristics of thin SiO_2 films on silicon, has been presented in the above study. Modification to the FNT equation has been sought in order to correct for the temperature dependent effective mass and barrier height. Calculations of the current-voltage characteristics have been shown to be in accord with experimental data. The influence of the presence of protrusions at the Si-SiO_2 interface on the current and breakdown fields has been discussed.

Acknowledgements

The authors are thankful to the New Jersey Commission of Science and Technology for partial financial support. We thank H. J. Wang for providing SIMION simulations of E-field distributions.

References

[1]. M. Lenzinger & E. H. Snow, J. Appl. Phys., 40, 278 (1969)

[2]. N. M. Ravindra, et al. Mal. Res. Soc. Symp. Proc., 105, 169 (1988)

[3]. Z. A. Weinberg, J. Appl. Phys., 52, 5052 (1982).

[4]. G. Krieger & R. M. Swanson, J. Appl. Phys., 52, 5710 (1981)

[5]. C. M. Obsburn & E. J. Weitzman, J. Electrochemical Soc., 119, 603 (1972)

[6]. N. M. Ravindra, et al, Phys. Stat. Sol., 70, 623 (1982)

[7]. A. C. Sharma, et al. Phys. Stat. Sol., 120, 715 (1983)

[8]. N. M. Ravindra, et al, J. Mat. Res., 2, 216 (1987)

[9]. A. H. Carim & R. Sinclair, Mat. Lett., 5, 94 (1987)

[10]. Z. A. Weinberg, Solid-State Elec., 20, 11 (1977)

[11]. R. Williams, J. Appl. Phys., 40, 278 (1969)

[12]. C. M. Osburn, MCNC Memorandum, March (1986)

PART V

Clean Surfaces and
Chemisorbed Layers

LOW ENERGY ELECTRON MICROSCOPY OF SURFACE PROCESSES ON CLEAN SI(111) and SI (100)

E. BAUER, M. MUNDSCHAU, W. SWIECH AND W. TELIEPS
Physikalisches Institut, Technische Universität Clausthal,
D-3392 Clausthal-Zellerfeld and SFB 126

ABSTRACT

Low energy electron microscopy (LEEM) is briefly introduced and its application to the study of surface defects, surface phase transitions on Si(111), crystal growth and sublimation on Si(100) is illustrated.

FUNDAMENTALS OF LEEM

LEEM is a non-scanning surface imaging technique which uses elastically backscattered slow electrons ($E_0 \approx 1.5 - 150$ eV corresponding to $\lambda \approx 10 - 1$ A) in (nearly) normal incidence conditions. The light-optical analogon of a LEEM instrument is a metallurgical microscope with an immersion objective: fast electrons ($E \approx 15$ KeV corresponding to $\lambda \approx 0.1$ A) enter the objective lens on one side and are decelerated in it to the desired low energy E_0. This is achieved by making the specimen the cathode in a so-called cathode lens which is a combination of a more or less homogeneous field along the optical axis, which decelerates the incident and accelerates the reflected electrons, and an electrostatic or magnetic lens. This lens produces a real image of the virtual image which is created behind the specimen by the homogeneous field. In order to obtain (nearly) parallel illumination the incident beam is focussed by the condensor into the back focal plane of the objective in which also the diffraction pattern formed by the reflected electrons is located. The large aberrations of electron lenses make it necessary to use only a small angular cone for imaging. This cone is selected by an aperture in the back focal plane.

Except for the necessity to separate incident and reflected beams, which is done by a magnetic prism, the rest of the optics is like in a conventional transmission electron microscope: two additional lenses or lens combinations ("intermediate" and projective lens) enlarge the intermediate image produced by the objective to the desired final magnification; alternately they can image the diffraction pattern in the back focal plane of the objective onto the detector. The detector is a micro channel plate image intensifier followed by a fluorescent screen.

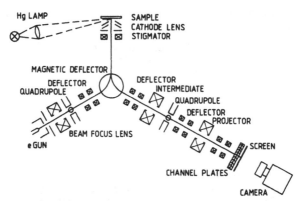

Fig. 1. Schematic of low energy electron microscope

Fig. 1 shows a schematic of the instrument. In addition to
the components already mentioned it is equipped with stigmators
to correct the astigmatism of the objective lens and of the
magnetic prism, with deflection systems for beam alignment and
with several accessories mounted at side entry ports which point
at the specimen: a UV lamp which is used to generate
photoelectrons for photoemission electron microscopy (PEEM), an
auxilary electron gun for secondary electron emission microscopy
(SEEM), an ion gun which can also be used for SEEM but serves
mainly for specimen cleaning, evaporators for in situ film
growth studies and directed gas sources for adsorptean and
surface reactions. The complete microscope is bakeable and can
reach a base pressure in the high 10^{-11} to low 10^{-10} torr range.

The resolution of the system is limited by the aberrations
of the homogeneous acceleration field. If the optimum aperture
is used which is determined by diffraction at the aperture,
chromatic and spherical aberration, then the resolution is given
approximately by

$$d \approx 4 \times 10^{-4} \ (\Delta V_0)^{\frac{1}{4}} V_0^{-\frac{1}{4}} F^{-\frac{1}{4}} \qquad (1)$$

were d is in m, $e\Delta V_0$ ist the energy spread in eV, $eV_0 = E_0$ is
the start energy of the electrons at the surface in eV and F is
the field strength in Vm^{-1}. For F = 10 $kVmm^{-1}$ and an energy
spread of 0.6 eV one obtains a resolution range from 8.5 mm to
2.7 mm for E_0 varying from 1.5 eV to 150 eV. The resolution of
the electrostatic triode lens which is used in the present
system is by a factor of two worse but its replacement by a
magnetic lens should allow to approach the resolution values
just mentioned.

Although the resolution increases with increasing energy,
intensity considerations make it preferable to work at the lower
end of the energy range, nearly down to the mirror microscopy
mode ($E_0 < 0$) which is useful too, for example if field
distributions in front of the surface are of interest. In the
energy range from 0 to about 20 eV most surfaces have high
specular reflectivities in normal incidence, reaching values
above 50 % at certain energies, e.g. in band gaps. Adsorbates

can change the reflectivity of the clean surface dramatically, either by direct backscattering from the adsorbate at energies at which the reflectivity of the substrate is low, by interference between adsorbate and substrate waves, by adsorbate-induced reconstruction or deconstruction or by other surface effects.

As a consequence strong diffraction contrast occurs between differently oriented grains, reconstructed and unreconstructed regions or clean and adsorbate-covered areas. In the absence of differences in crystal structure there is still another contrast mechanism, interference contrast, due to optical path differences, e.g. between waves reflected from the terraces adjoining a surface step ("geometric phase contrast") or between the waves reflected from the front and back surface of a thin parallel slab ("quantum size contrast"). In addition, topography contrast may occur, e.g. on rough surfaces due to field distortions. In the following applications of LEEM several of these contrast mechanisms will be used. More information on the instrumental aspects, resolution, intensity and contrast in LEEM can be found in several review papers [1-4] and the references given in them.

THE DEFECT STRUCTURE OF CLEAN SILICON SURFACES

While geometric phase contrast produces excellent contrast of monoatomic steps on metal surfaces, it is too weak on Si surfaces to be useful, presumably because of a compensating effect of the strain fields around the steps. Other contrast mechanisms have to be found, therefore. On Si(111) one can use step decorateon by (7x7) structure nuclei close to the (7x7) ↔ (1x1) transition [5].

As already known from earlier reflection electron microscopy studies [6], these nuclei form only on the upper side of the step so that up and down steps can be clearly distinguished (Fig. 2a). Fig. 2b shows a typical decoration pattern.

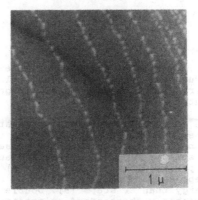

Fig. 2. Si(111) surface. a) Schematic of (7x7)nuclei. b) Normal bright field image of decorated steps (E_0 = 10.5 eV) [5].

On Si(100) monoatomic steps can be imaged indirectly by making use of the fact that the azimuthal orientation of the (2x1) superstructure differs on both sides of the step (Fig. 3a). When the incident beam is tilted somewhat into one of the superperiodicity directions then the diffraction conditions for the two orientations differ and one obtains the socalled tilted bright field contrast [3] between (2x1) and (1x2) regions. The boundaries between these regions are the steps. Fig. 3b shows LEEM images obtained by tilting the beam towards the [011] and the [01$\bar{1}$] directions. The terrace distribution in Fig. 3b with

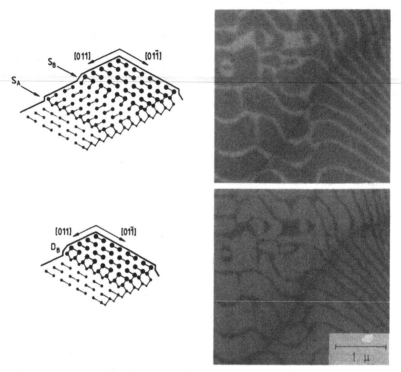

Fig. 3. Si(100) surface. a) Schematic of domain and step structure. b) Tilted bright field LEEM images ($E_0 \approx 6$ eV) [7]. The sharp dark line seen on all images is a crack in the channel plate.

one terrace type dominating is atypical. In general both terrace types have comparable width and the steps differ in roughness, in particular when oriented close to one of the <011> directions as seen in Fig. 4a. A detailed analysis shows that the smooth steps are S_A steps (see Fig. 2a) while the rough steps are S_B steps which, if straight, would have a 15 times larger formation energy than S_A steps [9]. Double steps are never seen on surfaces with standard orientation ([100]±0.5°) when cooled from high temperatures. According to numerous studies with other techniques misorientations >2° are necessary for their stability, or a terrace width between double steps of 8nm which is below the present resolution (15 nm) of LEEM. Even for a misorientation of 0.5° the terrace width between monoatomic steps is only 15.5 nm.

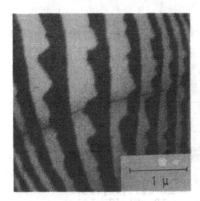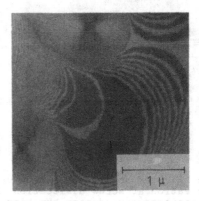

Fig. 4. Steps and terrace structures on Si(100). a) Smooth and rough steps (E_0 = 6 eV) [8]. b) Hillock and large terrace formation (E_0 = 7 eV) [11].

Larger terraces can be obtained not only by orienting the surface better but also by appropriate crystal treatment. Straining the crystal leads to the preferred sublimation of one terrace type causing the formation of a quasi-single domain surface (Fig. 3b) [10]. Step pile-up at sublimation-depressing localized impurities is the most effective way of forming large terraces. Terraces exceeding 6 μm in diameter can be obtained in this maner, of course at the cost of increased surface roughness ("sublimation hillocks") in other areas. Fig. 4b shows this phenomenon [11]. Details can be found in refs. [10, 11], in particular about the interactions between sublimation and slip steps.

THE (7x7) ⟷ (1x1) PHASE TRANSITION ON THE SILICON (111) SURFACE

The first microscopic studies of this phase transition were made by reflection electron microscopy [6, 12]. They gave ambiguous results concerning the order of the transition which was caused by the difficulty of image interpretation due to the strong image foreshortening. The early LEEM studies [5] settled this question in favor of a first order transition but also showed clearly a strong dependence of the microstructure of the surface and of the details of the phase transition upon the surface pretreatment. This has already been reviewed briefly [7, 13]. Here only two remarks should be added to the video movie illustrating the kinetics of the phase transition:
i) Screw dislocations are preferred nucleation sites of the (7x7) structure as shown in Fig. 5a by the larger size of the (7x7) island at the end of the step, the emergence point of the screw dislocation [10]. ii) Even strongly contaminated surfaces can be treated in such as manner that they show an excellent

(7x7) reconstruction. In Fig. 5b the impurities have segregated
into precipitates (black particles) which have acted during
subsequent sublimation as pinning centers for the advancing

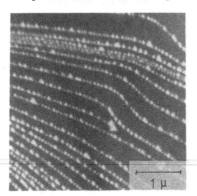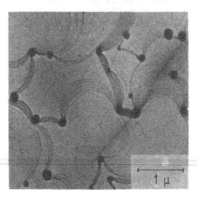

Fig. 5. Step and terrace structures on Si (111). a) (7x7)
nucleation at screw dislocation (E_0 = 10 eV) [10]. b)
Precipitate-caused step bunching (E_0 = 10 eV) [7, 13].

steps. This caused the formation of large clean, atomically flat
terraces on which the (7x7) ↔ (1x1) transition proceeds in the
same manner as on the cleanest surface studied. In the image
shown, the surface is completely covered by (7x7) structure.

KINETICS OF THE SUBLIMATION OF THE SILICON (100) SURFACE

In the previous section the structure and properties of
surfaces after sublimation have been described. Now the kinetics
of the sublimation process will be discussed briefly. This
process can be observed in real time up to about 1350K. Above
this temperature thermionic emission, which is independent of
superstructure orientation, becomes so strong that it overwhelms
the diffraction contrast. The rapid increase of thermionic
emission above 1350K is probably caused by a strong increase of
the Si adatom concentration which lowers the work function. The
high adatom concentration follows from annealing experiments of
surfaces quenchend from T $\gtrsim$ 1350K: many small islands form on
large terraces which is possible if there were many adatoms or
small clusters below the resolution limit. At 1350K sublimation
is so rapid that the surface makes the impression as if it were
boiling. At lower temperatures sublimation proceeds slowly
enough so that several processes can be identified: i)
"Lochkeim" formation in large terraces with subsequent lateral
growth of the monolayer thick circular hole to μm dimensions –
on sufficiently large terraces – before the next "Lochkeim"
forms in the monolayer below. ii) Width oscillations of narrow
terraces caused by the high adlayer concentration. iii)
Simultaneous sublimation from hills and valleys. iv) Growth of
large terraces and hillocks at sublimation-suppressing
impurities. These phenomena can be best seen in real time video
recordings. Details will be reported elsewhere [14].

HOMOEPITAXY OF SILICON ON THE SILICON (100) SURFACE

This much-studied process has been investigated with LEEM at very low supersaturations mainly with the goal to determine the equilibrium shape of two-dimensional Si(100) crystals [15]. For this purpose the substrate temperature was decreased and the deposition rate increased until slow step flow growth could be observed, with occasional nucleation only at surface imperfections. Fig. 6a shows a snapshot of the surface under these growth conditions. The elongated cusped shape of the topmost dark terrace on the large bright terrace which is still little influenced by edge effects indicates already a strong anisotropy of the step energy γ, the long side of the steps being bounded by

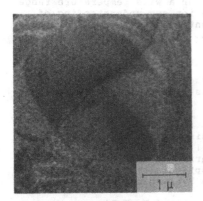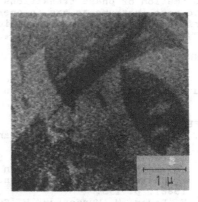

Fig. 6. Homoepitaxy of Si on Si(100). a) Terrace shape during step flow growth. b) Two-dimensional islands nucleated at somewhat higher supersaturation ($E_0 = 5$ eV). Videoreproductions [15]

S_A steps while the S_B steps are replaced by the cusps as one would expect on the basis of Wulffs law [16] and the anisotropy of γ [9]. The growth rate in the directions of the long sides of a terraces is much slower than in the cusp direction, so that the tip of the upper dark terrace catches up with the long side of the lower terrace and forms a double step. Although LEEM cannot decide whether or not there is a narrow band of the lower terrace left at the apparent double step, step energy considerations leave little double that a D_B double step (Fig. 3a) formed because $\gamma D_B << \gamma S_A + \gamma S_B$ [9].

If the supersaturation is carefully increased then nuclei form occasionally on the larger terraces and grow slowly so that their growth can be easily monitored (Fig. 6b). The islands grow in the same cusped elongated shape seen before (Fig. 6a) in the step flow growth and are rotated 90° to each other on adjoining terraces as required by the anisotropy of the step energy. With further increase of the supersaturation the nucleatein rate increases further until the islands become so small that LEEM no longer can resolve them. The absence of any step anisotropy at high temperatures mentioned in the previous section is not in contradiction to the growth observations because at high temperatures the entropy contribution to the step free energy destroys the anisotropy. Details of the growth studies will be reported elsewhere [15].

CONCLUSIONS

Low energy electron microscopy is still far from being developed to its full potential - at the time of writing of this article only one instrument is fully operational - but has already demonstrated its power in the study of surfaces. Its main virtue is the high intensity available which allows real time studies of surface processes such as crystal growth and sublimation or phase transitions over a wide temperature range. Once more instruments are in operation our understanding of surface processes should significantly accelerate and deepen.

ACKNOWLEDGEMENTS

Financial support by the Deutsche Forschungsgemeinschaft and by the Volkswagen-Foundation is gratefully acknowledged.

REFERENCES

1. E. Bauer and W. Telieps, Scanning Microscopy Suppl. 1, 99 (1987).
2. E. Bauer and W. Telieps, in Surface and Interface Characterization by Electron Optical Methods, ed.by A. Howie and V. Valdre (Plenum Publishing Corp., New York, 1988), p. 195.
3. E. Bauer, M. Mundschau, W. Swiech and W. Telieps, Ultramicroscopy 31, 49 (1989).
4. E. Bauer, in The Physics and Chemistry of Solid Surfaces, edited by R. Vanselow an R. Howe (Springer, Berlin, 1990), to be published.
5. W. Telieps and E. Bauer, Surface Sci. 162, 163 (1985); Ber. Bunsenges. Phys. Chem. 90, 197 (1986).
6. N. Osakabe, Y. Tanishiro, K. Yagi and G. Honjo, Surface Sci. 109, 353 (1981).
7. E. Bauer, M. Mundschau, W. Swiech and W. Telieps, in Evaluation of Advanced Semiconductor Materials by Electron Microscopy, ed. by D. Cherns (Plenum Publishing Corp., New York, 1989), p. 283.
8. W. Telieps, Appl. Phys. A 44, 55 (1987).
9. D.J. Chadi, Phys. Rev. Lett. 59, 1691 (1987).
10. M. Mundschau, E. Bauer, W. Telieps and W. Swiech, Phil. Mag., in print.
11. M. Mundschau, E. Bauer, W. Telieps and W. Swiech, Surface Sci., in print.
12. Y. Tanishiro, K. Takayanagi and K. Yagi, Ultramicroscopy 11, 95 (1983).
13. E. Bauer, M. Mundschau, W. Swiech and W. Telieps, Inst. Phys. Conf. Ser. No 93ʲ (1988), Vol. 1, p. 213.
14. M. Mundschau, W. Swiech and E. Bauer, to be published.

15. W. Swiech, M. Mundschau and E. Bauer, to be published.
16. C. Herring, in Structure and Properties of Solid Surfaces, ed. by R. Gomer and C.S. Smith (Univ. Chicago Press, Chicago, 1953), p. 5.

The Atomic Geometry of the Reconstructed (001)
Surface of the Rutile Phase of SnO_2

Charles B. Duke and Michael R. Thompson
Molecular Science Research Center
Battelle Pacific Northwest Laboratory
Box 999
Richland, Washington

Abstract.

The tight-binding total energy formalism developed for tetrahedrally coordinated compound semiconductors has been extended to
rutile-structure oxides and applied to calculate the surface atomic
geometry and electronic structure of SnO_2 (001). Two stable
structures, separated by an energy barrier, are found. The first
consists of slightly relaxed surface geometry with the top layer
oxygen atoms relaxed outward by approximately 0.12A, and cations
inward by 0.25A. The second geometry is a more massively reconstructed surface in which the four-coordinate surface Sn atoms
attain highly distorted tetrahedral coordination.

Introduction.

One of the most significant observations in semiconductor surface
science is that the cleavage surfaces of tetrahedrally-coordinated
compound semiconductors exhibit "universal" mechanisms for surface
reconstruction [1]. For a specific crystallographic surface, the
various materials exhibit reconstructions which are independent of
the identity of the material, are characterized by bond-length
conserving processes, and result in coordinate shifts which are
identical in direction for the various materials with magnitudes
scaling with the bulk lattice constants. This raises the intriguing possibility that broad classes of materials which exhibit
common crystallographic habit, will also possess common surface
structures of their cleavage faces, depending on the character of
the chemical bonding as reflected by the bulk material. Motivated
by this concept, we have extended our studies to the common rutile-
phase oxides, in particular, SnO_2 (001), to explore surface
reconstruction in octahedrally coordinated systems.

The Model.

Metal oxide structures are commonly described in the context of the
eutactic or close-packed arrangements of the oxide-ions into
layers, with some subset of interstices occupied by the metal atoms
[2]. These "hard-sphere" models allow the rationalization of bulk
structural behavior from simple intuitive principles. However,
they do not allow us to probe more complex problems such as the
interplay of electronic and ionic factors in the control of
structure, or surface rehybridization and reconstruction. The work
outlined here utilizes the tight-binding ansatz which treats the
description of the structure electronically, without explicitly
treating Coulombic interactions. We expect that the consequences of
surface-bond rehybridization and subsequent surface reconstruction
would give results consistent with the simpler intuitive concepts,
and will also allow us to quantify the driving force for surface
relaxation.

Total-Energy Tight-Binding Formalism. The sp^3s^* empirical tight-binding model used here is an extension of that of Harrison, Vogl, Chadi and others [3], and involves only the first-near neighbor interactions about each of the Sn and O atoms using an sp^3 basis. The inclusion of the excited s-like state s^* to the usual sp^3 basis set has the effect of reducing the energy of the indirect conduction band minimum by coupling to antibonding p-like conduction states, but has no effect in the present case of the insulator SnO_2. Since this model formally treats only bonding interactions, each of the oxygen atoms possess a non-interacting (non-bonding) p-state orthogonal to the plane of its three Sn-O bonds. While the general features of the band structure are in agreement with others, some models include higher order coordination shells which results in dispersion of the non-bonding oxygen bands by approximately +/-0.5 eV as a function of k. However, this treatment has had little effect on the outcome of surface reconstruction in this system. The total-energy, E_{tot} of an electron-ion system can be expressed as

$$E_{tot} = E_{ee} + E_{ei} + E_{ii} \qquad (1)$$

where Eee, Eei, and Eii are the electron-electron, electron-ion, and ion-ion interaction energies, respectively. In a practical sense, the variation of the total energy associated with atomic displacements is given by

$$E_{tot} = E_{bs} + \sum_{i,j} (U_1 \, \varepsilon_{ij} + U_2 \, \varepsilon^2_{ij}) \qquad (2)$$

in which the first term is the contribution of the band structure energy and the second term represents an empirical correction for the double counting of the electron-electron interactions in the band structure calculation, and includes the ion-ion interaction energy. The quantity ε_{ij} is the fractional change in the bond length between atoms labeled i and j with respect to the equilibrium value. The empirical elastic constants ε_{ij} are determined by fitting to bulk elastic moduli. We retain only quadratic terms in the fractional bond length changes.

Given U_1 and U_2, we can calculate the surface atomic geometries by minimizing the total energy given by Eq.(2) or, equivalently, by iteratively reducing for each atom the Hellmann-Feynman forces calculated with the tight-binding model. These total energy calculations are performed using a finite slab, periodic in two-dimensions and discontinuous in the third, composed of layers parallel to the surface under study. Diagonalization of the slab Hamiltonian yields the one-electron energy eigenvalue spectrum $E_n(k)$, which appears in Eq(2). Surface states are identified by studying the localization of the resulting one-electron states, and their changes during the minimization process. In this manner, it is possible to ascertain the driving force for surface relaxation and an understanding for the subsequent stability which results.

Results.

The energy minimization of seven and fourteen atomic-layer models for the (001) surface of rutile-SnO_2 have been completed. For a seven layer model, we have allowed the first six atomic-layers to undergo shifts in their atomic positions proportional to the Hellmann-Feynman force computed for each of the atoms using the tight-binding formalism. The seventh layer is held fixed as a condition of convergence to the bulk structure. Convergence to a minimum energy structure is judged by an incrementally small change in the total-energy (less the 10^{-5} eV), and vanishing forces. We have observed two energy minima resulting from these calculations. The first corresponds to a mild, somewhat predictable reconstruction which involves the relaxation of the first oxygen layer outward into the vacuum, and movement of the four-coordinate surface Sn atom inward. The second minima represents a more massive reconstruction which drives the four coordinate surface Sn atom toward highly distorted tetrahedral coordination.

As illustrated in Fig. 1 for the first reconstruction, the shifts in the atomic geometry are all perpendicular to the cleavage surface, along the slab c-axis. The two independent surface oxygens

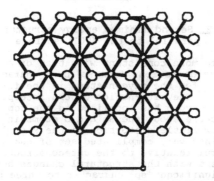

Figure 1. Reconstructed surface looking parallel to the (001) surface representing first minima.

move approximately 0.13A above the former (001) plane, the four-coordinate surface Sn draws inward by 0.25A, and the surface bond angle around Sn changes from 180° to 158°. Alternating sub-surface layers of Sn and O atoms move inward or outward in varying degrees. Changes in Sn-O bond lengths are moderate, ranging from 2.00A to 2.09A, and bond angle changes are apparent deep into the slab. The driving force for the reconstruction can be inferred by inspection of the eigenvalue spectrum before and after reconstruction. Two bands which lie near -1.5eV in the band structure of the unreconstructed model stabilize by lowering approximately 0.5eV during reconstruction, have full pi-symmetry in the bulk, and represent several layer-wise bonding/antibonding combinations of Sn and O p_z orbitals. The lowering of the energy is associated with an increase in the pi-bonding character between the Sn/O layers 2-3, 4-5 and, 6-7, while increasing the antibonding character between interleaving layers. This effect diminishes as a function of depth in the slab. Several bands lying in the range of -5.0eV to -7.5eV,

typical of Sn-s and O-p orbital interactions, indicate marked
increases in sigma bonding between Sn and O in the first two
layers, indicative of the shortening of the Sn-O bonds relative to
the bulk.

The second minimum energy structure, as illustrated in Fig.2,
results in a more massive reconstruction of both surface and sub-
surface structure. The geometry about the four-coordinate surface-

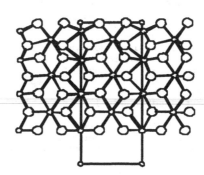

Figure 2. Second minimum energy structure viewed
parallel to the (001) surface.

Sn atom has become more consistent with that of a highly distorted
tetrahedron. Large deviations from sp^2 bond character is apparent
in the bond angles for the oxygen atoms in the first through third
layers of the slab. The alternation of the interplanar spacing
between planes of oxygen has localized in the top four layers and
the overall structure about the Sn and O atoms in layers four
through seven show little distortion in the x-y plane. These
features result in a very complicated set of changes in the
eigenvalue spectrum relative to the unreconstructed surface.
Features consistent with the structural changes here include; the
development of significant sp^3 character to three bands at approximately -4.2eV, -4.5eV and -5.2eV, which involve the Sn and O atoms
in the top three layers of the slab, and the continued development
of the pi-like bands described in the former reconstruction into a
set of bands with greater pi-character for Sn and O atoms in the
lower layers. However, since this reconstruction results in several
critical contacts between oxygen atoms in the second and third
layers of the slab, the validity of this model as an energy minima
is somewhat uncertain.

Synopsis.

We have generated two models for surface reconstruction of the
(001) surface of the rutile-phase of SnO_2. These minima are
separated by a relatively low energy barrier. The major features
of the structure representative of the first minima are consistent
with those which might be predicted via simple hard-sphere models,
ie. the mild relaxation of the outer oxygen surface into the
vacuum and a concomitant inward movement of the four coordinate Sn
atom in response to conserve coordination. Our calculations have
given an rough estimation of the magnitude of the atomic position

shifts, and have indicated the mechanism under which surface bond rehybridization takes place, namely, the preservation of approximate pi-bonding in the presence of the discontinuity. However, the second minima would not be easily predicted by simple considerations due to the complexity of the structural transition from a tetrahedral surface structure to the octahedral bulk. While this latter result is intriguing, the stability of this structure is in doubt due to the critical non-bonded oxygen-oxygen contacts which result in sub-surface layers. The calculation of the total energy is done under the tight-binding formalism without second near-neighbor interactions. Calculations which utilize a more complete Hamiltonian are indicated.

References.

1. C.B. Duke, in Surface Properties of Electronic Materials, D.A. King and D.P. Woodruff, Eds. (Elsevier, Amsterdam, 1988), pp. 69-118; C.B. Duke and Y.R. Wang, J. Vac. Sci. C.B. Duke and Y.R. Wang, ibid, 6, 1440 (1988).

2. Jerremy Burdett, Inorg. Chem., 24, p. 2244-2253 (1985); Michael O'Keeffe, Acta Crystallogr., Sect. A., A33, 924 (1977).

3. D.J. Chadi, Phy. Rev B, 9, 2074 (1979); C. Mailhiot, C.B. Duke, and D.J. Chadi, Surf. Sci., 149, 366 (1985); S. Munnix and M. Schmeits, Surf. Sci., 126, 20 (1983); W.A. Harrison, Phys Rev., B8, 4487 (1973); W.A. Harrison in, Electronic structure and Properties of Solids (Freeman, San Francisco), 1980; P. Vogl, H.P. Hjalmmarson, and John D. Dow, J. Phys. Chem. Solids, 44, 365 (1983).

A NEW DEFECT ON THE RECONSTRUCTED Si(100) SURFACE :
AN *AB INITIO* MOLECULAR-DYNAMICS STUDY

SIGEO IHARA*, SHI LUN HO,* TSUYOSHI UDA**, and MASAHIKO HIRAO**
*Central Research Laboratory, Hitachi, Ltd., Kokubunji, Tokyo 185, Japan
**Advanced Research Laboratory, Hitachi, Ltd., Kokubunji, Tokyo 185, Japan

ABSTRACT

An interstitial dimer, which is symmetric and recessed from the surface, is proposed as a new defect structure on reconstructed Si(100) surfaces. The structure of the interstitial dimer is studied using *ab initio* molecular-dynamics simulations, and shown to be stable. Possible arrangements of the new defects are discussed.

1. INTRODUCTION

As a consequence of its importance in device technology, the Si(100) surface has been one of the most intensively studied of all semiconductor surfaces[1]. Nevertheless, the detailed atomic structure of this surface remains controversial. Among the models proposed to explain the experimental data, the dimer model is the most widely accepted The original dimer model, in which the dimer is symmetric, was proposed in order to explain the 2X1 symmetric surface pattern observed by early diffraction studies[2]. In this model the 2X1 symmetric structure is obtained by dimer reconstruction of the surface atoms; pairs of atoms on the ideal bulk sites on the surface, and which have two dangling bonds per atom, move towards each other to form bonds or dimers, thereby reducing the total energy. The higher order c(4X2), p(2X2), c(2X2) patterns, which have been reported in later experimental studies, were classified according to the arrangements of the buckling dimers or asymmetric dimers, in which one atom buckles away from the surface while the other buckles towards the surface[3,4]. The change in the LEED pattern at 200K was attributed to a phase transition in buckling dimer orientations. Although the symmetric dimer model was supported in some reports, the buckling dimer model have generally accepted as being the most successful in consistently explaining the wealth of experimental data (See Ref.1, and references therein).

The situation changed when Pandey[5], recognizing the importance of the surface defects, proposed the missing dimer or dimer vacancy model. Using total energy calculations, he showed that a surface consisting of symmetric dimers with dimer vacancies is energetically more stable than one composed of buckling dimers. Scanning tunneling microscopy (STM) images[6] show that the surface at room temperature has many areas with defects, and that they are randomly distributed over the surface. It has been thought that these defects are vacancies. Most of these defects on the surface form small clusters, and the surface dimers near this region are asymmetric, giving c(4X2) and p(2X2) symmetric areas. Sometimes defects appear in symmetric dimer rows giving 2X1 symmetry, and are considered to be individual dimer vacancies, similar to the dimer vacancy defects proposed by Pandey[5]. Thus isolated dimer vacancy and dimer vacancy clusters are associated with symmetric and asymmetric dimer structures, respectively. Moreover, a number of recent experiments[7] have indicated that the larger periodicities, such as c(4X4), c(8X8), and (2Xn) structures (where n lies in the range 6-11), may also be explained in terms of the ordering of dimer vacancies. However, recent STM experiments[5] indicate that there is yet another type defect which cannot be explained by a dimer vacancy model. It is clear, therefore, that the nature of surface defects remains one of central problems in the study of Si(100) surfaces

2. AB INITIO MOLECULAR DYNAMICS

In applying the molecular-dynamics technique to the study of the Si(100) surface, the calculation of the forces experienced by atoms by quantum mechanical means is essential, since bonding properties strongly depend on quantum effects which are not accurately modeled by empirical potentials. Molecular dynamics using empirical potentials is known to yield unrealistic results for isolated atoms on the surface[8]. Accordingly, density functional theory, incorporating a pseudopotential scheme, was employed. In order to compute the total energy of the system, the local density approximation was used for the exchange correlation energy. Local pseudopotentials[9] rather than non-local ones were employed, since the former retain a satisfactory picture of reconstruction while being computationally less demanding. For numerical reasons, the electronic states were calculated using first order equations of motion in the dynamically simulated annealing, instead of the second order ones used by Car and Parrinello [10,11]:

$$\zeta\, \partial|m,t> /\partial t = - (H - e_m)\, |m,t> , \qquad (1)$$

where H is the total Hamiltonian of the system, ζ is the convergence factor, $|m,t>$ is an eigen vector of state m at time t and $e_m = <m,t|H|m,t>$. Equation (1) was evaluated by numerical integration, except for the diagonal components, which were integrated analytically to promote rapid convergence. Gram-Schmidt orthogonalization was then used to prevent all the wave functions from falling into the ground state. In each time step, self-consistency was checked. The Hellmann-Feynman forces were calculated in momentum-space.

In all the simulations the surface was modeled by 6-layers, with each layer containing 4 atoms. The atoms in the two lowest layers were fixed in their ideal bulk positions, while those in the upper 4 layers were allowed to move freely. Periodic boundary conditions were applied in all three dimensions with a 6.5 Å deep vacuum region. The energy cut-off was 6 Ry (1750 plane waves), and the charge density was calculated by Γ point sampling for dynamical simulations. The integration of the equations of motion for atomic coordinates was performed using the leap-frog algorithm with a time step of 0.5 fs. The initial atomic coordinates and the temperature of the system were varied. In some simulations, the ideal lattice positions were used as the initial configuration in order to observe dimer formation, while in others, symmetric and asymmetric dimer configurations were used to check initial condition dependency. The target temperature in all simulations was 300K. The excess energy which arises from relaxation from the initial states was removed by rescaling the velocities. The simulations comprised of 3000 - 6000 steps. In order to enable the accurate comparison of the energies of various atomic configurations obtained from dynamical simulations, static total energy calculations, with four-k-point sampling, were also performed. All simulations were performed on the supercomputers HITAC S-810 and S-820.

3. RESULTS AND DISCUSSION

Formation of the Asymmetric Dimer and the Symmetric Dimer

The formation of two (011) dimers in the free Si(100) surface was observed in every run started from ideal lattice positions in our simulations. In most simulations (at about 300K), 2X1 structures, which stem from asymmetric dimers, were obtained within 0.5 ps. In order to compare the total energies, we also performed simulations starting from symmetric dimer configurations. The energy difference (in the 2X1 structure), which was obtained from static calculations, between the symmetric and asymmetric dimers, is small (the former are 0.01 eV/dimer more stable than the latter).

Structural Properties of the Interstitial Dimer

In some of the simulations, started from ideal lattice positions, a new defect structure (Fig.1) was formed during the first 0.5 ps: a pair of atoms, which form a dimer, are recessed from the surface and are located at the interstitial position. We refer to this dimer as an interstitial dimer. The interstitial dimer is found adjacent to a surface dimer (Fig.1b). Furthermore, both the interstitial dimer and the surface dimer are symmetric. The average bond lengths of the interstitial and surface dimers are 2.33 Å and 2.24 Å, respectively. These two values are close to the corresponding values of previous dimer models[1-8]. Note that values of the former and the latter are close to the nearest neighbor separation in the crystalline solid and the Si_2 molecular bond length, respectively.

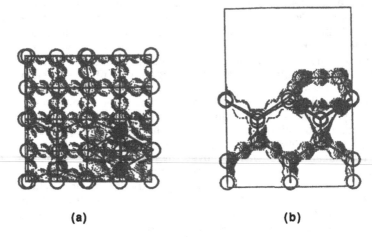

(a)   (b)

Fig.1   Interstitial dimer configuration: To aid interpretation the interstitial dimer atoms are displayed as closed circles. The remaining Si atoms are represented by open circles and electron charge density by dots. (a) View of surface from above. The surface and interstitial dimers (both symmetric) may be seen in the lower left and lower right quarters respectively. (b) Side view of the simulation box. The periodic boundary is indicated by the rectangular box and the deep vacuum region above the surface is readily apparent. Both dimers are on the right hand side of the figure.

Interstitial Dimer versus Dimer Vacancy

It is useful to compare the interstitial dimer to the dimer vacancy proposed by Pandey (see Fig.2). In the Pandey model, the dimer vacancies play a twofold role in lowering the surface energy. Firstly, they allow the formation of a dimer in the second layer, and reduce the net number of dangling bonds by two. Secondly, they make the surface flatter and increase the degree of π bonding. Thus, the distance along the $[0\bar{1}1]$ direction between the two surface dimers on either side of the dimer vacancy decreases (Fig.2a). By contrast, in the interstitial dimer model, the distance between the two surface dimers next to the interstitial dimer will increase (Fig.2b), because the displacements of the second layer atoms bonded to the interstitial dimer are expansive. Recent calculations[12], show that the creation energy of a dimer vacancy is 0.28 eV/defect. However, static total energy calculations by us indicate that the interstitial dimer configuration is at most 0.1 eV/dimer (0.2 eV in total) less stable than two asymmetric dimers also obtained by us. (Note that there are 2 dimers in our system.) The creation energy of an interstitial dimer, at a concentration of 0.5 defects/(lattice unit)^2, is therefore estimated as 0.2 eV. Thus our calculations suggest that the interstitial dimer would be observed by experiment.

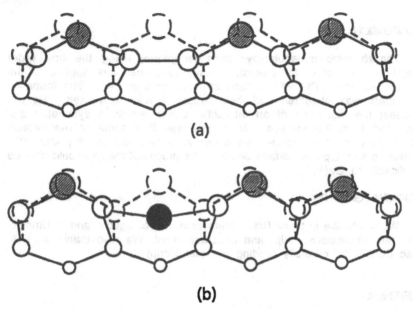

Fig.2 Dimer vacancy model (a) and interstitial dimer model(b). Solid, shaded, open , and broken circles represent Si atoms at the positions in the interstitial dimer, the relaxed surface dimers, the relaxed inner layers, and the ideal surface dimers, respectively.

Relation of Interstitial Dimer to the STM Images

As was stated earlier, one may observe in the STM images obtained by Hamers *et al.* [6], what appear to be individual dimer vacancies in the symmetric dimer rows. However, an interstitial dimer in the dimer rows would give the same image as a dimer vacancy. Consequently, the defects observed by Hamers *et al.* which have been interpreted as individual dimer vacancies, may in fact consist in part of interstitial dimers. High resolution STM, would distinguish between the interstitial dimer proposed here and the established dimer vacancy.

Strain considerations would also favor the existence of the proposed interstitial dimer in the defect complexes. For example, the interstitial dimers have the possibility of creating clusters with themselves or of forming defect complexes with dimer vacancies. The double dimer vacancy, which is apparently observed by the STM study of Hamers and Köhler[6] as a pair of dimer vacancies, may in fact be a pair of interstitial dimers or a vacancy-interstitial dimer complex, since the strain of the latter two is smaller than that of the former. On the basis of our calculations, we propose the interstitial dimer. It is possible that this dimer can be observed by STM.

CONCLUSION

The *ab initio* molecular-dynamics technique, using the first order equations of motion for electronic wave functions, was applied to the Si(100) surface. The 2X1 surface structure was studied. The formation of dimers was observed and their structures were examined. Results suggest the existence of an interstitial dimer, which is symmetric and recessed from the surface. It is proposed that some of the defects which have been interpreted as dimer vacancies are actually interstitial dimers, a new type of surface defect. The proposed model should also be applicable to Ge(100).

ACKNOWLEGEMENTS

We would like to thank Drs. Y.Murayama, K. Yamaguchi, and Y. Umetani for their discussions, help, and encouragement. We also thank Dr. D.A. Mac Donaill for carefully reading our manuscript.

REFERENCES

[1] D.R. Haneman, Adv. Phys. **31**,165 (1982).
[2] R.E. Schlier and H.E. Farnsworth, J. Chem. Phys. **30**, 917 (1959).
[3] D.J. Chadi, J. Vac. Sci. Technol. **16**, 1290 (1979); D.J. Chadi, Phys. Rev. Lett. **43**, 43 (1979); M.T. Yin and M.L. Cohen, Phys. Rev. **B24**, 2303 (1981).
[4] J. Ihm, D.H. Lee, J.D. Joannopoulos, and A.N. Berker, J. Vac. Sci. Technol. **B1**, 705 (1983).
[5] K.C. Pandey, in Proceedings of the Seventeenth International. Conference on the Physics of Semiconductors, edited by D.J. Chadi and W.H. Harrison (Springer-Verlag, New York, 1985), pp.55-58.
[6] R.J. Hamers, R.M. Tromp, and J.E. Demuth, Phys. Rev. **B34**, 5343 (1986) and references cited therein; R.J. Hamers and Köhler, J. Vac. Sci. Technol. **A7**, 2854 (1989).
[7] D.M. Rohlfing, J.Ellis, B.J. Hinch, W. Allison, and R.F. Willis, Surf. Sci. **207**, L955 (1989) and references therein.
[8] F.F. Abraham and I.P. Batra, Surf. Sci. **163**, L752 (1985).
[9] J. Ihm and M. L. Cohen, Phys. Rev. **B21**, 1527 (1980).
[10] R. Car and M. Parrinello, Phys. Rev. Lett. **55**, 2471 (1985).
[11] M.C. Payne, J.D. Joannopoulos, D.C. Allan, M.P. Teter, and D.H. Vanderbilt, Phys. Rev. Lett. **56**, 2656 (1986).
[12] N. Roberts and R.J.Needs, J. Phys. Condens. Matter 1,3139 (1989).

SI(001) MOLECULAR BEAM EPITAXY: ENHANCED DIFFUSION OR BONDING?

S. CLARKE, M.R. WILBY, and D.D. VVEDENSKY
The Blackett Laboratory and Semiconductor Interdisciplinary Research Centre,Imperial College, Prince Consort Road, London, SW7 2BZ, United Kingdom

ABSTRACT

Through application of a lattice model of the Si(001) surface, implemented in a Monte Carlo growth simulation we investigate the structural evolution of the Si(001) surface during molecular beam epitaxy. Particular emphasis is placed upon identifying the role of both enhanced diffusion and directional bonding.

INTRODUCTION

Homoepitaxial growth on the Si(001) surface provides an ideal system for studying fundamental aspects of molecular beam epitaxy (MBE). Absent are complications associated with binary systems, such as GaAs, but present is a growth mode very strongly dependent upon local changes induced by the reconstruction. The Si(001) surface exhibits a 2x1 reconstruction comprising ordered rows of paired atoms (dimers). When a surface is prepared it inevitably includes some misorientation from the desired plane, manifesting in steps on the surface. A monatomic a step forms a domain boundary between 2x1 and 1x2 reconstructions, the direction of dimer formation rotating through 90° at the step. In contrast a surface comprising biatomic steps forms a single domain surface. Both cases lead to two types of steps, those with the upper terrace dimer axis normal to the step (type-A) and those in which the dimer axis is parallel (type-B) (Fig. 1). These steps are labeled S_A and S_B or D_A and D_B for single or double height steps, respectively. Similarly we label domains in which the

Steps Descending in [110] Direction ⟶

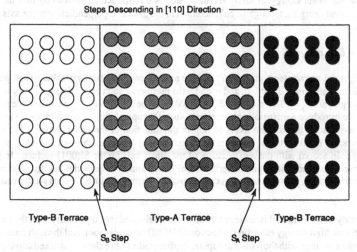

Type-B Terrace   Type-A Terrace   Type-B Terrace

S_B Step S_A Step

Figure 1. A schematic illustration of a monatomically stepped Si(001) surface.

Mat. Res. Soc. Symp. Proc. Vol. 159. ©1990 Materials Research Society

dimer axis is normal to the steps as type-*A* and parallel as type-*B*.

A prominent feature of Si(001) homoepitaxy is the formation of elongated clusters on the surface with the elongation in the direction perpendicular to the dimer axis of the atoms comprising the cluster. Revealed by the scanning tunneling microscope (STM) [1], elongated clusters have been ascribed to either enhanced diffusion parallel to the dimer axis of the under-layer [1] or to a difference in residence times at the S_A and S_B steps forming the edges of these clusters [1,2]. In this paper we investigate the origin of elongated cluster formation, testing each hypothesis in a simulation of Si(001) MBE. We then analyse the implications of our findings for step stability and propagation at high temperatures.

THE SOLID-ON-SOLID MODEL OF Si(001) MBE

We adopt as the basis of our model a simple cubic lattice in which neither overhangs nor vacancies are permitted, the solid-on-solid criterion. In this framework we include two kinetic processes, the deposition and migration of surface adatoms, the long residence time of Si atoms on the surface permitting the neglect of evaporation [3]. Atoms are deposited by randomly generating surface sites with a rate FA, where F is the flux and A the surface area. Atoms are allowed to maximize the number of nearest neighbor-bonds formed within a square of dimension $2S+1$, centred upon the arrival point [4], with $S = 3$. Each and every surface atom is ascribed an Arrhenius inter-site hopping rate:

$$k(E,T) = k_o \exp(-E/k_B T) \qquad (1)$$

where $k_o = 10^{13}\text{s}^{-1}$ is the vibrational frequency of a surface atom, k_B is Boltzmann's constant, T is the substrate temperature, and E is a configurational energy barrier to hopping.

The energy barrier to migration comprises two elements, a site independent surface barrier, plus a term representing the number of nearest-neighbor bonds in the plane of the surface. To account for the lower energy of an S_A versus S_B step, we introduce an anisotropy into the nearest-neighbor bond energies, through an enhancement in the direction perpendicular to the axis of dimer formation:

$$E = E_s + mE_{//} + nE_{\perp}, \qquad (2)$$

where E_s is the surface term, $E_{//}$ and $E_{\perp}$ are the nearest-neighbor bonds parallel to and perpendicular to the enhancement direction, and m and n the number of such bonds formed. Values for these bond energies are $E_s = 1.3\text{eV}$ and $E_{//} + E_{\perp} = 0.5\text{eV}$, with the ratio $B = E_{//}/E_{\perp}$ an adjustable parameter. Finally we introduce a diffusion enhancement, such that upon being selected to hop to an adjacent site, an atom is D times more likely to hop in the direction of the enhancement. Each of these enhancements is taken to be in the same direction and to rotate through 90° each successive monolayer in accord with the reorientation of the dimer axis on the Si(001) surface. In previous studies we have found that a parameterization of $B = 10$ and $D = 1$ provides results most consistent with experiment [5], a result verified by comparison of the model with STM data on stepped surfaces [6].

Analysis of the model is achieved through comparison of the step density with the intensity of the reflection high-energy electron-diffraction (RHEED) specular spot, and through comparison of the domain coverage with the half-order spots. Earlier studies have shown a marked correspondence between the step-density and specular RHEED [2,5], reproducing the azimuthal and domain dependence of the RHEED oscillation period [7] and post growth recovery of the intensity [8].

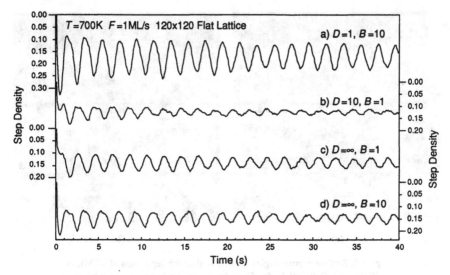

Figure 2. The sensitivity of the step-density oscillations upon the inclusion
of an anisotropy in surface diffusion, with or without an anisotropy in surface
nearest-neighbor bonding.

DIFFUSION VERSUS BOND ENHANCEMENT

Presented in Fig. 2 are step density evolutions for the growth of 40 monolayers on a 120x120
site lattice at $T=700K$ and $F=1ML/s$. The step-density has been resolved along the x-axis of the lattice
maximizing sensitivity to the formation of elongated clusters. As in previous studies we find that
case (a) $B=10 D=1$, is most consistent with experiment exhibiting prolonged biatomic oscillations.
Conversely case (b) $B=1 D=10$, shows rapidly decaying oscillations, and even setting $D=\infty$ (c) does
not substantially increase the stability of the oscillations. An important feature to note here, is the
180° phase difference between cases (a) and (b), evidently the the direction of elongation due to
diffusion enhancement is orthogonal to that due to directional bonding. In Fig. 3 are surface
morphologies after the growth of 0.25ML for cases (a) and (c). Obvious is the more elongated nature
of islands generated by a bonding enhancement versus a diffusion enhancement. Finally in Fig. 2(d)
is the combined case in which $B=10$ and $D=\infty$, the oscillations having the same phase as in case (a),
but with lower stability.

In order to understand this behavior in the case of diffusion enhancement, consider a square
cluster on a surface in which diffusion in the layer of the cluster is in the x-direction. An atom
migrating to the cluster forms a single bond with the cluster, however the lifetime of the atom in this
configuration although much longer than for a free atom, still allows substantial mobility. Upon
hopping the atom must move either away from or onto the cluster. If it moves onto the cluster it must
diffuse along the y-direction until arriving at an edge of the cluster parallel to the x-axis, forming
another single bond. In this case, however, if the atom hops again there is a good chance it will move
into a site still possessing a single nearest-neighbor bond. Subsequently its lifetime at an edge parallel
to the diffusion enhancement is longer than at an edge normal to it and the island will grow
preferentially normal to the enhancement.

This study illustrates that although diffusion enhancement may lead to cluster elongation it will

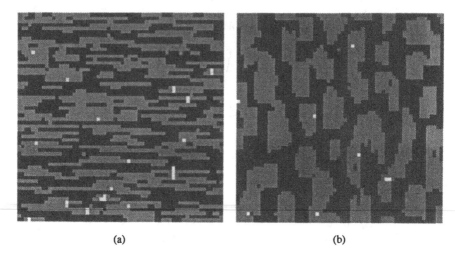

(a)   (b)

Figure 3. Surface morphologies generated after the deposition of 0.5ML at $T=700K$ and $F=0.5ML/s$. (a) Corresponds to the bonding and diffusion anisotropies detailed in Fig. 2(a), and (b) to Fig. 2(c).

be perpendicular to the direction of diffusion and create clusters in which the aspect ratio is very low. Bond enhancement, on the other hand yields surface morphologies more in accord with STM data and even when combined with an infinite diffusion enhancement, dominates the growth mode.

STEP STABILITY AND DOMAIN COVERAGE

In Fig. 4 we show the evolution of the even-layer coverage, corresponding to the area of the surface covered by a particular domain (type-B), during high-temperature annealing, growth and recovery on a 120x120 lattice comprising 12 monatomic steps. Initially the surface is allowed to relax to remove *memory* of the artificially imposed starting configuration. Upon commencing growth we observe a transition to a dominance of type-B. At the lower temperature ($T=825K$) we observe oscillations in the "coverage" as growth proceeds by the formation and coalescence of islands on the steps, however at the higher temperature ($T=850K$) all atoms are incorporated directly into the step edge in "step-flow" mode. Terminating the beam flux leads to a return to the steady state observed before growth. Experimentally the same trend is observed, with a half-order RHEED intensity corresponding to a mixed 2x1/1x2 domain before and after growth, but upon opening the Si shutter the intensity rapidly attains a value indicative of a type-B 2x1 surface [9].

To gain insight into this behavior we analyze the surface morphology (Fig. 5) at $T=850K$, after 10s (*i.e.* the equilibrated surface) and after 20s (*i.e.* the surface during growth). In each case we observe a pronounced difference in the structure of the S_A versus S_B steps. Steps in which the bond enhancement is parallel to the steps are very smooth (S_A), whilst for those in which the bond anisotropy is normal (S_B) are very rough. The origin of the domain change is revealed by the relative energies of the steps. If the bond enhancement is parallel to the steps, the step is very stable and atoms adjacent to the step have a low lifetime in that site. Conversely, if the bond enhancement is normal to the step edge then the step is unstable, but atoms adjacent to it will have a very high lifetime. During growth, the longer residence time of atoms at an S_B step ensures that it grows in preference to the S_A

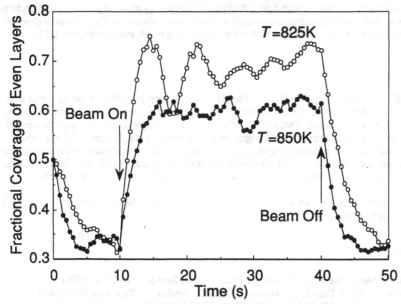

Figure 4. Fractional coverage of even numbered layers (type-B terraces) during annealing growth and subsequent recovery on a monatomically stepped vicinal substrate. The surface used in this simulations was a 120x120 array, with 12 monatomic steps, exposed to a flux F=0.25ML/s.

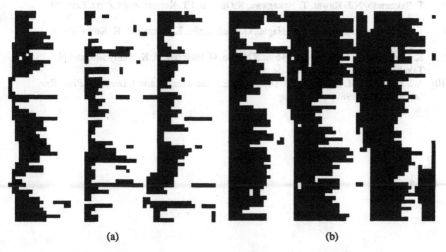

Figure 5. "Domain maps" of (a) an equilibrated surface after 10s annealing at T=850K and (b) the same surface following 10s growth. These are 60x60 subsections of the lattices used in Fig. 4. The steps descend from left to right and the black shading corresponds to to even layers (type-B).

step, as atoms do not have enough time to find their lowest energy environment, before interacting with atoms newly arrived from the incoming flux. These observations are in direct accord with recent STM studies of the Si(001) surface, with evident alternating rough and smooth steps [10].

SUMMARY

We have shown that the inclusion of an anisotropy in nearest-neighbor bonding within a lattice model of the Si(001) surface may account for many observed morphological features during MBE. Conversely inclusion of anisotropic diffusion in our model yields surface morphologies with little correspondence to experiment. Of importance to our understanding of growth is the observation that it is the lifetime of an atom in a given site that determines the growth mode, not the mechanism of transport to or from that site. This is vividly demonstrated by growth on highly vicinal surfaces, in which we observe dramatically different step structures during growth and at equilibrium.

SC would like to thank the U.K. Science and Engineering Research Council for support through a fellowship.

REFERENCES

[1]   Y.-W Mo, R. Kariotis, D.E. Savage, M.G. Lagally, *Surf. Sci.* **219** L551 (1989).
[2]   M.R. Wilby, S. Clarke, T. Kawamura, and D.D. Vvedensky, Phys. Rev. B (in press).
[3]   M. Ichikawa and T. Doi, *Appl. Phys. Lett.* **50**, 1141 (1987).
[4]   S. Clarke and D.D. Vvedensky, *Phys. Rev. B* **37**, 6559 (1988).
[5]   S. Clarke, M.R. Wilby, D.D. Vvedensky, and T. Kawamura, *Phys. Rev. B* **40**, 1369 (1989).
[6]   E.J. Van Loenen, Private Communication.
[7]   T. Sakamoto, N.J. Kawai, T. Nakagawa, K.Ohta, and T. Kojima, *Appl. Phys. Lett.* **47**, 617 (1985).
[8]   T. Sakamoto, T. Kawamura, S. Nagao, G. Hashiguchi, K. Sakamoto, K. Kuniyoshi, *J. Crystal Growth* **81**, 59 (1987).
[9]   K. Sakamoto, T. Sakamoto, K. Miki, S Nagao, G. Hashiguchi, K. Kuniyoshi, and N. Takahashi, *J. Electrochem Soc.* (in press).
[10] A.J. Hoeven, J.M. Lenssinck, D. Dijkkamp, E.J. van Loenen, and J. Dielman, *Phys. Rev. Lett.* **63**, 1830 (1989)

XPS ANALYSIS OF THE SAPPHIRE SURFACE AS A FUNCTION OF HIGH TEMPERATURE VACUUM ANNEALING

E.D. Richmond
Naval Research Laboratory, Code 6816, Washington D.C. 20375-5000

ABSTRACT:

For the first time the ($1\bar{1}02$) surface of sapphire has been investigated by X-ray photoelectron spectroscopy to ascertain chemical changes resulting from annealing in vacuum at 1300° C and 1450° C. As received substrates had a substantial surface C contaminant. For substrates that were chemically cleaned before inserting them into the MBE system no trace of carbon is detected. A residual flourine contaminant results from the cleaning procedure and is desorbed by the vacuum annealing. Spectra of annealed substrates are compared to the unannealed chemically cleaned substrates. The annealed substrates exhibit 0.4 to 0.5 eV shift to higher binding energy of the Al peak and a 0.3 eV shift to higher binding energy of the O peak. In addition, a 2% depletion of oxygen from the surface occurs.

INTRODUCTION:

For the past few years NRL has been investigating the effect of high temperature annealing of the ($1\bar{1}02$) sapphire substrates annealed at 900° C, 1100° C, 1300° C, and 1450° C prior to deposition of a silicon epilayer, because of the importance of silicon on sapphire (SOS) for radiation-hard microelectronics. The films were investigated for strain properties[1] and their growth kinetics[2]. In a comparison with chemical vapor deposited (CVD) SOS, the Molecular Beam Epitaxy (MBE) SOS contained an order of magnitude less microtwin volume, less strain, better crystalline quality, higher mobilities, and lower interface states[3] One potential source of the differences between CVD SOS and MBE SOS may be due to the chemical changes of the surface due to the high temperature vacuum annealing.

EXPERIMENT:

The 3" ($1\bar{1}02$) sapphire wafers were chemically cleaned using the procedure shown in Table 1. They were then inserted immediately into the NRL Si MBE/Surface Analytical System[4]. The substrates were analyzed in the analysis chamber which consists of a VG ESCALAB MkII with a dual anode Mg/Al source. The sources are operated at 300W and 600W for the Mg and Al sources respectively. For each source, a widescan was taken of each sample with a pass energy of 50 eV, and then high resolution scans at steps of 0.1 eV and a pass energy of 10eV were taken of the photoelectron peaks Al(2p), Al(2s), O(1s), C(1s), F(1s). Also the Auger spectra for the transitions O(KVV) and Al(KLL) were obtained. C1 slits were used throughout. The angle between the sample normal and the detector is 15° .

RESULTS AND DISCUSSIONS

Figure 1 shows a widescan of a chemically cleaned but unannealed sample. The only peaks which are discernible are the Al(2p), Al(2s), O(2s), O(1s), and F(1s) photoelectron peaks and the O(KVV) and F(KVV) Auger peaks. The F(KVV) peak and the O(2s) peak are not labelled and occur around 610 eV and 35 eV respectively. The C(1s) peak at about 295 eV and the C(KVV) Auger peak around 1000 eV are obviously absent. This is confirmed by high resolution spectra, which, in one case evidences a small signal, < 0.6%, almost indiscernible from the noise, and in another sample no signal at all. After an anneal at 1300° C or 1450° C there is no trace of carbon or flourine on the surface. To provide a measure of the relative surface sensitivity, the mean free path calculated by Penn[11] for a binary solid such as Al_2O_3 is shorter than you would find in an elemental solid such as Al beccause of the larger density of valence band electrons and core electrons. For a Mg source the mean free path in Al_2O_3 for Al(2s) is 15.7Å , for C(1s) is 13.9Å , and for O(1s) is 11.1Å . For comparison, the mean free path for Al(2s) from the metal is 21.6Å . Before chemical cleaning, a significant C photoelectron and

Auger peaks are found[4].

To extract the charging of the sample, the valence band maximum (VBM) is determined as suggested by Gignac et al.[5] The VBM is determined by the intersection of the slope of the upper valence band edge with the background. I have found that if one does the extrapolation manually then an error is introduced ranging from tens of millivolts to several tenths of a volt, or possibly even more. To avoid this, both the background and slope of the valence band are fitted by a linear least square fit. For the valence band slope data is included ranging approximated from 16% to 84% above the background in order to avoid the nonlinear portions of the slope.

The upper valence band is shown in Fig. 2 for a clean but unannealed specimen of sapphire. The linear least square parameters for the equation $y = aE + b$ are

Bkgrd: $a = 0.388$ cps/eV $b = 82.3$ eV

$E_l = 6.9$ eV $E_r = 13.8$ eV

VB: $a = 185.60$ cps/eV $b = -2541$ eV

$E_l = 14.4$ eV $E_r = 15.6$ eV

The intersection of these two lines give a valence band maximum of 14.17 eV. This is to be compared to a manual determination of 13.75 eV. The importance of this is more readily seen in Table 2, error of this magnitude would invalidate any conclusions about a shift in peak energy resulting from the in-situ anneal. In addition, an error of this size would cause a random shift in he unannealed peak energy, which is seen to be onstant from sample to sam le.

TABLE 1
SAPPHIRE CHEMICAL CLEANING PROCEDURE
1. HOT ACETONE
2. HOT TRICLOROETHYLENE
3. HOT ACETONE
4. HOT METHANOL
5. RINSE DI H_2O
6. 1:1 H_2O_2:H_2SO_4
7. RINSE DI H_2O - 5 MIN.
8. 4:1:1 H_2O:HCl:H_2O_2-15 MIN. BOILING
9. REPEAT STEP 7
10. 10:1 H_2O:HF 30 MIN.
11. REPEAT STEP 7
12. REPEAT STEP 8
13. REPEAT STEP 7
14. BLOW DRY WITH N_2
15. TRANSFER TO LOADLOCK
16. ANNEAL 30 MIN. IN VACUUM AT SPECIFIED TEMPERATURE

The shift in the Al(2s) peak energy due to the annealing is shown in Table 2. First notice that

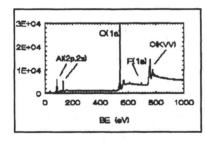

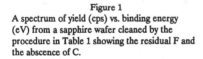

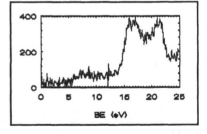

Figure 1
A spectrum of yield (cps) vs. binding energy (eV) from a sapphire wafer cleaned by the procedure in Table 1 showing the residual F and the abscence of C.

Figure 2
A spectrum of the yield (cps) of the upper valence band of sapphire vs binding energy (eV).

the unannealed peak energies, after correcting for charging, are the same sample to sample, as suggested above. After an anneal, the peak moves to higher binding energies. For a 1450° C anneal the shift is on the order of 0.5 eV. For a 1300° C anneal, the shift is slightly less, around 0.4 eV. The difference between these two shifts is just outside the error bars of ± 0.05 eV. The error bars are derived from the fact that the spectra are sampled every 0.1eV. Thus, the Al(2s) binding energy shift appears to be effected by the in-situ anneal temperature.

Table 2 also shows similar data for the shift of the O(1s) peak binding energy. Here there is a 0.33 eV shift from the unannealed value of 531.68 eV. For the O(1s) there is little sensitivity to the in-situ anneal temperature.

In order to quantitatively determine the percentage of oxygen, two basic approaches are pursued. One is to use the form factors derived by Wagner et al.[6]. The other is to use the equation[8]

$$I_a(E_a) = \sigma_a(h\nu)\, D(E_a) \iint L_a(\gamma) \iint J_O(x,y) T(x,y;\gamma\,\phi,E_a) \int N_a(x,y,z) e^{-z/(\lambda\cos\theta)}\,dz dx dy d\phi\, d\gamma \quad (1)$$

here I_a equals the flux intensity for state 'a', D is the detection efficiency for energy E_a, L_a is the angular asymmetry of the intensity of the photoemission from each atom 'a', J_O is the x-ray flux intensity, T is the analyzer transmission at energy E_a; N_a is the density of atoms 'a'; and, λ is the mean free path. Assuming that N_a is not a function of z, then the last integral simplifies to $N_a\lambda\,(E_a)\cos\theta$. Further if I_a is referenced to some standard, and we concern ourselves with only the photoelectron lines of the symmetric 's' level, in which case L_a is constant, then we find that

$$N_a = I_a/\sigma_a T\,\lambda \quad (2)$$

where I_a is given by the peak height or the peak intensity, σ_a is given by Scofield's cross sections, $T \propto E^{-1}$, and $\lambda \propto E^n$ where $n = 0.5$ (Seah and Dench[10]), ~ 1 (Penn[11]), or .75 (Powell[12]). In the case of using form factors, the denominator in eq. (2) is replaced with a parameter S as given by Wagner et al.[6]

Making the assumption that N_a is constant as a function of 'z', I find that the most consistent and realistic results occur by using a peak area for I_a and Wagner's form factors. The form factors are based on an empirical based mean escape depth with an energy dependency of $\lambda \propto E^n$, with $n = 0.66$.

The peak area is determined by synthesizing the peak using the equations described in Seah and Briggs[8], which is provided with the ESCALAB software. The area of the synthesized peak is then calculated for the Al(2s) peak and the O(1s) peak. Then the quantity N_O/N_{Al} is computed and the oxygen concentration is found from $N_O = 1/\{1 + (N_O/N_{Al})^{-1}\}$. The unannealed oxygen concentration was normalized to 60% in all cases. The oxygen concentration for the annealed samples was similarly scaled so that Fig. 3

TABLE 2		
BINDING ENERGY (eV)		
	O(1s)	Al(2s)
UNANNEALED	531.8	119.58
REFERENCE[14]	531.6	---
ANNEALED 1300°C	532.01	120.01
ANNEALED 1450°C	532.05	120.12

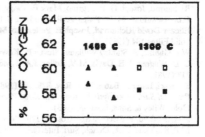

Figure 3
The oxygen concentration in percent in the surface layer for both Al and Mg x-ray sources for unannealed (Δ, $\square$) samples and samples annealed at 1300° C ($\blacksquare$) and 1450° C ($\blacktriangle$).

represents a true difference in oxygen concentration due to the in-situ high temperature annealing.

Examining Fig. 3, the oxygen concentration in the annealed samples varies from 57.6% to 58.7%, giving about a 2% oxygen decrease in the surface layer after either a 1300° C or a 1450° C anneal. Oxygen depletion due to high temperature annealing in vacuum has been previously found for the (0001) plane, but no quantitative analysis could be obtained from that experimental data.[13]

Chang[13] finds from LEED studies that the unannealed sapphire surface has a pseudo 1 x 1 surface structure. This changes to an equilibrium 2 x 1 surface structure after high temperature annealing. The binding energy shifts reported here are not inconsistent with these LEED results. The result that both atoms increase their binding energy may also be an indication of a surface relaxation as well as a reconstruction.

The fact that the Al binding energy increases by 0.1 to 0.2 eV more than the O binding energy is difficult to interpret without a better understanding of the exact surface structure. Following Ciraci[7], the Al surface atoms become less ionic. This might result in a strengthening of the Al backbonds, which would increase their binding energy. The loss of O from the surface could create surface vacancies as well as potentially drive the reconstruction.

SUMMARY:

The surface of (1̄102) sapphire has been investigated by XPS as a function of high temperature in-situ annealing. The results indicate that the chemical cleaning process used before inserting the wafers in the MBE system eliminates any C contamination, but leaves a residual surface concentration of flourine. After an anneal at 1300° C or 1450° C, the flourine is removed from the surface. The change in the binding energy of both the Al and O peaks is not inconsistent with both a surface relaxation and reconstruction. The binding energy shift of the Al is about 0.1 eV greater than the O binding energy shift for an anneal at 1300° C and about 0.2eV at 1450° C. An O loss of about 2% from the surface is measured after the high temperature anneal.

REFERENCES

1. Joseph Pellegrino, Syed Qadri, Eliezer Richmond, Mark Twigg, Carl Vold, MRS Symp. Proc.,(Materials Research Society, Pittsburgh, 1988) v. 116, p. 395.
2. Joeseph G. Pellegrino, Eliezer D. Richmond, Mark E. Twigg, MRS Sym. Proc. (Materials Research Society, Pittsburgh, 1988), v. 116, p. 389;Joe Pellegrino, Mark Twigg, Eliezer Richmond, MRS Sym. Proc. (Materials Research Society, Pittsburgh, 1988), v. 107, p. 383; Eliezer Dovid Richmond, Joseph G. Pellegrino, Mark E. Twigg, 1989 Fall MRS Symposium, Presentation D2.4.
3. E.D. Richmond, M. E. Twigg, A. R. Knudson, S. Qadri, N. Green, Thin Solid Films 184 (1989).
4. Eliezer Dovid Richmond, Joseph E. Pellegrino, Mark E. Twigg, Syed Qadri, Michael T. Duffy, Thin Solid Films 184 (1989).
5. W.J. Gignac, R. Stanley Williams, Steven P. Kowalczyk, Phys. Rev. B32, 1237 (1985).
6. C. D. Wagner, L.E. Davis, M.V. Zeller, J.A. Taylor, R.H. Raymond, L.H. Gale, Surf. Interface Anal. 3, 211 (1981).
7. S. Ciraci, Inder P. Batra, Phys. Rev. B28, 982 (1983).
8. Practical Surface Analysis by Auger and X-ray Photoelectron Spectroscopy, eds. D. Briggs and M.P. Seah (John Wiley & Sons, London, 1983).
9. J. H. Scofield, J. Electron Spectrosc. 8, 129 (1976).
10. M.P. Seah and W.A. Dench, Surf. Interface Anal. 1, 2 (1979).
11. David R. Penn, J. Electron Spectrosc. 9, 29 (1976).
12. C.J. Powell, Surface Science 44, 29 (1974).
13. Chuan C. Chang, J. Vac. Sci. and Technol. 8, 500 (1971).
14. Handbook of X-ray Photoelectron Spectroscopy, C.D. Wagner, W.M. Briggs,L.E. Davis, J.F. Moulder, G.E. Mullenberg (Editor) (Perkin Elmer Corporation, Eden Prairie, MN., 1979).

ADSORPTION AND TRIBOCHEMICAL REACTIONS

W. M. MULLINS† AND T. E. FISCHER‡
† Purdue University, West Lafayette, IN 47907
‡ Stevens Institute of Technology, Hoboken, NJ 07030

ABSTRACT

A model for chemisorption based on the Lewis theory is proposed and developed. The model reproduces the observed relationship between aqueous solubility and band gap for covalent oxides. A relationship is shown between the calculated free-energy of the surface reactions and the observed effect of water on the wear of these materials. This relationship is explained in terms of surface reaction rate and equilibrium. Similar calculations for the adsorption of $-CH_3$ functional groups are presented. The results are compared to water adsorption calculations and related to experimental wear test results for model fluids on covalent oxides. The effect of water on lubricant adsorption is discussed.

1. Introduction

Ceramic materials are attracting much interest in friction and wear applications because of their high hardness and apparent chemical inertness. Significant chemical effects on wear, tribochemistry, have been demonstrated in the literature. Relative humidity has been demonstrated to dramatically effect the wear rate of Si_3N_4 [1] and in certain conditions produce surface lubricity [2]. Water has also been shown to produce chemisorption embrittlement (CE) in silica and glasses [3], Al_2O_3 [4] and partially stabilized zirconia (psz) [5]. In addition, highly polar organics such as stearic acid, traditional boundary layer lubricants for metals, have been shown to increase wear in Al_2O_3 and psz. Non-polar organics such as paraffin that are inert to metals have been shown to be effective boundary layer lubricants for many ceramics [5,6].

In general, surface reactions during wear are very important to wear behavior of ceramics. The surface chemistries of these materials are very different from metals so that little of the traditional lubrication technology is applicable. In addition, surface chemistries vary significantly in these materials so that an empirical understanding of one or two is not sufficient to design effective lubricants for general application. A systematic approach to lubricant molecule design must start with the fundamentals of the surface adsorption reaction.

Surface chemistry is determined by the electronic structure of the constituents through the Lewis acid-base theory. The adsorption reaction is modelled as a simple chemical bonding problem between a single molecule and a surface, approximated by the linear combination of orbitals (LCO) method [7]. Thermodynamic equilibrium behavior for a large ensemble are determined by using Fermi statistics in the single molecule model to determine relative probabilities of orbital occupations.

The model was used to calculate the extent of the surface adsorption reaction as a function of band gap for H_2O and $-CH_3$ on a series of covalent oxides. Regions for lubricity, CE and boundary layer lubrication are suggested for the results. Criteria are suggested for synergistic effects between dissimilar adsorbing species that can be used to predict the effects of ambient of lubricant performance and possibly to control the adsorption effects of complex lubricant systems.

2. Macroscopic Model Results

The model used [7] is based on a theory originally described by Mulliken [8] and extended later by Salem [9] and Klopman [10] to describe acid-base interactions for individual molecules. This model assumes that acid-base adduct formation reaction can be described by an LCO and uses the overlap and resonance integrals of the involved orbitals, along with their relative energies, as parameters. The semi-empirical extension to surface reactions used here has included the Fermi energy of the species as a parameter to predict average interactions for an ensemble [7]. The model has been used to successfully show an experimentally determined relationship between Fermi-energy and the aqueous acidity of surfaces for aluminum and silicon oxides.

During the surface reaction, a net charge is transferred from the base to the acid to form a net charge dipole at the adduct. Assuming a constant conformation for the water molecule and no effect of defect states in the substrate Fig. 1 shows a plot of the square of the calculated charge dipole formed as a function of band gap for water on a covalent ceramic surface. The curve is seen to have a relative minimum at $E_g \cong 8.0$ that corresponds to zero charge transfer of a neutral reaction. Variation from this case increases the net charge transfer of the reaction.

Also plotted on Fig. 1 are the reported solubilities of various covalent oxides [11] as a function of band gap [12]. With the scales suitably adjusted, an obvious relationship between solubility and band gap are observed. Since water is highly polar, solubility of any species is related to the magnitude of its charge and the dielectric constant of the solvent [13].

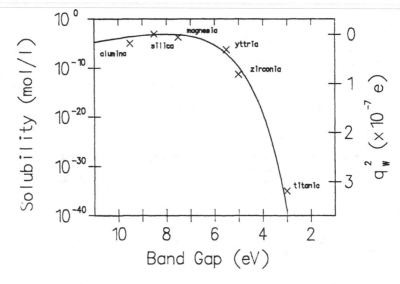

Fig. 1.   Literature reported aqueous solubility as a function or band gap for "covalent" oxides. Calculated Huckel charge (solid line) is shown as a function of band gap.

A maximum interaction is predicted for SiO_2 which has the highest solubility. The solubility of this materials is sufficient to produce complete dissolution of the surface in a tribochemical wear situation. This is believed to be the cause of lubricity. Strong chemisorption is noted for alumina, magnesia, yttria and potentially zirconia, but the magnitude of the adduct dipole limits the extent of any possible surface reaction. Sufficient chemisorption is possible for CE to be observed but the solubility is to low to allow for the healing of surface cracks [14] and smoothing of the surface [2] associated with lubricity. For titania, the adduct dipole is so large that no extensive reaction is predicted physisorption is expected to dominate chemisorption. Neither CE nor lubricity are expected for titania.

The model interaction for a $-CH_3$ functional group as a a function of band gap can be described in the same manner as for the water molecule. Assuming a similar configuration to the water molecule, Fig. 2 shows a the predicted relationship between band gap and surface adduct dipole. The relationship observed is similar to that of Fig. 1, but shifted to a lower band gap region due to the lower energy gap of the $-CH_3$ functional. Based on the results for water, significant adsorption is expected in the entire range of ceramics. A maximum for adsorption is expected for Y_2O_3 but solubility in a non-polar media is expected to by small due to the low dielectric constant. For this reason, CE is considered possible for Y_2O_3 and possibly ZrO_2.

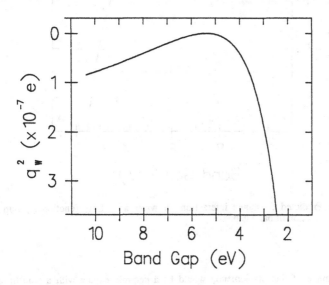

Fig. 2.   Calculated Huckel charge on the $-CH_3$ functional group as a function of substrate band gap using the same orbital parameters as for water.

3. Discussion

A maximum chemical interaction is expected at the point of minimum net charge transfer. At this point, little electrostatic hinderance to reaction are expected at surface sites. There is an equal likelihood for species dissociation and little complexation is required so that the dissolution reaction is not hindered. At this point, also, chemisorption embrittlement is predicted to be at a maximum. Changing the band gap of the substrate increases the magnitude of the charge dipole formed by added formation. The polarity of the dipole is determined by the relative band gaps of the lubricant in the substrate if we neglect the effects of Fermi energy, surface states and doping. As a rule of thumb, the wider the band gap, the more acidic the material.

For two different adsorbing species their different electronic structures will allow for different relative minima in the charge transfer and different relative maxima for the interactions. Between the individual relative maxima, opposite polarities of dipoles are generally formed. The adsorption of the two species on alternating surface sites allows the dipoles formed during the adduct formation to compensate one another. The effective interaction curve for the two species is shown in Fig. 3 and is much wider than the curve shown in Figs. 1 and 2. Correspondingly, chemisorption embrittlement is expected to extend throughout the entire region.

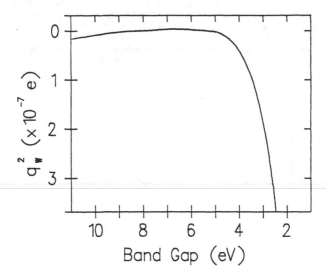

Fig. 3.   Predicted combined interaction for water and -CH$_3$ functional group as a function of band gap.

An example of this phenomenon would be a ceramic engine with a paraffin-based boundary layer lubricant. The water of combustion would react with the ceramic surface at the sites where the paraffin is specifically adsorbed. This would increase the adsorption curve for the paraffin to the point of chemisorption embrittlement for the material. The result would be a dramatic increase in the wear rate of the ceramic at all wear points.

4. Future Work

Molecular orbital calculations are currently underway to better characterize the surface interaction chemistry. In the next phase, fluorocarbon-based lubricants will be studied to determine adsorption chemistry and the effects of adventitious water on chemisorption embrittlement. The possible application of organophosphates and organosulphates in blocking the coadsorption of water will also be studied.

5. References

1.   H. Shimura and Y. Tsuya, *Wear of Materials,* ed. V. Ludema, ASME, New York, p. 452 (1977).

2.   T. E. Fischer, H. Liang and W. M. Mullins, Mat. Res. Soc. Symp. Proc. **140**, 339 (1989).

3.   S. M. Wiederhorn, S. W. Freiman, E. R. Fuller and C. J. Simmonr, J. Materials Sci. **17**, 3460 (1982); T. A. Michalske, and B. C. Bunker, J. Appl. Phys. **56**, 2686 (1984).

4.   N. Wallbridge, D. Dowson and E. W. Roberts, *Wear of Materials,* K. C. Ludema (ed.), ASME, New York (1983).

5. T. E. Fischer, M. P. Anderson, S. Jahanmir and R. Salher, *Wear of Materials,* K. Luedma (ed.), ASME, New York, p. 257-266 (1987).

6. S. Johanmir and T. E. Fischer, STLE Trans. **31,** 32 (1988).

7. W. M. Mullins, Surface Sci. **217,** 459 (1989).

8. R. S. Mulliken, J. Am. Chem. Soc. **74,** 811 (1952).

9. L. Salem, J. Am. Chem. Soc. **90,** 543 (1968).

10. G. Klopman, J. Am. Chem. Soc. **90,** 223 (1968).

11. W. F. Linke, *Solubilities: Inorganic and Metal-Organic Compounds, Vol. 1* (4th ed.), American Chemical Society, Washington DC, (1958) and W. F. Linke, *Solubilities: Inorganic and Metal-Organic Compounds, Vol. 2* (4th ed.), American Chemical Society, Washington DC, (1965).

12. W. H. Strehlow and E. L. Cook, J. Phys. Chem. Ref. Data **2,** 163 (1973).

13. M. Maroncelli and J. MacInnis, G. R. Fleming, Science, **243,** 1674 (1989).

14. Y. Bando, S. Ito. and M. Tomozawa, J. Am. Cer. Soc. **67,** C36 (1984).

PART VI

Scanning Tunneling Microscopy

ATOMICALLY-RESOLVED SURFACE PHOTOVOLTAGE PROBED BY OPTICALLY-EXCITED SCANNING TUNNELING MICROSCOPY

R.J. HAMERS AND K. MARKERT
IBM T.J. Watson Research Center, Yorktown Heights, N.Y. 10598

ABSTRACT

Scanning Tunneling Microscopy combined with optical excitation techniques is used to study non-equilibrium electronic properties of clean
silicon surfaces with high spatial resolution. Tunneling potentiometry is
performed to measure the excess carrier distributions via the surface
photovoltage effect. Well-ordered regions of Si(111)-(7X7) show a uniform
surface photovoltage effect, while strong decreases are observed near defects. The decreases in the photovoltage are attributed to an increase in
the rate of recombination of electron-hole pairs in the vicinity of the defects.

INTRODUCTION

Since its invention in the early 1980's, the scanning tunneling microscope has proven itself as a versatile tool capable of addressing questions
in a wide variety of areas. An increasing number of studies have used STM
to explore the *electronic* properties of surfaces on an atomic scale, through
measurements of the tunneling I-V characteristics. While such I-V curves
provide information about the *equilibrium* electronic properties of the surface, in many applications dynamic processes such as carrier transport and
recombination are of equally great importance.

In this paper, we report on recent investigations combining optical
excitation with scanning tunneling microscopy to probe the non-equilibrium
electronic properties of silicon surfaces. We accomplish this by utilizing
the well-known surface photovoltage (SPV) effect. [1-3] Unlike conventional
SPV measurements which probe macroscopically large areas, by using an STM
tip as our measuring probe we are able to measure the SPV on regions of atomic
dimensions.

As a result of the high density of mid-gap surface states on silicon
surfaces, charge transfer between surface and bulk states occurs, with the
concurrent formation of a sub-surface space-charge layer. The electric field
in this space-charge layer is accompanied by band-bending which, on silicon

surfaces, places the Fermi level midway between the conduction and valence band edges at the surface, irrespective of the bulk doping. As a consequence, the band-bending is downward on p-type material and upward on n-type material.

Illumination of this surface with light having energy greater than the bulk bandgap will excite electrons from the valence to the conduction band, thereby creating an electron-hole pair. Since the absorption depth of light at 488 nm is $\simeq 10^4$ Å, electron-hole pairs are created not only near the surface, but also comparatively deep into the bulk of the sample. In most of the bulk there is no electric field, so that the concentration of electrons is the same as the concentration of holes. In the space-charge region, however, the electric field separates the electron-hole pairs. On p-type sample, with downward band-bending, the electric field drives the electrons toward the surface and the holes into the bulk. On n-type samples, the electric field drives the holes toward the surface and the electrons into the bulk. As a result, there will be a non-equilibrium accumulation of electrons at the surface of a p-type sample, and a depletion of electrons at the surface of an n-type samples. The accumulation of carriers of one charge and the depletion of those of the other charge in the near-surface region generates a voltage, referred to as the "Surface Photovoltage", or SPV. In our experiments, we probe the magnitude and sign of the SPV using the STM tip as a potentiometer.

EXPERIMENTAL

Figure 1 shows the overall experimental system. Illumination was provided with either a 10 mW He-Ne laser at 632.8 nm or a multiline 100 mW Argon-ion laser at 488-514 nm. As shown in figure 1, the linearly-polarized laser light passed through an electro-optic modulator (EOM) followed by a polarizing prism (PP), allowing the laser intensity to be controlled under computer control by application of a voltage to the modulator. A small fraction of the beam was split off using a beamsplitter (BS) so that the intensity could be monitored using a photodiode (PD). The remainder of the beam was directed with two steering mirrors into the ultrahigh vacuum system and onto a 1-inch focal length lens. The lens focussed the light to a spot of approximately 100 microns diameter at the tunneling junction.

The ultrahigh vacuum system is a two-chamber system with a base pressure of $\simeq 7 \times 10^{-11}$ Torr. A preparation and analysis chamber, equipped with Low Energy Electron Diffraction, Auger Electron Spectroscopy, and a mass spectrometer, is used for degassing and annealing of the samples. After cleaning as described previously to create large, flat terraces of Si(111)-(7X7), the samples are transferred to the second chamber containing the STM. The STM is similar in principle to that described in an earlier

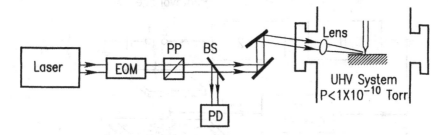

Figure 1. Diagram of optical system. EOM: Electro-optic modulator; PP: Polarizing prism; BS: Beamsplitter; PD: Photodiode.

publication [4], with some modifications. The microscope used here utilizes a different piezoelectric scanner consisting of a single plate for the X-Y scanner and a short tube for the Z-scanner, providing a 9.6 kHz resonant frequency (compared with 3.1 kHz for the earlier piezoelectric tripod design). The increased resonant frequency and the smaller dimensions of the microscope provide greatly improved immunity against external vibrations and thermal drifts.

The excess carriers produced by optical excitation can be probed in several ways. One way of observing the SPV effect is by monitoring the tunneling current I as a function of the applied voltage V. In the absence of illumination, zero applied voltage always results in zero current. However, when the sample is illuminated a photoelectric current is observed when the sample bias is zero, and a finite voltage must be applied to the sample in order to prevent any tunneling current from flowing. This difference in potential between the surface of the semiconductor and the bulk is the surface photovoltage.

In order to monitor the SPV in a continuous manner as the tip is scanned on the sample, we utilize a potentiometry system as shown in figure 2. The control electronics are gated using two sample-and-hold amplifiers (S/H1 and S/H2) such that two types of measurement are performed. These measurements are performed at a rate of $\simeq$ 2 kHz, which is much faster than the scan speed of the tip across the surface. Consequently, they can be considered to be simultaneous. During the first part of the gating cycle, a constant voltage is applied to the sample (represented by the battery in figure 2). Sample-and-hold SH1 is in the "sample" state, and the amplified tunneling current is monitored using feedback-loop controller FB1, which compares the actual tunneling current with a pre-set value (usually 1 nA), and applies a voltage to the Z-piezo in order to keep the tunneling current constant. This part of the cycle is functionally equivalent to the conventional constant-current STM topography measurement. During the second part of the cycle, the STM is operated as a potentiometer. This is accomplished by deactivating S/H1

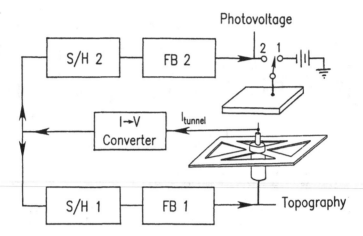

Figure 2. Schematic of the electronics used for photovoltage measurements. Electronics denoted "1" are active during the topography part of the cycle, those denoted "2" are active during the potentiometry part of the cycle.

so that the sample-tip separation remains constant; simultaneously, sample-and-hold S/H2 is activated, and the actual tunneling current is measured with a second feedback controller FB2. This controller *nulls* the tunneling current by applying a voltage to the sample bulk in order to keep the tunneling current equal to zero. In the absence of illumination, this should be identically equal to zero, while when the sample is illuminated this should be equal to the surface photovoltage (except inverted in sign).

RESULTS

Figure 3 shows some results obtained on Si(111)-(7X7) using the potentiometry system described above. On the left (3a) is shown the conventional STM "topography", given by the output of feedback controller FB1. The atomic structure of the (7X7) unit cell is clearly visible. Differences between the various adatoms with the unit cells and between the two halves of each unit cell arise from electronic structure effects, as described in previous publications. [5,6] Near the top of the image, a boundary between two different (7X7) domains can be observed. This appears as a shift in the otherwise straight line joining the deep corner holes of the (7X7) unit cells. At the domain boundary, it is impossible to satisfy the normal coordination of all silicon atoms, so an irregular arrangement of adatoms is observed.

Figure 3. Atomically-resolved measurements of topography and surface photovoltage on Si(111)-(7X7) surfaces. a) Topography, with bias of -1 V; b) Surface photovoltage, average value=140 mV. c,d) 3-dimensional representation of same data as in a and b, except including portion where laser was blocked.

Figure 3b shows the local surface photovoltage measured in the second part of the feedback cycle (i.e., the output of feedback controller FB2), essentially simultaneously with the topography image shown in figure 3a. On this P-type material, the illumination produces excess electrons (a negative voltage) at the surface. In fig. 3b, the grayscale is adjusted such that surface locations showing a large, negative SPV are white, while those showing zero photovoltage are black; no regions of positive photovoltage are observed on p-type material. A comparison of the topography (3a) with the local surface photovoltage (3b) shows that in regions where the topography shows a relatively well-ordered (7X7) reconstruction, the SPV shows a relatively uniform photopotential of -140 mV. At locations where the left image shows defects, the photovoltage decreases. At relatively large defects, as in the upper left corner, the photovoltage drops to zero.

A stringent test of the experimental system is to block the laser and to observe what happens to the SPV signal. This is shown as a three-dimensional representation shown in figures 3c and 3d. Note that this is the same region shown in figures 3a and 3b, except that the three-dimensional view in 3c,d includes an additional portion where the laser was blocked. The forward left corner of figures 3c and 3d correspond to the top left corner of figures 3a and 3b. Toward the right of figures 3c,3d (and at the end of 3a,3b) the laser was blocked. When the laser is blocked, the "topography" in 3c shows an immediate decrease in the apparent surface height. This is photothermal effect - when there is no absorption in the tunnel junction region the sample and/or tip cool and contract; in constant-current STM, this cooling appears as a decrease in the apparent height of the surface. A close inspection of the declining portion of figure 3c shows that despite this thermal cooling effect, atomic resolution is maintained and the atomic structure of the (7X7) reconstruction is still visible. In figure 3d, the simultaneously-measured surface photovoltage is shown. This image shows that when the laser is blocked, the photovoltage drops to zero within the noise limit of 5 mV. There are two significant observations to be made regarding the noise when the laser is blocked. First, the data shows that that after blocking the laser, the residual noise on the SPV is extremely small- much smaller than the corrugations observed when the laser was on. Secondly, the residual noise shows no correlation with the sample topography. Taken together, these results demonstrate that even the atomic-scale corrugation of the photovoltage observed in figure 3 is a real effect, and that the SPV signal is derived *entirely* from the non-equilibrium electronic state populations induced by the optical excitation.

From figure 3, and by statistical analysis of a large number of such images, we find that more than 95% of all defects strongly decrease the photovoltage over distances of $\simeq$ 20Å. This distance is somewhat smaller than the electrostatic screening length of $\simeq$ 80Å calculated from the bulk doping. However, we note that surface segregation of the boron dopant together with the excess carriers generated by our photoexcitation are both expected to reduce this screening length.

The SPV is also dependent on the illumination intensity. In figure 4, we show that dependence of the SPV on the illumination intensity, measured on a well-ordered region of Si(111)-(7X7). Also shown is a two-parameter fit to a function of the form $V = A \log(1 + BI)$, where I is the illumination intensity and A and B are constants of the fit. The experimental data is

clearly in excellent agreement with this functional form; the rationale for
the function will be discussed below.

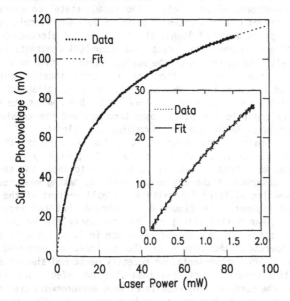

Figure 4. Dependence of surface photovoltage on illumination intensity, on
a well-ordered region of Si(111)-(7X7).

DISCUSSION

We note that the observed photovoltage is a steady-state voltage determined
both by the rate at which carriers drift toward the surface (which is determined in turn by the band-bending) and the rate at which the electron-hole
pairs recombine at the surface. As a result, the reduced photovoltage observed near defects can have two origins. First, it could arise from a
surface location where the band-bending is small, so that the electric field
in the subsurface space-charge region is small. Second, it could arise from
an increased recombination rate in the vicinity of the defect.

Information regarding the origin of this contrast can be obtained by
measuring the functional dependence of the SPV on the illumination intensity.
Previous studies[7, 8] of surface recombination in macroscopic samples have
shown that in the low-injection limit (i.e., low laser intensities), the
photovoltage on p-type material can be expressed as: $V_{SPV} = A \log(1 + \delta n/n)$,

where the constant A is determined by the local band-bending and δn and n represent the concentration of photo-excited electrons and the total concentration of electrons, respectively. The steady-state concentration of photoexcited electrons can be written in terms of bulk macroscopic constants such as the absorption depth of light, the efficiency of electron-hole pair production, the local recombination rate, and other bulk constants.[7, 8] Since most of these (with the exception of the surface recombination velocity) are expected to be spatially uniform, they can be grouped together into a new constant, so that the functional dependence of the photovoltage on the illumination intensity can be written as: $V = A \log(1 + BI)$, where the constant A is essentially equal to the static band-bending, and the constant B is inversely proportional to the surface recombination velocity.

In the above equation, the constant "A" primarily reflects the static band-bending on the surface, and the constant "B" primarily reflects the local rate of recombination. Thus, it is possible to determine the origin of the contrast observed in the SPV measurements by making measurements of the SPV as a function of laser intensity in "ideal" regions of the surface and in defected regions. In figure 4, we showed that our experimental measurements on clean Si(111)-(7X7) can be accurately fit by this simple function over more than three order of magnitude in illumination intensity. In order to understand the nature of the contrast we observed in the spatially-resolved SPV measurements, we recently performed similar measurements with spatial resolution of less than 15 Å on defects and on nearly ideal regions of the surface. The results of such measurements are presented in another publication, [9], but are summarized here. Briefly, those measurements show that on on the Si(111)-(7X7) region, the SPV varies according to the log function at low intensities, but at higher intensities a saturation of the SPV is evident. This is in stark contrast to the behavior observed at defects; here, the SPV is typically much lower, and, more importantly, shows no sign of saturating as the illumination intensity is increased further.

The observation that the defect shows only a small SPV which does not saturate with increased intensity is only consistent with the conclusion that the defect acts as an efficient recombination center for electron-hole pairs. If the defect was simply characterized by a smaller band-bending, we would expect that it would show a smaller SPV effect, but would also saturate at comparatively low intensities.

One of the rather unique features of the clean Si(111)-(7X7) surface is that the density of surface states is essentially continuous between the valence and conduction band edges. In general, such a metallic surface would be expected to give rise to a high recombination rate, since electrons in the conduction band can easily cascade down through the surface states, back into the valence band. Recombination mediated by a metallic surface is therefore expected to be very efficient, as multiphonon processes are not required. Our results show that defects such as (7X7) domain boundaries, as well as oxidation-induced defects, which *reduce* the density of states within the gap, actually lead to an *increase* in the recombination rate. Clearly, a microscopic understanding of recombination processes requires a greater understanding of the detailed nature of the surface states than has been available previously.

In the specific case of the Si(111)-(7X7) surface, previous studies with the scanning tunneling microscope [5] and with photoemission spectroscopy [10] find that the state lying closest to mid-gap on Si(111)-(7X7) is accurately described as a two-dimensional Tamm state. The photoexcited carriers, on the other hand, are primarily in the three-dimensional bulk conduction band. For recombination to take place via the surface states, the electron must somehow get from the three-dimensional conduction band into this two-dimensional surface state. In order to conserve momentum, this must involve a scattering process. At localized surface defects, the two-dimensional nature of the surface states is destroyed, effectively coupling surface and bulk states. Our results indicate that the role of atomic-sized defects in the recombination dynamics is primarily to act as scattering centers, allowing the electrons to effectively couple from the bulk states into the continuum of two-dimensional surface states.

CONCLUSIONS

The combination of optical excitation with scanning tunneling microscopy provides a new method for probing the *non-equilibrium* electronic properties of surfaces with high spatial resolution. Using the scanning tunneling microscope tip as an atomic-scale potentiometer, as it possible to probe the steady-state buildup of carriers in the space-charge region, thereby obtaining new information on recombination dynamics in surface states. On clean Si(111)-(7X7) surfaces, our results demonstrate that defects increase the rate of electron recombination by providing a means for coupling between bulk and surface states.

REFERENCES

[1] W.Kuhlmann and M. Henzler, Surf. Sci. **99**, 45 (1980).

[2] J. Lagowski, C.L. Balestra, and H.C. Gatos, Surf. Sci. **29**, 213 (1972).

[3] L.J. Brillson and D.W. Kruger, Surf. Sci. **102**, 518 (1981).

OBSERVATIONS OF SURFACES BY ELECTRON MICROSCOPY DURING STM OPERATION.

J.C.H.Spence , U. Knipping, R. Norton, W. Lo and M. Kuwabara*.
Dept. of Physics
Arizona State University
Tempe AZ 85287

*I.B.M. James Watson Lab.
Yorktown Heights
New York 10598.

ABSTRACT.

A scanning tunnelling microscope is described which operates inside a transmission electron microscope in the reflection mode. The device is used to study the mechanism of STM contrast in graphite and semiconductors. It allows for the observation of any strain during tunnelling, using the reflection electron diffraction contrast mechanism. The first results of our new transputer-based digital imaging system for STM are reported.

INTRODUCTION.

Our understanding of the STM imaging mechanism would be greatly improved if an independent imaging method were available to observe the sample surface during STM operation. An STM has therefore been constructed for operation inside a transmission electron microscope, as shown in figure 1. The STM uses a tube type scanner [1,2] and is in the

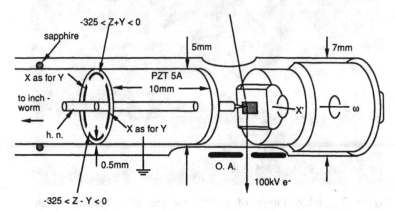

Figure 1. Diagram of the apparatus, perhaps the smallest STM yet built. Here O.A. is the objective aperature of the electron microscope, h.n. is a hypodermic needle. The electron beam path is also shown.

form of a side-entry holder for the Philips EM400 TEM. It is controlled by a digital electronics system [3]. Reflection electron microscope (REM) images may be obtained of the region under the tip during tunnelling from cleaved bulk samples in a vacuum of about 10^{-7} Torr. The holder offers control of one tilt axis during operation while the second axis must be preset for a particular diffraction setting since the diffracted beam used for imaging must be aligned with the optic axis of the electron microscope. The high mechanical and thermal stability of the electron microscope stage contributes to the STM performance. The spatial resolution of the REM technique is limited to about 1nm in the direction transverse to the beam and the images are foreshortened by a factor of about 40 in the beam direction. RHEED patterns can also be obtained from the same sub-micron region. A dark-field (non-scanning) diffraction-contrast reflection electron microscope (REM) image may be obtained (as shown in figure 2) simultaneously with any STM images. These REM images are extremely sensitive to elastic or plastic strain [4]. Video recording of the REM images is used to provide fast recording. In this way the sequence of events involved in tunnelling as well as any resultant elastic or inelastic deformation of the surface may be recorded.

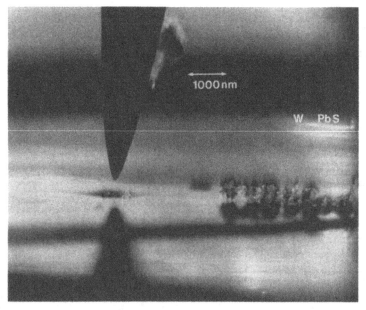

Figure 2. REM image of a PbS surface after collision with a tungsten tip. The shadow image of the tip is at top while the mirror image is at bottom. Surface steps formed at the time of cleavage are also visible.

This system is currently in the final stages of development and evaluation. In this paper we summarise earlier work, describe our digital imaging system and report the first results from it, and present some preliminary results of experiments aimed at detecting strain due to surface forces between tip and substrate.

THE DIGITAL IMAGING SYSTEM.

Our image aquisition and processing system has been designed and built at ASU together with all necessary software. The machine-language code for the transputers used was written in OCCAM by U. Knipping. The input analog servo loop is under digital control, and an OPA128 pre-amplifier has been used as the first stage. The tunnelling signal is digitised by a 16 bit A/D converter with a 20 microsecond minimum sampling time, providing a 50 kHz bandwidth. Image aquisition and processing is handled by four 32 bit Inmos Transputers which allow parallel processing and operate independently at 20 MHz. These are linked by fast serial connections. Data aquisition is handled by a T414 transputer with 1 Mb of DRAM memory. Data processing, filtering, three-dimensional image rotations, etc. are handled by a 2.5 Megaflop T800 transputer, also with 1 Mb of DRAM. The images are stored on an 800 Mbyte write once, read many times optical disc (WORM (Maxstore)), with another T414 and 1 Mb DRAM responsible for data storage. The final T414 is used to control the graphics display on a flat screen color monitor. Input commands are supplied from a keyboard under the control of a 6809 8-bit microprocessor, using menus on a second black and white monitor. Grey-scale and three-dimensional image displays are available, together with various filtering and smoothing functions. A logarithmic amplifier is used, and images are obtained in the constant current mode

Figure 3 shows two successive STM images obtained with the new digital electronics of an indentation resulting from a tip collision with a PbS[100] surface, recorded in air. Each recording occupied four minutes. The main features of the images are reproduced and the drift rate is found to be about 0.7 nm per minute.

RESULTS.

The instrument has been built for two purposes (i) to study the elastic deformation processes which may accompany tunnelling in STM [5,6], and (ii) to study the strain-field associated with van der Waals surface forces from a tip in the absence of bias or tunnelling [7, 8]. In previous work on PbS using a tungsten tip, an REM image of the gap during tunnelling had been obtained [9], and on InP using a Pt/Ir tip, images of the strain-field associated with tunnelling, probably through a thin insulating layer of hydrocarbon contamination, were taken [10]. The elastic strain-field associated with tunnelling is seen to extend over hundreds of nanometers. These REM images are sensitive to atomic displacements of a small fraction of an Angstrom. This strain is not

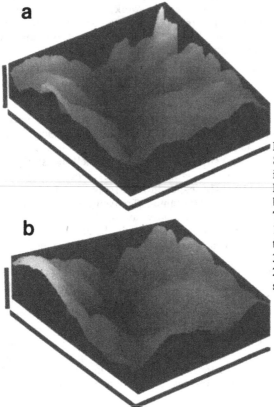

a

b

Figure 3. Successive STM images of an indentation on a PbS[100] surface caused by a W-tip collision. The scanned area is 110nm X 110nm, while the height bar denotes 37nm. Successive 256 X 256 point, four minute scans are shown.

always seen, however, perhaps due to differences in surface cleanliness in the electron microscope (the electron microscope vacuum is about 10^{-7} torr). The REM images are not sensitive to the presence of a few atomic layers of amorphous material. The observed dependence of strain on bias setting (higher bias presumably resulting in a larger "gap" and, hence, less strain) is consistent with this picture [10]. A tight-binding theory of pressure-dependent resonant tunnelling through an insulating layer has recently been proposed [11] which may account for these observations. Comparisons of the observed strain-field with calculated REM images are planned in order to determine the magnitude and sign of the strain (dilation or compression). If tunnelling is assumed to occur through a thin insulating layer, rough estimates of the pressure in the gap may be made using a modification to Hertz's solution for the point indentation problem in elasticity [12]. These pressures may approach a few GPa, and so modify the electronic structure [11].

We have not yet suceeded in obtaining REM images of graphite during tunnelling. The inelastic indentation damage caused by tip crashes has been imaged simultaneously by REM and STM, and the effect of the tip in generating crystallographic surface steps on [110] surfaces of InP and [100] surfaces of PbS after gentler crashes is commonly observed. A frequent observation is the attachment of a fragment of the sample to the tip so that tunnelling actually occurs between like materials, with obvious implications for work-function measurements and spectroscopy. Figure 4 shows two successive REM images of a tungsten tip to which debris is attached (following an earlier crash) during tunnelling at 1 nA on PbS (tip bias -200mV, tip negative). The lower tip image is a mirror image resulting from the Bragg reflection process. Estimates of the size of the gap from REM images such as this must allow for the Goos-Hanchen effect which typically produces a gap of about 2nm.[9]. It may be estimated from any small opaque object lying on the surface near the tip. The contribution to the tunnelling current from the illuminating 100kV electron microscope beam must also be considered, and may be allowed for (it is typically about one tenth of the tunnelling current and extremely stable, independent of tip position) [9]. The contribution to tip-surface forces of charge deposited by the electron beam on the tip and surface may be isolated by varying the tip bias and the electron beam intensity.

Our most recent work has concerned comparisons of work-function measurements from regions showing strain during tunnelling with those that do not. We are also attempting to obtain REM images of graphite and and to image the surface strain resulting from van der Waals forces between tip and sample without bias or tunnelling current. Figure 5 shows an REM image obtained at 100 kV using the (022) specular beam reflected from the [01-1] surface of GaAs. A W tip is seen at a large distance from the surface in figure 5(a). The tip and sample are electrically grounded. In figure 5(b), the tip has been brought to within 16 nm of the surface, however electrical measurements show that no contact has been made. Despite no other changes in experimental conditions, the image contrast is seen to change in the region under the tip, suggesting strain. Experimental and theoretical work [13] indicates

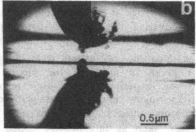

0.5μm

Figure 4. Fragments of a PbS substrate attached to a tungsten tip. Tunnelling is occuring from the fragments to the substrate. The fragments are loosely attached causing the tunnel current to be unstable

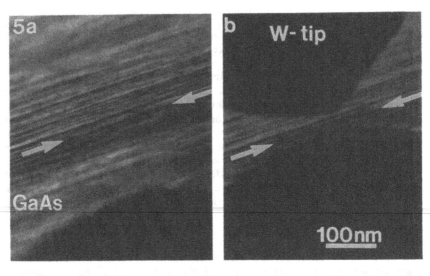

Figure 5(a). This shows a fresh tungsten tip approaching a heavily stepped cleaved GaAs (01-1) surface. In figure 5(b) the tip-surface gap has been reduced to 16 nm, without prior contact. The arrows indicate the same region on both pictures.

that appreciable forces (in the nN range) can be expected over distances of a few nanometers (for gold) or larger (for graphite) [8] between an AFM tip and a surface. In our experiments, at least three types of forces must be considered which may produce surface tractions. These are (a) Van der Waals forces. (b) Forces due to any tip bias and tip-surface capacitance. (c) Forces between charge deposited by the electron beam in any hydrocarbon contamination layers on the tip and/or sample. Image forces due to any induced charge and accounting for the dielectric constants of the materials must be considered. Any surface free charge σ generates a surface field whose normal component is σ/ε_o, to which the material will respond elastically. For example, if it were assumed that a layer of charge q is deposited by the electron beam onto a hydrocarbon layer covering the tip, then the total attractive electrostatic image force between a spherical tip and an uncharged conducting plane can be shown to be $5.8q^2/d^2$ nN, where d is the tip-surface distance in Angstroms and q is given in units of the electronic charge (q=1 for one electron). However, experience from high resolution electron microscopy indicates that specimen charging produces clearly observable effects on electron microscope images which were not seen in this work. For conducting samples, such as this heavily doped GaAs (10^{19} dopant atoms/cc) and a W tip, REM images are not visibly affected by charging effects from the beam. This suggests that electrostatic forces are small in our experiments, and that the strain observed while tunnelling occurs because of tip contact through a layer of contamination. To estimate the importance of beam-induced charge, experiments are in progress to measure the tip charge directly using an electrometer. The sign (dilation or compression) of any strain in the surface layer has yet to be determined. Detailed contrast calculations and further experiments will be required to determine the dependence (if any) of the REM image features on electron beam current, tip bias and distance between tip and surface. We emphasise the effects of foreshortening on these images, so that very small changes in the images may correspond to large effects along the beam direction. Foreshortening can be reduced through the use of higher order beams for imaging at a cost of image intensity.

CONCLUSIONS

These preliminary results show that the REM method, combined with STM, provides a very useful combination of techniques for the study of strain caused by tunnelling through insulating surface layers. STM imaging in air, liquids, or conventional vacuum is believed to occur by this mechanism. This technique may also be useful for clarifying the imaging mechanism in graphite. Quantitative results await detailed calculations of REM contrast changes due to sharp tips, which are planned. These can then be correlated with experimental measurements from REM images. We find that in some cases an observable elastic strain-field is seen in the electron microscope accompanying tunnelling through contamination. This may extend over many hundreds of nanometers [10]. The most important experimental problems to be resolved are the crystallographic alignments of the sample with the optic axis and beam of the electron microscope. These are needed to obtain greater resolution and intensity in the REM images.

ACKNOWLEDGMENT.

This work was supported by the N.S.F. National center for High Resolution Electron Microscopy at Arizona State University, and by N.S.F. grant number DMR 88-13879

REFERENCES.

1. G. Binnig and D.P.E. Smith. Rev. Sci. Instruments. **57**, 1688 (1986)
2. J.C.H.Spence, Ultramicros. **25**, 165 (1988).
3. Designed and constructed by U. Knipping and R. Norton.
4. T. Hsu and J.M. Cowley, Ultramicros. **11**, p.293 (1983).
5. R.J.Colton, J.Vac. Sci. Tech. **A6**, p.349 (1988)
6. I.P. Batra and S. Ciraci. J. Vac. Sci. Tech. **A6**, p.313 (1988)
7. D. Tabor and R.H.S. Winterton. Proc. Roy. Soc. **A312**, p.435 (1969).
8. N. Burnham and R. J. Colton. J. Vac. Sci. Tech. **A7**, p.2906 (1989).
9. M. Kuwabara, W. Lo and J.C.H.Spence . J. Vac. Sci Tech **A7**, p. 2745 (1989)

10. M. Kuwabara, W. Lo and J.C.H.Spence. Proc. 47th Ann. EMSA meeting. 1989. G. Bailey Ed. San Francisco Press, San Francisco.
11. S.M.Lindsay, O.F. Sankey, Y. Li , C. Herbst and A. Rupprecht. (1989) J. Phys. Chem. submitted.
12. I.F. Sneddon. Int. J. Eng. Sci. **3**, 47 (1965).
13. Intermolecular and surface forces. J.N.Israelachvili. Academic Press. 1985.

IMAGING BY SLIDING PLANES IN SCANNING TUNNELLING MICROSCOPY

John D. Todd and John B. Pethica
Department of Metallurgy and Science of Materials, Oxford University,
Parks Road, Oxford OX1 3PH, UK

ABSTRACT

Scanning tunnelling microscope images of layered materials in a non-uhv environment exhibit various anomalous phenomena, including enhanced corrugation heights, periodicity over large areas and a marked absence of point defects. We have modified a precision indentation device to allow STM rastering of a tip across a surface, while simultaneously monitoring mechanical contact. Images we have obtained from this apparatus on an HOPG sample exhibit atomic scale resolution with contact areas much larger than a single atom. Contrast in the image results from periodic conductance fluctuations as the layers of the sample undergo shear in the region of the tip. We provide a model for this process, which explains a variety of curious, and otherwise unrelated phenomena occurring during STM imaging of these materials.

INTRODUCTION

In the past few years the scanning tunnelling microscope (STM) has yielded atomic resolution images of surfaces in air and aqueous solution with astonishing ease. Published data on HOPG and other layered materials seldom fail to exhibit features on the scale of only a few Ångströms, despite the high probability of surface contamination in a non-uhv environment. Moreover, such images display curious phenomena, which the conventional picture - that of tunnelling from a single tip atom a nanometre or so above the surface [1] - fails to explain. These include an absence of point defects, the blurring of any non-periodic features or visibility of the lattice through the feature, anomalously large amplitudes of atomic-scale structure, and barrier heights much lower than the work function [2,3]. A common feature of the materials investigated is an easy shear plane parallel to the surface. In striking contrast to these results, careful STM of HOPG cleaved and heated in uhv shows no atomic-scale features [4].

The nature of these anomalies, and the context in which they occur, strongly suggest that definite contact is occurring between tip and sample. A recent study establishes the plausibility of this conclusion, by demonstrating 'atomic-scale' imaging even when the tip has been moved more than 20Å towards the surface[5]. Moreover, indirect measurement of the forces occurring during atomic-scale imaging, inferred from the displacement of a cantilever, suggests that a force of greater than 1µN is needed in order to establish a tunnelling current [6]. The important implication for STM and atomic force microscopy (AFM) is that the probe resolution in these cases is no longer atomic. A proposal that tunnelling from a single atom still occurs, despite contact over a larger area mediated by contamination [7], raises the further problem of how the single atom geometry can be stable during tip rastering .

We describe here an experiment in which contact was uneqivocally demonstrated during imaging, and develop an alternative model, suggested earlier [8], in which the conductance through a constricted region of lattice shear will fluctuate with the period of the lattice. Besides being immediately relevant to STM, the mechanism illuminates other interesting processes at this scale, such as block shear and theoretical lattice strength [9].

'ATOMIC' IMAGING BY CONDUCTION THROUGH SLIDING LATTICE PLANES

Contact between tip and sample explains several of the otherwise unrelated anomalies seen in STM. A scanning tip touching the surface must be accompanied by shear in the region of constricted, point contact, geometry. This could be accomplished, for example, by dislocation diffusion. In order to understand the geometry throughout the process, we have made a computer simulation, in which we relax a slab of atoms interacting via the Finnis-Sinclair potential for gold [10], after giving the top atomic layer a rigid shift. Figure 1 shows the results. For a small shift, the strain is elastic. Increasing the shift causes further strain until suddenly inelastic block shear occurs at one plane. Further elastic strain follows until again the upper lattice jumps across to the next lattice site.

A full treatment of the imaging process for constant height scanning would then involve an iterative solution of the Boltzmann equation, describing the total current through the shearing aperture as a sum

$$I = I_0 + I_1 + I_2 + ... \tag{1}$$

where the terms involve no scattering, one phonon processes and so on (see figure 2). I_0 is equivalent to the Sharvin resistance $4\lambda\rho/3\pi a^2$, where the mean free path, λ, is greater than the contact radius, a (the ballistic limit). The second term arises in the diffusive regime, and gives a correction of order a/λ, the Maxwell resistance, $\rho/4a$. We use a simplifying assertion to estimate the fluctuating resistance during scanning: As the upper lattice moves to the next lattice site, the scattering interface can be treated as a grain boundary which appears then vanishes again. The resistivity due to grain boundaries/boundary area per unit volume, S say, has been measured for copper [11], agreeing with a recent pseudopotential calculation [12]. The additional resistance due to a circle of grain boundary of radius a is then $S/\pi a^2$, and the contrast C in STM images will be simply $(S/\pi a^2)/I$, i.e.

$$C = 3S / 4\lambda\rho \quad \text{(ballistic regime), and} \tag{2}$$

$$4S / \pi\rho a \quad \text{(diffusive regime)} \tag{3}$$

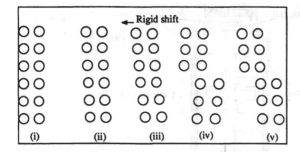
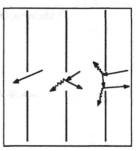

Figure 1. Circles represent atomic positions in two planes of a slab of infinite area. The slab has been relaxed from a rigid shift of the top layer. As the shift increases, shear is initially elastic (i-iii), then the upper part of the slab undergoes block shear (iv), before further elastic shear (v).

Figure 2. Ballistic transmission, single phonon scattering and two phonon scattering of an electron in the region of shear.

In the ballistic case C is independent of a and so for very small contacts the contrast should be stable. The result is a 10% fluctuation for copper. In the absence of data for HOPG, which may be influenced by intercalation, we remark that the value is close to that typically found in our constant height scanning. Equation (3) implies a contact radius of 80nm for copper to give 10% contrast. This is a considerable size, and the calculation shows that scattering is sufficient for the mechanism we propose. Next we describe an experiment to measure mechanical properties while imaging.

EXPERIMENTAL METHOD

In order to obtain unequivocal measurements of forces, compliance and hence contact area during scanning in air, we have combined a load-controlled nanoindenter together with STM sample translator and electronics. The indenter imparts forces to a precision of 100nN and is sensitive to a displacement of 0.3 Å. It is described in detail elsewhere [13]. Figure 3 shows a diagram of the apparatus used.

An electrochemically etched tungsten tip was mounted in place of the standard diamond indenter , and the rigid sample stage was replaced by a piezo-electric tube scanner, allowing horizontal movement of the sample. By means of an electromagnetic coil, fixed to the shaft below which the indenter tip was mounted, and a concentric magnet in the head unit, a force could be applied to the shaft, bringing tip and sample into contact. The shaft is connected to two leaf springs in the indenter head unit , with total stiffness of 48N/m. Movement towards the surface was halted when the instrument controller sensed a stiffness 50% greater than this value. By applying a small sinusoidal modulation to the tip loading and measuring the resulting amplitude of oscillation in displacement, typically 0.5Å, the contact compliance and hence contact area could be determined immediately [14].

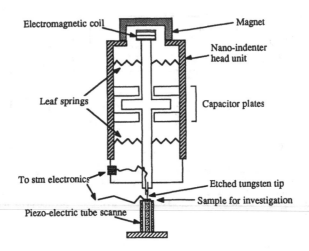

Figure 3. Nanoindenter as modified for STM imaging

A bias voltage of 1V or less was applied between the tip and sample, and the resistance of the tip-sample junction was recorded. Continuous scanning of the sample stage under the tip was achieved by applying sawtooth waveforms directly from a function generator, thus reducing noise from high voltage amplifiers and allowing continuous display of images on an oscilloscope. The entire apparatus was housed within an insulated chamber in a temperature-controlled environment, in order to reduce thermal drifts and acoustic disturbance. With these precautions it was possible to achieve stability sufficient to operate the equipment as a scanning microscope, without feedback in the vertical motion of the sample. Thus compliance could be measured simultaneously with imaging.

Initial investigations were carried out using HOPG, being the most ubiquitous material for multifarious non-uhv STM studies, and yet source of a disturbing variety of unexplained phenomena. In order to investigate the wider applicability of the model we have also begun to inspect an electropolished a slab of Ruthenium, a relatively inert metal whose hcp structure ensures a preferred direction of lattice shear.

RESULTS AND DISCUSSION

After contact the tip remained stable vertically to within a nanometre. Sometimes current appeared first (see figure 4). Figure 5 shows a typical image exhibiting resistance contrast varying on the atomic scale from an HOPG sample. The familiar six-fold symmetry is clearly visible with only 3-point smoothing to reduce 1/f noise. The ambient resistance is 1MΩ, with approximately 10% fluctuation. This image was obtained at 10Hz scan rate, and in definite contact; the phase

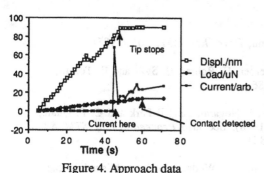

Figure 4. Approach data

Figure 5. Image of HOPG
from the Nanoindenter.

difference between load and displacement modulations had changed by a value corresponding to a stiffness greater than 40kN/m. The spacing of the periodic structure on the atomic scale is not dependent on scan speed.

Conductance over an extended area as we propose has been shown by Colton et al. [15] to be capable of giving the variety of periodic features observed on layered materials . The periodic features of the image depend on the angle of rotation of the planes. A similar explanation of different image types has been provided by Batra et al. for AFM [16]. Our results suggest that quite adequate contrast and stability are available from the sliding mechanism. Determining the corresponding height modulation from current contrast depends on the effective barrier height, but for 10% fluctuations even high barriers will give Ångström movement. The model also suggests that the lateral extent of defective features will give an idea of the contact area. Images of Au(111) in uhv suggest that contact may be only a few atoms across. Thus the process reveals novel and interesting mechanical properties at this scale.

ACKNOWLEDGEMENTS

We thank Dr. Adrian Sutton for providing the code for the Finnis-Sinclair relaxation, and Professor Sir Peter Hirsh F.R.S. for laboratory facilities.

REFERENCES

1. J. Tersoff and D. R. Hamann, *Phys. Rev.* **B31**, 805 (1985)

2. G. Binnig, H. Fuchs, Ch. Gerber, H. Rohrer, E. Stoll and E. Tosatti, *Europhys. Lett.* **1**, 31 (1986)

3. T. R. Albrecht, H. A. Mizes, J. Nogami, Sang-il Park and C. F. Quate, *Appl. Phys. Lett.* **52**, 362(1988). See also *Proceedings of STM 88* (Oxford) Microsc.**152**(1988)

4. D. Lawunmi, M. C. Payne and M. Welland, to be published in the *Proceedings of the1989 European Conference on Advanced Materials and Processes.*

5. D. P. E. Smith, G. Binnig and C. F. Quate, *Appl. Phys. Lett.* **49**, 1166 (1986)

6. C. M. Mate, R. Erlandsson, G. McClelland and S. Chiang, *Surf. Sci.* **208**, 473 (1989)

7. H. J. Mamin, E. Ganz, D. W. Abraham, R. E. Thompson and J. Clarke *Phys. Rev.* **B34**, 9015 (1986)

8. J. B. Pethica, *Phys. Rev. Lett.* **57**, 3235 (1986)

9. A. Kelly, *Strong Solids* , Clarendon Press, Oxford, 1973

10. G. J. Ackland, G. Tichy, V. Vitek and M. W. Finnis, *Phil. Mag.* **56**, 735 (1987)

11. P. V. Andrews, M. B. West and C. R. Robeson, *Phil. Mag.* **19**, 887 (1969)

12. G. Lormand, *J. Physique Colloque* **C6**, 283 (1982)

13. J. B. Pethica and W. C. Oliver, *Physica Scripta* **T19**, 61 (1987)

14. J. B. Pethica and W. C. Oliver, *Mat. Res. Soc. Symp. Proc.* **130**, 13 (1989)

15. R. J. Colton, S. M. Baker, R. J. Driscoll, M. G. Youngquist, J. D. Baldeschwieler and W. J. Kaiser, *J. Vac. Sci. Technol.* A**6**(2), 349 (1988)

16. F. F. Abraham and I. P. Batra, *Surf. Sci.* **209**, L125 (1989)

ROLE OF TIP MATERIAL IN SCANNING
TUNNELING MICROSCOPY

C. JULIAN CHEN

IBM T. J. Watson Research Center, Yorktown Heights, NY 10598, USA

ABSTRACT

In this paper, we show that atomic resolution in scanning tunneling microscopy (STM) originates from p_z or d_{z^2} states on the tip. Consequently, only a limited selection of tip materials can provide atomic resolution: d-band metals, for example, Pt, Ir, Pd, Rh, W, Mo; semiconductors that tend to form p-like dangling bonds, for example, Si.

In the early models of STM, the tip is conceived as a macroscopic body of metal, whose material nature, or elemental composition, is not important.[1] For example, the s-wave model describes the tip as a potential well with a hemispherical end of radius R. On metal samples, this leads to the notion that the tunneling current is proportional to the charge density of the sample surface at the center of curvature of the tip.[1] As a consequence, it predicts that individual atoms on low Miller index surfaces are not detectable. Instead, only profiles of superstructures with length scale greater than 6 Å can be resolved.[1] On the other hand, STM has repeatedly resolved individual atoms on many metal surfaces, including the "smoothest" ones, such as Au(111), Al(111), and Cu(100), which have nearest-neighbor atomic distances 2.5-2.9 Å (Table 1). Those experimental results are in sharp contradiction to the predictions of the s-wave theory. Some authors propose models of STM imaging based on *mechanical interactions* between tip and sample.[4,6] None of these are able to provide a consistent explanation of all the observed phenomena.[6,7]

Table 1. Atomic resolution observed on metal surfaces by STM

surface	atomic spacing	method	corrugation (current-variation)	reference
Au(111)	2.87 Å	current	(10 %)	2
Au(111)	2.87 Å	topographic	0.15 Å	3
Al(111)	2.88 Å	topographic	0.1 - 0.8 Å	4
Cu(100)	2.55 Å	topographic	0.2 Å	5

The s-wave model predicts no atomic resolution on metals because it conceives the tip as a *macroscopic continuum*.[1] Consequently, the image predicted with such tip resembles a macroscopic continuum. Clearly, to account for the observed atomic resolution, the conception of the tip should be "exact", i. e., atomic, as well. In other words, the *actual* tip states have to be considered.[8 – 11] Experimentally, there are only a few metals have been used as tip material,[9] notably, W, Pt, Ir. These metals are well-known d-band metals, i. e., more than 90% of their density of states near the Fermi level is from the d-electrons.[12] Thus, under these experimental conditions, *there is virtually no atomic s-wave states on the tip.* Interestingly enough, these d-states provide an adequate explanation of the observed atomic resolution.

Take an example, tungsten. It has a well-known tendency to form d_{z^2} dangling bonds on surfaces.[12] To study the imaging process with a tungsten tip, Ohnishi *et al.* performed first-principle calculations of the electronic states of tungsten clusters. [10] They found that at the apex atom of either W_4 or W_5, there is a localized metallic d_{z^2} state very close to the Fermi level. Using Green's function methods, they also show that the tunneling conductance is predominately contributed by that d_{z^2} tip state. [10] We will show that such d_{z^2} tip states can allow greater atomic detail on the sample to be observed than that from the charge-density contour.

First, we make a qualitative explanation in terms of the *reciprocity principle* in STM: upon interchanging the tip state and the sample state, the image should be identical.[11] For free-electron metals, it is well known that the surface charge-density, as well as the tunneling current for an s-wave tip, can be represented with reasonable accuracy as a sum of the charge densities of individual atoms, each made of s-states (*atomic charge superposition*).[1] As shown in Fig. 1, with a d_{z^2} tip state, the center of the apex atom no longer traces the charge-density contour of the sample. Instead, it traces the charge-density contour of a *fictitious surface* with d_{z^2} states on each atom. Since the charge-density distribution of a d_{z^2} state is much narrower than that of an s-state, the peaks at individual atoms become much more pronounced.

Figure 1. An intuitive picture of the enhanced atomic resolution due to a d_{z^2} tip state, in light of the reciprocity principle of STM.[11]

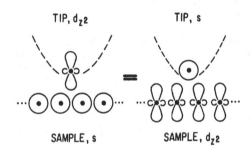

TIP, d_{z^2} TIP, s

SAMPLE, s   SAMPLE, d_{z^2}

To make the argument more quantitative, we will use the *derivative rule[11]* to show that a d_{z^2} state acts as a *quartic high-pass filter* to the fine features of the image. It has been proved that for a p_z tip state, the tunneling matrix element is proportional to the first derivative of the sample wavefunction, $M \propto |\partial\psi/\partial z|$; and for a d_{z^2} tip state, the tunneling matrix element is proportional to the second derivative of the sample wavefunction,[11]

$$M \propto |\partial^2\psi/\partial z^2| .$$

The p and d tip states generate substantial tunneling-current enhancements to atomic-size features. Consider a Fourier component of surface wavefunction[1] with transverse wavevector $\mathbf{k}$,

$$\psi(\mathbf{k}) = C(\mathbf{k}) \exp\{ - (\kappa^2 + \mathbf{k}^2)^{1/2} z\} \exp(i\mathbf{k} \bullet \mathbf{x}),$$

where $\mathbf{x} = (x,y)$, and κ is the decay constant,[1] related to the workfunction ϕ by $\kappa = (2m_e\phi)^{1/2}/\hbar$. A typical value is $\kappa \simeq 1$ Å^{-1}. It follows immediately that a p_z tip state generates a matrix element proportional to $(1 + \mathbf{k}^2/\kappa^2)^{1/2} \psi(\mathbf{k})$, and a d_{z^2} tip state generates $(1 + \mathbf{k}^2/\kappa^2) \, \psi(\mathbf{k})$. The enhancement of tunneling matrix elements due to the *prefactors* is shown in Fig. 2. The tunneling current is proportional to the square of the matrix element. Therefore, for small features, i. e., for Fourier components with $|\mathbf{k}| > \kappa$, the p_z tip state acts as a *quadratic* high-pass filter, whereas a d_{z^2} tip state acts as a *quartic* high-pass filter. Notice that the condition $|\mathbf{k}| > \kappa$ means the length scale of the feature is smaller than $2\pi/\kappa \simeq 6$ Å, which is just the criterion of being truly atomic. With this derivative rule, quantitative predictions to the images can be made.

Figure 2. Enhance-ment of tunneling matrix elements by p and d tip states. The enhancement of tunneling current is the square of the enhancement of the matrix element.

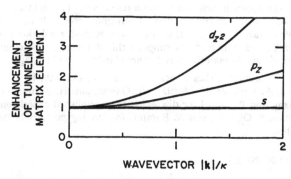

Using the atomic-charge superposition method,[1] the charge distribution of metal surfaces with different atomic spacings can be calculated, as well as the theore-tical STM image. Figure 3 shows some typical results with s, p_z, and d_{z^2} tip states. The variation of *corrugation amplitude* with tip-sample distance is shown. The details of numerical calculations will be published elsewhere.

Figure 3. Theoretical corrugation amplitude for different tip states. For metals, with atomic spacing 3 Å, with a d_{z^2} tip state, large corrugation should be observed at a short tip-sample separation, whereas the corrugation of charge density , i. e., of the image from an s-wave tip, is too small to be detectable. For superstructures with periodicity 7.5 Å, the p- and d-tip states generate corrugation enhance-ment as well, although smaller.

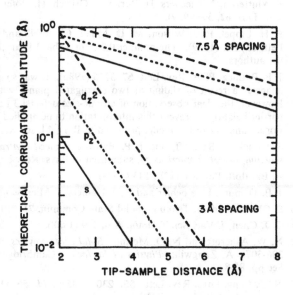

It is not surprising that Lang's numerical simulation of STM images predicts no true atomic resolution.[13] In that model, a Na atom is used to represent the tip, and the simulated image of a metal atom on sample surface appears to be a pancakelike bump with diameter 9 Å and maximum height 1.6 Å. Consequently, for surfaces with nearest-neighbor atomic distance smaller than 9 Å, that model predicts no atomic resolution. Actually, the Na atom has s-valence electrons only, and in reality has never been used as tip material.

In conclusion, the material nature or the elemental composition of the STM tip is a crucial factor in achieving atomic resolution. The right tip element should be either a d-band metal or certain element semiconductor. If the end of the tip is made of noble metals (Cu, Ag, Au) or alkali metals, high resolution cannot be achieved. Although all existing evidence supports this statement, a systematic study of STM resolution versus tip material may further clarify the STM imaging mechanism.

The author wishes to acknowledge J. E. Demuth, R. J. Hamers, J. Tersoff, I. P. Batra, A. Baratoff, V. Moruzzi, N. Garcia, and H. Rohrer for helpful discussions, S. Chiang and P. Lippel for discussion of unpublished experimental details as well as J. Pethica, S. Ohnishi, and A. Baratoff for sending me unpublished material and manuscripts.

REFERENCES

1. J. Tersoff and D. R. Hamann, Phys. Rev. Lett. **50,** 1998 (1983); Phys. Rev. B **31,** 805 (1985); J. Tersoff, Phys. Rev. B **39,** 1052 (1989).

2. V. M. Hallmark, S. Chiang, J. F. Rabolt, J. D. Swalen, and R. J. Wilson, Phys. Rev. Lett. **59,** 2879 (1987);

3. Ch. Wöll, S. Chiang, R. J. Wilson, and P. H. Lippel, Phys. Rev. B **39,** 7988 (1989).

4. J. Wintterlin, J. Wiechers, H. Burne, T. Gritsch, H. Höfer, and R. J. Behm, Phys. Rev. Lett. **62,** 59 (1989).

5. P. H. Lippel, R. J. Wilson, M. D. Miller, Ch. Wöll, and S. Chiang, Phys. Rev. Lett. **62,** 171 (1989). The number 0.2 Å in Table 1 is an private information from the authors.

6. J. B. Pathica, Phys. Rev. Lett. **57,** 3235 (1986). It was suggested that the observed image is a result of sliding of two corrugated planes with the same periodicity. However, the clear observation of atomic-size defects[3] and the 3.5 eV apparent barrier height[4] prevent this interpretation to be applied on those metal surfaces. For an analysis of the model proposed in Ref [4], see Ref. [7].

7. S. Ciraci, A. Baratoff, and I. P. Batra, *Enhanced contrast and atomic forces in scanning tunneling microscopy,* submitted to Phys. Rev. Lett.

8. A. Baratoff, Physica **127B,** 143 (1984).

9. J. E. Demuth, U. Koehler and R. J. Hamers, J. Microscopy **151,** 299 (1988).

10. S. Ohnishi and M. Tsukuda, Solid State Commun. **71,** 391 (1989).

11. C. J. Chen, J. Vac. Sci. Technol. A **6,** 319 (1988).

12. N. W. Ashcroft and N. D. Mermin, *Solid State Physics,* Saunders, 1976. See pp. 306-307. A. Zangwill, *Physics at Surfaces,* Cambridge University Press, 1988. See pp. 82-83.

13. N. D. Lang, Phys. Rev. Lett. **55,** 230 (1985); *ibid.,* **56,** 1164 (1986).

STM Of Gold On Graphite

P. A. THOMAS[+], W. H. LEE, R. I. MASEL[*]
University of Illinois, 1209 W. California St. Urbana Il, 61801

ABSTRACT

Scanning Tunnelling Microscopy was used to examine the surface structure of gold films and supported gold particles. It was found that freshly evaporated gold films have a relatively smooth surface morphology. However, after 1 hr of annealing, a series of dendrites were clearly visible on the surface of the films. With additional annealing, the dendrites got larger, until the film broke into particles. STM images of annealed gold particles showed that annealed gold particles larger than 150 Å have a polyhederal shape and a smooth surface morphology. However, the surfaces of the particles smaller than 60 Å remained highly corrugated even after a long anneal. We believe that the difference in morphology of large and small particles may have some profound implications in supported metal catalysis.

INTRODUCTION

The surface structure of supported metal catalysts is quite important to the design of the catalysts. At present, there is considerable information about the structure of the metal particles in supported metal catalysts from high resolution transmission electron microscopy (TEM). However, many of the details of the surface structure are missing. The objective of the work reported here is to try to use scanning tunnelling microscopy (STM) to provide structural information which compliments the structural information obtained via TEM.

EXPERIMENTAL

Our experiment was to evaporate or sputter a small amount of gold onto a cleaved graphite or mica substrate, anneal the sample at high temperature, rapidly cool, then examine the sample with the STM.

Gold on graphite samples were prepared in a small turbopumped vacuum system with a base pressure of about 1×10^{-9} torr. Freshly cleaved HOPG graphite substrates were loaded onto a transfer rod, and then transferred into the vacuum system. A small amount of gold was then evaporated onto the substrate. The substrate was then heated to 600° C for 18 hours in vacuum, then rapidly cooled. The sample was then removed from the vacuum system and examined with the STM.

We have also done work with gold/palladium films. A cleaved mica substrate was sputter coated with a gold/palladium alloy using a Polaron EM5400 SEM coating system. They were then either used as is, or heated at 300 C in air for various times, and then cooled to room temperature and loaded in the STM.

All of the STM work was done with a homemade STM based on the previous design of Lyding, et al.[1]. One should refer to Thomas[2] for details.

[+] Current Address: Zip 40 E, IBM, Rt 52, Hopewell Junction Ny 12533

[*] Send Correspondence to this author

RESULTS

Figures 1-4 show some of the results we have obtained. Unfortunately, the length limitations of this paper make it difficult for us to discuss these images in detail. However, basically we find that when we first deposit the films, they look as shown in figure 1. The films are relatively smooth, with large scale undulations, but few features on the 20-100 Å length scale. When we anneal the samples, we observe irregular oscillations in the surface structure as though dendrites are growing on the surface as shown in figure 2.

The films break into particles upon further annealing. We do not have good images of the particles at the metal loadings in the film data. However, at lower metal loadings, we can image individual particles. Generally, we have found that when we deposit gold onto cleaved HOPG graphite substrates then anneal, we get rows of particles which line up along steps on the graphite surface. The areas between the steps appear to be devoid of particles. The rows of particles are also visible with a scanning electron microscope (SEM).

Figures 3 and 4 show images of some of the individual particles which we have observed. We have found that after annealing, the particles larger than 150 Å seem to have a relatively smooth surface structure shown in figure 3. Contour plots of the particles show that they are polyhedral. They also appear to have rounded edges. However, this may be an artifact of the STM. As the particles get smaller, however, the particle shapes change. We observe what appears to be a series of waves in the particles surface and the particle no longer has a polyhedral shape. The smallest particles look as shown in figure 4. All of our images of gold particles smaller than about 60 Å look very rough as though the particles are composed of a series of individual gold clusters which are only loosely held together. Note however, that these highly corrugated structures grow spontaneously upon annealing. We have not observed them in our images of freshly deposited gold.

DISCUSSION

When we first obtained the images in figure 4, we wondered whether the corrugated structures in our images were some sort of artifact. One could produce images like those in figure 4 because of ghost images associated with tunneling from multiple points on our STM tip. One also has to consider the possibility that the corrugations were somehow associated with a feature on the graphite substrate.

Fortunately, ghost images associated with tunneling from more than one point on the tip are easy to detect in experiments like ours. A ghost is essentially an image of the tip projected onto the particle. As a result, it will appear as a feature which is seen in the same place on the images of many equivalent sized particles in the same scan. We do not see such characteristics in the images used to generate the data in figure 4. We also have obtained similar images with multiple tips; the images do not change appreciably with changing tip bias or direction. From theory, we know that still, our images must be a convolution of the tip geometry and the surface geometry of the particles. However, at present the evidence suggests that the corrugations in figure 4 are associated with some real feature on our samples rather than a feature on the tip.

The influence of the graphite substrate is more difficult to quantify. We do not observe features like those in figures 3 and 4 on our graphite substrate in the absence of gold, and the particles we observe seem to correspond to those expected from SEM images of the same samples. Thus, while we have not completed the I/V measurements we need to prove that we are observing gold particles, it seems that the images in figures 3 and 4 are associated with gold particles and not a defect on the graphite.

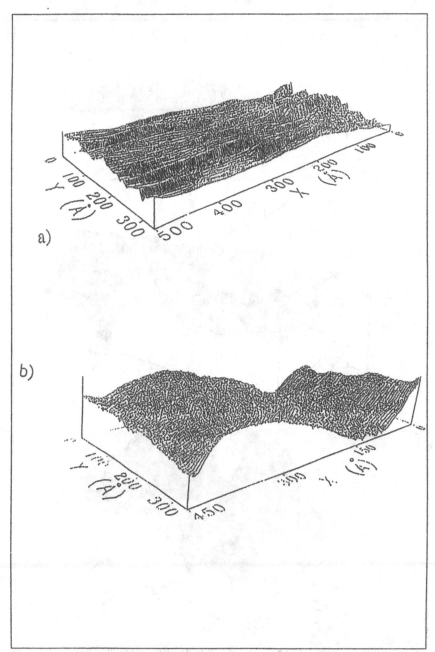

Figure 1   Constant Current STM images of a cleaved mica substrate which had been sputter coated with a gold/palladium film with a nominal thickness of a) 50 Å b) 1600 Å.

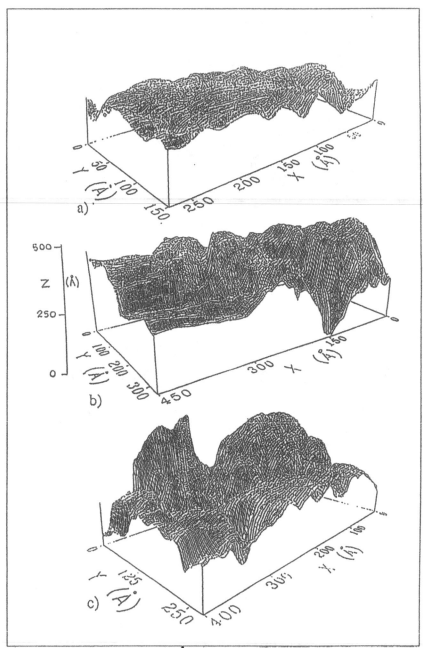

Figure 2   STM images of a 200 Å gold/palladium film which had been annealed at 300 C in air for a) 0, b) 3 hrs, c)6 hrs.

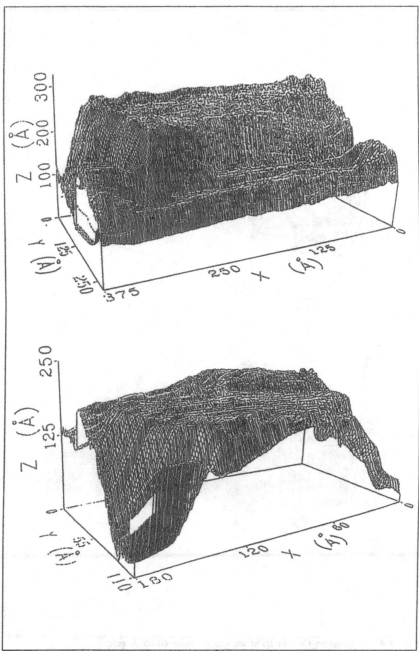

Figure 3 STM images of some 200-400 Å gold particles on graphite
substrates.

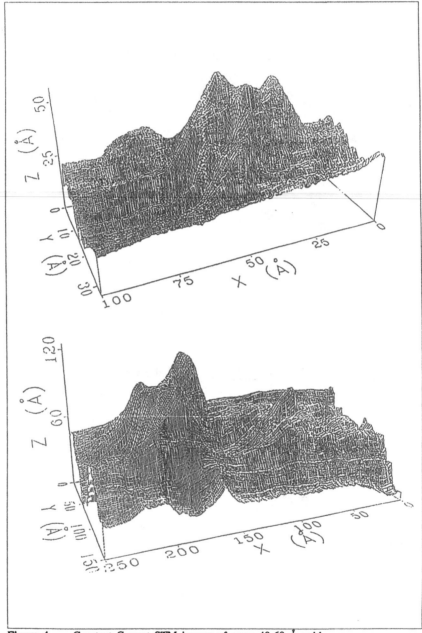

Figure 4   Constant Current STM images of some 40-60 Å gold
particles supported on graphite. Note that there are two
particles in the lower image.

Previous TEM studies of gold on graphite also support this assignment. Previous studies have shown that gold particles larger than 200 Å have a polyhederal shape similar to that in figure 3. However, polyhederal particles are unusual when the particle size is less than 100 Å. Instead, the particles are usually multiply twinned. Smith and Marks[4] have found that some of the multiply twinned particles have simple stacking faults. However, many 10-100 Å gold particles actually consist of small gold clusters strung together. Smith and Marks call such structures "poly-particles". Our images in figure 4 are consistant with Smith and Mark's atomic resolution TEM images of poly-particles. We see more of the details of the surface morphology than did Smith and Marks. However, it is not unreasonable that a poly-particle would have a surface morphology like that in figure 4.

We do not know to what extent the graphite substrate is affecting the shapes of the gold particles. We only observe gold particles which are pinned at steps on the graphite substrate. Although we do not know why the particles are pinned, it is certainly possible that the pinning process could cause gold agglomerates to form. The steps in the graphite surface could also influence the particle shape. Thus, it is unclear whether the shapes we observe are characteristic of gold, on any support, or merely of gold on these particular substrates.

Still, even if the particle shapes were influenced by the substrates it would still be important to heterogeneous catalysis. Afterall, typical catalyst supports are corrugated substrate. If the smallest metal particles in a working supported metal catalyst were like those in figure 4 the particles would be expected[3] to behave much differently than particles like those in figure 3. Thus, if the particle shapes in real catalysts were like those in figure 4, it would be important, even if the shapes were dictated by the support.

CONCLUSIONS

In summary, then, we find that freshly evaporated or sputtered films look relatively flat with the STM. A series of dendrites grow upon annealing which are easy to image with the STM. Large annealed gold particles have a polyhederal shape, with rounded edges. However, particles smaller than about 60 Å appear highly corrugated. One does have to consider image artifacts in the later pictures. However, it appears that there is a real difference between the surface morphology of gold particles smaller than 60 Å and that of gold particles larger than 150 Å.

ACKNOWLEDGMENT

This work was supported by the National Science Foundation under Grant DMR 86-12860. The SEM work was done at the University of Illinois Center for Microanalysis of Materials which is supported as a national facility, under National Science Foundation Grant DMR 86-12860.

LITERATURE CITED

1. J. W. Lyding, S. Skala, J. S. Hubacek, R. Brockenborough, G. Gammie, Rev. Sci. Instruments. 59, 1897 (1988).
2. P. A. Thomas, PhD Dissertation, University of Il., Urbana Il. (1989).
3. See for example, M. Bordart, G. Djega-Mariadassou, Kinetics of Heterogeneous Catalytic Reactions, Princeton Univ. Press, (1984).
4. D. J. Smith, L. D. Marks, J. Crystal Growth, 54, 433 (1981).

PART VII

Encapsulated Surfaces

INTERFACIAL SUPERSTRUCTURES STUDIED BY GRAZING INCIDENCE X-RAY DIFFRACTION

KOICHI AKIMOTO, JUN'ICHIRO MIZUKI, ICHIRO HIROSAWA AND
JUNJI MATSUI

Fundamental Research Laboratories, NEC Corporation
34, Miyukigaoka, Tsukuba, Ibaraki 305, Japan

ABSTRACT

Surface superstructures (reconstructed structures) have been observed by many authors. However, it is not easy to confirm that a superstructure does exist at an interface between two solid layers. The present paper reports a direct observation, by a grazing incidence x-ray diffraction technique with use of   synchrotron radiation, of superstructures at the interface. Firstly, the boron-induced $\sqrt{3} \times \sqrt{3}$ R30° reconstruction at the Si interface has been investigated. At the a-Si/Si(111) interface, boron atoms at 1/3 ML are substituted for silicon atoms, thus forming a $\sqrt{3} \times \sqrt{3}$ R30° lattice. Even at the interface between a solid phase epitaxial Si(111) layer and a Si(111) substrate, the boron-induced $\sqrt{3} \times \sqrt{3}$ R30° reconstruction has been also observed. Secondly, SiO_2/Si(100)-2x1 interfacial superstructures have been investigated. Interfacial superstructures have been only observed in the samples of which SiO_2 layers have been deposited with a molecular beam deposition method. Finally, the interfaces of MOCVD-grown AlN/GaAs(100) have been shown to have 1x4 and 1x6 superstructures.

Introduction

There are many observations of superstructures (reconstructed structures) on semiconductor or metal surfaces. These superstructures are mostly believed to exist only on a clean surface fabricated under the condition of ultra-high vacuum. However, it is very recently that buried interface reconstruction has been studied [1-10]. The grazing incidence x-ray diffraction technique [11] is particularly suited to the study of interfacial structures, though techniques for surface studies, such as low-energy electron diffraction (LEED), reflected high-energy electron diffraction (RHEED) and scanning tunneling microscopy (STM) cannot study buried interfaces. From technological and scientific points of view, the investigation on the interface structure is very important for understanding of electronic properties of semiconducting materials.

In this paper, firstly, the boron-induced $\sqrt{3} \times \sqrt{3}$ R30° reconstruction at the Si interface has been investigated. Secondly, the 2x1 superstructure at the SiO_2/Si(100) interface has been studied. Finally,

Mat. Res. Soc. Symp. Proc. Vol. 159. ©1990 Materials Research Society

the interfaces of MOCVD-grown AlN/GaAs(100) have been shown to have 1x4 and 1x6 superstructures.

The present experiment was performed with using the synchrotron radiation at BL-9C, which is installed by NEC Corporation at the Photon Factory (PF) in Tsukuba. Incident x-rays with a wavelength of 1.5Å impinged on the sample at a grazing incidence angle of 0.2° for Si systems or 0.3° for GaAs systems. Diffracted x-rays were collected by a scintillation counter through 0.17° Soller slits. The normal of the scattering plane was nearly parallel to the surface normal of the sample.

Boron induced interfacial superstructure

A variety of different metals are known to induce a $\sqrt{3} \times \sqrt{3}$ R30° reconstruction on a Si(111) surface. In the case of column III metals (Al, Ga, In), it is generally admitted that a $\sqrt{3} \times \sqrt{3}$ R30° reconstruction involves metal adatom geometry in a three-fold hollow site on the Si(111) surface, in such a way that substrate dangling bonds are eliminated. Although a boron, which is a most important column III element as an acceptor impurity in Si device technology, also induces the $\sqrt{3} \times \sqrt{3}$ R30° reconstructed structure on the Si(111) surface [12-14], there is no consensus on the real space structure of $\sqrt{3} \times \sqrt{3}$ R30° -B reconstruction. Recently, boron-induced $\sqrt{3} \times \sqrt{3}$ R30° reconstruction was seen at the a-Si/Si(111) interface[4] and also seen between an epitaxial Si(111) layer and a Si(111) substrate[4] by a grazing incidence x-ray diffraction. And more recently, a difference in the stability of the boron-induced and gallium-induced $\sqrt{3} \times \sqrt{3}$ R30° structures at the a-Si/Si(111) interface was reported[15].

Firstly, the results on the $\sqrt{3} \times \sqrt{3}$ R30° -B structure at the a-Si/Si(111) interface are presented. The sample preparation procedure is as follows. After a Si(111)-7x7 RHEED pattern was obtained, boron from an HBO_2 crucible cell was deposited on the substrate at 750°C. The $\sqrt{3} \times \sqrt{3}$ R30° RHEED pattern was observed at the coverage of 1/3 ML of boron. After that, an a-Si layer of 100Å in thickness was deposited on the $\sqrt{3} \times \sqrt{3}$ R30° -B surface structure at room temperature. A RHEED pattern showed an amorphous structure.

Figure 1 shows the observed $\sqrt{3} \times \sqrt{3}$ R30° structure factors. The radius of each solid circle is proportional to the observed structure factor with the correction for the polarization, Lorentz factor and the variation of the active sample area. The star marks show the bulk fundamental lattice points. We refer to the $1/3(\bar{2}\bar{2}4)$ reflection in the three-dimensional lattice space as (10) in the two-dimensional one.

The Patterson function was calculated as shown in Fig. 2 using the observed structure factors except the bulk reflections. We assume three-fold symmetry of the data of structure factors. The Patterson peaks were clearly found along the $[\bar{1}\bar{1}2]$ direction. The distance from an

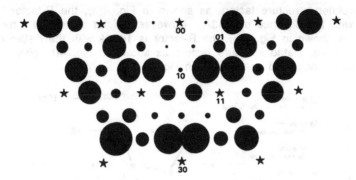

Fig. 1. Observed structure factors for the $\sqrt{3} \times \sqrt{3}$ R30° -B at the a-Si/Si(111) interface.

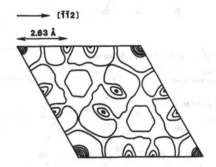

Fig. 2. Patterson function calculated from the observed structure factors in Fig. 1.

origin is 2.63Å (normally 2.22Å for a bulk Si crystal). Two simple interpretations are possible to explain the Patterson peaks. One is of Si lattice expansion caused by B atom adsorption, as shown in Fig. 3(a). The other interpretation is of Si lattice contraction, as shown in Fig. 3(b) For the latter case, it needs substitution of boron atoms for silicon atoms. If no substitution of boron atoms is assumed, the Patterson peaks should have appeared at shorter distance from the origin than the normal distance (2.22Å).

Figures 4(a) and (b) are structure factors maps calculated from the structure of Figs. 3(a) and (b), respectivley. By varying the distance of the silicon-boron bond (in-plane projected length), we obtained the minimum R-factor value. R-factor is defined as follows.

$$R = \frac{\Sigma |F_o - |F_c||}{\Sigma F_o},$$

where Fo is an observed structure factor and Fc is a calculated one. For the calculated structure factors as shown in Fig. 4(a), the R-factor is 35.4% with the distance of 2.43Å. However, for the calculated structure factors as shown in Fig. 4(b), the R-factor is 16.1% with the distance of 1.94Å. The calculated structure factors as shown in Fig. 4(b) give similar values to the observed ones shown in Fig. 1.

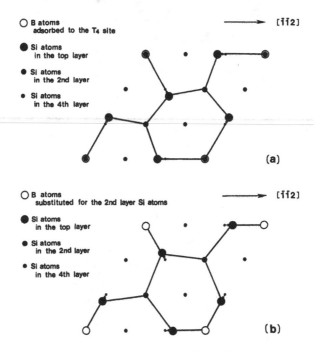

Fig. 3. In-Plane structure models for the $\sqrt{3} \times \sqrt{3}$ R30° -B at the a-Si/Si(111) interface. (a) shows the absorbed boron model and (b) shows the substitutional boron model.

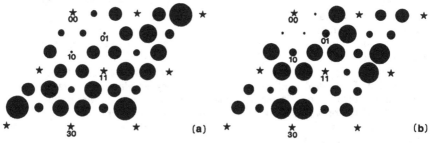

Fig. 4. Calculated structure factors. (a) calculated using the adsorbed boron model in Fig. 3(a), and (b) calculated using the substitutional boron model in Fig. 3(b).

As mentioned above, in the case of the in-plane projected structure, boron atoms are substituted for silicon atoms, thus forming a $\sqrt{3} \times \sqrt{3}$ R30° lattice and the distance of boron-silicon bond (in-plane projected length) is 1.94Å±0.08Å. By taking account of in-plane displacement of the fourth-layer silicon atoms, the R-factor value was not improved remarkably. By the present results, however, it is not clear that boron atoms are substituted for silicon atoms in the top layer or in the second layer or in both layers on the surface. On the surface Si(111) $\sqrt{3} \times \sqrt{3}$ R30° -B reconstruction, it was reported[14,16] that the boron atoms are substituted for second layer silicon atoms and silicon adatoms sit on a particular site. And quite recently, silicon adatoms were reported[16] to be displaced from its ordered site to the random site under room temperature Si deposition up to 2ML. Therefore, our interface structure model is consistent with the surface structure model. The silicon-boron bond distance in our analysis on the interface reconstruction is similar to that of the surface reconstruction. Since the present results give no information about the depth direction, further experiments are needed for precise structure analysis.

Secondly, the results on the $\sqrt{3} \times \sqrt{3}$ R30° -B structure at the interface between an epitaxial Si(111) layer and a Si(111) substrate are presented. The sample preparation procedure is as follows. After a Si(111)-7x7 RHEED pattern was obtained, boron from an HBO_2 crucible cell was deposited on the substrate at 750°C. The $\sqrt{3} \times \sqrt{3}$ R30° RHEED pattern was observed at the coverage of 1/3 ML of boron. After that, an a-Si layer of 100Å in thickness was deposited on the $\sqrt{3} \times \sqrt{3}$ R30° -B surface structure at room temperature. Finally, by raising substrate temperature up to 620°C, a solid phase epitaxial Si(111) layer, changing from an amorphous layer, was formed, which was confirmed by Kikuchi pattern of RHEED.

Figure 5 shows the observed $\sqrt{3} \times \sqrt{3}$ R30° -B structure factors. Large star marks show the bulk fundamental lattice points. At the

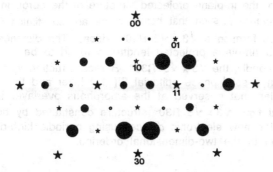

Fig. 5. Observed structure factors for the $\sqrt{3} \times \sqrt{3}$ R30° -B at the interface between a solid phase epitaxial Si(111) overlayer and a Si(111) substrate.

reciprocal lattice points indicated by small star marks, peak intensity reflects the epitaxial Si(111) overlayer. The diffraction intensity is about 1/10 of that observed at the a-Si/Si(111) interface. The observed structure factors map (Fig. 5) resembles the structure factor map (Fig. 1) observed at the a-Si/Si(111) interface. Therefore, the $\sqrt{3} \times \sqrt{3}$ R30° reconstructed structure at the epitaxial overlayer is about the same as that at the a-Si/Si(111) interface. By varying the distance of the silicon-boron bond, we obtained the minimum R-factor value of 36.0% with the distance of 1.96Å, where two atomic layers (a double layer) are assumed to be involved in stabilizing the $\sqrt{3} \times \sqrt{3}$ R30° reconstruction. By taking account of in-plane displacement of the fourth-layer silicon atoms, the R-factor value is improved to be 26.2%. In calculation, the in-plane boron-silicon bond length is assumed to be 2.03Å and the in-plane silicon-silicon bond-length containing the fourth-layer silicon atoms is assumed to be 2.13Å. From only the present results, it is not clear which model is acceptable, because diffraction intensity is very small. It is intuitively clear that the substitutional type $\sqrt{3} \times \sqrt{3}$ R30° structure as shown in Fig. 3 (b) allows the overlayer to grow epitaxially, although, other type $\sqrt{3} \times \sqrt{3}$ R30° structure, such as an adsorption to a T_4 site, does not hold the case.

In the previous report[4], a Si(111) layer was epitaxially grown on the $\sqrt{3} \times \sqrt{3}$ R30° -B surface structure at the substrate temperature of 600°C. In that case, the diffraction peak intensity is very weak (about 1/3) in comparison with that of a solid phase epitaxial Si(111) overlayer.

In the case of solid phase epitaxial growth, Sb "δ-doping" was reported[17]. Sb atoms are thought to be located at one atomic layer. The substitutional type $\sqrt{3} \times \sqrt{3}$ R30° structure constituted by boron atoms may realize high-density δ-doping with an additional two-dimensional ordering. The periodic δ-doping leads to enhancement in charge carrier mobility under suitable circumstances[18].

In conclusion, the in-plane projected structure of the boron induced $\sqrt{3} \times \sqrt{3}$ R30° -B structure is such that boron atoms are substituted for silicon atoms, thus forming a $\sqrt{3} \times \sqrt{3}$ R30° lattice. The distance of boron-silicon bond (in-plane projected length) is found to be 1.94Å±0.08Å. Secondly, the $\sqrt{3} \times \sqrt{3}$ R30° -B reconstruction was also observed between a solid phase epitaxial Si(111) layer and this structure resembles that observed at the amorphous overlayer interface. The substitutional type $\sqrt{3} \times \sqrt{3}$ R30° structure constituted by boron atoms may develop new electronic devices using periodic high-density δ-doping in addition to the two-dimensional ordering.

Superstructure at the SiO₂/Si(100) interface

Silicon is a dominant material in electronic devices, due partly to the high quality of its oxide. In this day, the thickness of oxide layer becomes thinner, the influence of interface structure upon the electronic devices becomes larger. The atomic structure at the interface must be understood to control electronic character. Although many atomic structure models [19-21] have been proposed, there has been no agreement with each other so far.

In this paper, the $SiO_2/Si(100)$ interface has been studied. Two types of sample were prepared. One of them have SiO_2 layers deposited with a molecular beam deposition method (MBD method) at room temperature. The MBD is a SiO_2 deposition method in a Si MBE system by co-deposition of Si and O_2. The thickness of the SiO_2 layer was about 50Å. The other type of sample was thermally oxidized sample. Sample was oxidized at 700°C, 750°C or 850°C in dry atmosphere, and also oxidized in wet atmosphere at each temperature. In addition, we prepared an a-Si/Si(100) sample for reference. A 100Å amorphous silicon layer was deposited in the MBE equipment at room temperature.

Several fractional-order reflections were observed with the samples in which SiO_2 layers were deposited with a MBD method. The superstructures of 2x1 were found to exist at the interface of the $SiO_2/Si(100)$. On the other hand, there was no fractional-order reflection, nor diffraction peak that indicates the existance of crystalline SiO_2 in thermally oxidized samples. Also in the a-Si/Si(100) system, no fractional-order reflection was observed. This result tells us that the observed 2x1 superstructure exists at the interface between SiO_2 and Si(100), not between a-Si and Si(100). In Fig. 6, the map of observed structure factors of the $SiO_2/Si(100)$-2x1 sample is shown. We refer to the $(0\bar{1}1)$ and (011) indices, as (-10) and (01), respectively. The radius of each solid circle is proportional to the observed structure factor, and the star marks show fundamental reciprocal lattice points. At the points of open circles, no clear diffraction peak was observed.

First of all, we considered the crystalline SiO_2 models. The observations of the cristobalite-like [20] and the tridymite-like [19]

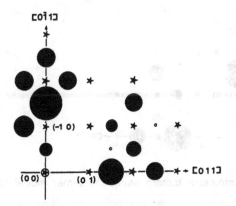

Fig. 6. Observed structure factors for the $SiO_2/Si(100)$-2x1 structure.

crystalline phases at SiO$_2$/Si(100) interface were reported, respectively. To consider the cristobalite-like case, the arrangements in which some layers of cristobalite (100) exist on Si(100), was adopted. The lattice mismatch between silicon and cristobalite is about 7% in this relation. But this cristobalite-like model is expected to give a $\sqrt{2}$ x $\sqrt{2}$ R45° diffraction pattern. Therefore, cristobalite-like model is not suitable for the observed 2x1 structure. As for the tridymite-like model, the orientational relationship between the Si substrate and the tridymite proposed by Ourmazd [19] is that Si(01$\bar{1}$), Si(011) and Si[100] are parallel to tridymite (100), tridymite (002) and tridymite [010], respectively. Throughout the fitting procedure, the arrangements that gave an R-factor of 16%, were obtained. However, one of the bond length between silicon and oxygen atoms is larger than that in usual crystalline or non-crystalline SiO$_2$ by 25%. Therefore, these arrangements should not be realistic.

It is believed that the 2x1 superstructure on a clean Si(100) is stabilized by dimers. Then, as a starting structural model, we adopted dimer models that were proposed by several authors [22-24]. Throughout the fitting procedure to decrease R-factor, the arrangements that gave a R-factor of 15%, were obtained. These optimum arrangements did not depend on the starting models. The calculated structure factors of this model are shown in Fig. 7. The difference between calculated and observed structure factors is rather large at (-2 1/2) and (-2 -1/2) reciprocal lattice points. To make allowance for experimental error, this dimer model cannot be denied. By taking account of one or two additive oxygen atoms and one additive silicon atom in the unit cell, the R-factor value is improved [25]. Further experiments are required in order to clarify the detail atomic arrangements at the interface.

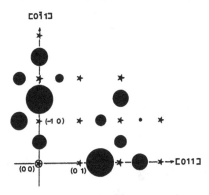

Fig. 7. Calculated structure factors using the dimer model for the 2x1 structure.

Superstructure at the AlN/GaAs interface

Recently, we showed for the first time that a superstructure exists at the insulator/semiconductor (AlN/n-GaAs(001)) interface[6]. This interface was fabricated under the condition of H_2 ambient gas at a pressure of 0.1 atm. The AlN/GaAs structure was fabricated in a MOCVD system. We reported that the H_2 petreatment leads to formation of 1x4 and 1x6 superstructures at the interface between GaAs and an AlN film deposited at 220°C. Figure 8 shows a map of structure factors of the 1x4 superstructure. The structure factors at the (0h) and ($\bar{2}$h) reciprocal lattice points are larger than those at the ($\bar{1}$h) lattice points. The calculated diffraction pattern, assuming the existence of one or three dimer atoms in the unit cell on the direction of [$\bar{1}$10], resembles the observed one. Therefore, it is suggested that As-site dimer atoms exist in the 1x4 superstructure. This 1x4 structure seems to be a disordered version of the 2x4 or c(2x8) structures which are considered to have As dimer atoms along the [$\bar{1}$10] direction [26].

However, in the course of re-examinations, the 1x6 superstructures proved to be only occcasionally formed by the H_2 pretreatment while the 1x4 structure was reproducibly observed. We suppose that this ambiguity arose from unoptimized conditions such as the duration and temperature of the H_2 pretreatment. Also, uncontrolled residual As pressure in the reactor may have had an influence. In order to confirm that the 1x6 structure actually exists, we have developed a TMG (trimethylgallium) pretreatment. The GaAs surface was capped with a 60Å thick AlN film deposited at 220°C. From the grazing incidence x-ray diffraction measurements on the sample, it was confirmed that a 1x6 superstructure is reproducibly obtained by the TMG pretreatment. This fact suggests that the 1x6 structure is Ga-stabilized one. We can now mention that preparation of As-dimer-stabilized (1x4 structure) and Ga-stabilized (1x6 structure) surfaces are realized by the H_2 and TMG pretreatment in the MOCVD system respectively.

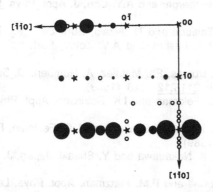

Fig. 8. Observed structure factors for the AlN/GaAs (100)-1x4 structure.

Acknowledgement

The authors would like to thank Drs. S. Fujieda, Y. Mochizuki, T. Tatsumi and H. Hirayama for the sample preparation and critical discussion. They also wish to express their appreciation to Drs. Y. Matsumoto and M. Ogawa for their encouragement.

References

1. J.M. Gibson, H.-J. Gossmann, J.C. Bean, R.T. Tung and L.C. Feldman, Phys. Rev. Lett. 56, 355 (1986).
2. I.K. Robinson, W.K. Waskiewicz and R.T. Tung, Phys. Rev. Lett. 57, 2714 (1986).
3. K. Akimoto, J. Mizuki, T. Tatsumi, N. Aizaki and J. Matsui, Surf. Sci. 183, L297 (1987).
4. K. Akimoto, J. Mizuki, I. Hirosawa, T. Tatsumi, H. Hirayama, N. Aizaki and J. Matsui, Extended Abstracts 19th Conf. on Solid State Devices and Materials (Business Center for Academic Societies Japan, Tokyo, 1987) p. 463-466.
5. J. Mizuki, K. Akimoto, I. Hirosawa, T. Mizutani, and J. Matsui, J. Vac. Soc. & Technol. B6, 31 (1988).
6. K. Akimoto, I. Hirosawa, J. Mizuki, S. Fujieda, Y. Matsumoto and J. Matsui, Jpan, J. Appl. Phys. 27, L1401 (1988).
7. S. Fujieda, K. Akimoto, I. Hiroswa, J. Mizuki, Y. Matsumoto and J. Matsui, Japan, J. Appl. Phys. 28, L16 (1989).
8. K. Akimoto, I. Hirosawa, J. Mizuki and J. Matsui, Proc. 2nd Int. Conf. on Formation of Semiconductor Interfaces, 1988, Takarazuka, Japan, to be published in Appl. Surf. Sci.
9. K. Hirose, K. Akimoto, I. Hirosawa, J. Mizuki, T. Mizutani and J. Matsui, Phys. Rev. B39, 8037 (1989).
10. D. Lorreto, J.M. Gibson and S.M. Yalisove, Phys. Rev. Lett. 63, 298 (1989).
11. W.C. Marra, P. Eisenberger and A.Y. Cho, J. Appl. Phys. 50, 6927 (1979).
12. H. Hirayama, T. Tatsumi and N. Aizaki, Surf. Sci. 193, L47 (1988).
13. V.V. Korobtsov, V.G. Lifshits and A.V. Zotov, Surf. Sci. 195, 466 (1988).
14. F. Thibaudau, Ph. Dumas, Ph. Mathiez, A. Humbert, D. Satti and F. Salvan, Surf. Sci. 211/212, 148 (1989).
15. R.L. Headrick, L.C. Feldman and I.K. Robinson, Appl. Phys. Lett. 55, 442 (1989).
16. R.L. Headrick, I.K. Robinson, E. Vlieg and L.C. Feldman, Phys. Rev. Lett. 63, 1253 (1989).
17. A.A. van Gorkum, K. Nakagawa and Y. Shiraki, Japan, J. Appl. Phys. 26. L1933 (1987).
18. A.F. Levi, S.L. McCall and P.M. Platzman, Appl. Phys. Lett. 54, 940 (1989).

19. A. Ourmazd, D. W. Taylor, J.A. Rentschler and J. Benk, Phys. Rev. Lett. 59, 213 (1987).

20. P.H. Fuoss, L.J. Norton, S. Brennan and A. Fisher Colbrie, Phys. Rev. Lett. 60, 600 (1988).

21. F.J. Himpsel, F.R. McFeely, A. Taleb-Ibrahimi, J.A. Yarmoff and G. Hollinger, Phys. Rev. B38, 6084 (1988).

22. W.S. Yang, F. Jona and P.M. Marcus, Solid State Commun. 43, 847 (1982).

23. J.A. Appelhaum and D.R. Hamann, Surf. Sci. 74, 21 (1978).

24. D.J. Chadi, Phys. Rev. Lett. 43, 43 (1979).

25. I. Hirosawa, K. Akimoto, T. Tatsumi, J. Mizuki and J. Matsui, Proc. of the Conf. on defect recognition and image processing for research and development of semiconductors, 1989, Tokyo, to be published in J. Cryst. Growth.

26. M.D. Pashley, K.W. Haberern, W. Friday, J.M. Woodall and P.D. Kirchner, Phys. Rev. Lett. 60, 2176 (1988).

315

DIRECT OBSERVATION OF A 7×7 SUPERSTRUCTURE
BURIED AT THE AMORPHOUS-Si / Si(111) INTERFACE

AKIRA SAKAI*, TORU TATSUMI** and KOICHI ISHIDA*

*Fundamental Research Laboratories, NEC Corporation,
34, Miyukigaoka, Tsukuba, Ibaraki 305, Japan
**Microelectronics Research Laboratories, NEC Corporation,
4-1-1, Miyazaki, Miyamae-ku, Kawasaki, Kanagawa 213, Japan

ABSTRACT

    Direct imaging of a 7×7 superstructure buried at the interface between
amorphous-Si and a Si(111) substrate was demonstrated by cross-sectional high
resolution transmission electron microscopy (HRTEM). The electron diffraction
pattern of the interface region in cross-section geometry showed diffuse
streaks of fractional order which suggested the existence of the
superstructure at the interface. The <110> cross-sectional HRTEM image showed
the atomic configuration of the projected interface superstructure, which had
a periodicity of 23 Å along the interface. Such periodicity corresponds to 7
times the 1/3{224} periodicity which is 3.3 Å at the interface. The interface
atomic model of the buried 7×7 superstructure was also proposed by using the
image matching technique between the experimental and computer simulated
images.

INTRODUCTION

    Surface and interface superstructures in Si materials have been widely
studied by large number of techniques. One of the most effective methods for
analyzing their atomic scale structures is transmission electron microscopy
(TEM). Contrary to the other techniques, for example scanning tunneling-
microscopy [1,2], low energy electron diffraction [3,4] and X-ray diffraction
[5,6], TEM provides us with both reciprocal and real space information of the
same area of samples. Transmission electron diffraction (TED) pattern
represents the obtainable reciprocal information and the analyses of the TED
pattern reveal the average atomic structures. Takayanagi et al. have used it
for analyzing a Si 7×7 surface reconstructed structure [7], and consensus
has been reached that a model proposed by them consisting of dimers, adatoms,
and stacking faults (DAS model) best describes the reconstructed surface.
    TED analysis was also applied to the study of a buried interface
structure and the preservation of a 7×7 periodicity at the interface between
amorphous-Si (a-Si) and a Si (111) substrate was first confirmed by Gibson et
al. [8].

The other characteristic performance of TEM is visualization of the superstructures in real space. High resolution (HR) TEM allows us to deduce their structures directly at the atomic level. In particular, cross-sectional HRTEM observation provides us with the atomic arrangement perpendicular to the superstructure plane. However, since the projection image of the superstructure is obtained by this observation, the image may be unclear unless only one domain of the superstructure is projected [8].

In this paper, we will demonstrate the direct imaging of the 7×7 superstructure buried at the a-Si/Si(111) interface by cross-sectional HRTEM. Furthermore, we predict the interface atomic structure by using image matching technique between the experimental and computer simulated images.

EXPERIMENTAL

In order to form a large domain of a 7×7 superstructure layer, a thick Si buffer layer of 3000 Å was epitaxially grown on a Si (111) substrate at 750 ℃ after the surface cleaning [9] in a molecular-beam-epitaxy chamber with base pressure 1.2×10^{-10} Torr. The thick buffer-layer produces an atomically flat surface. The validity of this procedure was proved elsewhere [10]. After the confirmation of a clean surface on the buffer layer showing a characteristic 7×7 reflection high energy electron diffraction pattern, a-Si was subsequently deposited by electron-beam evaporation to the thickness of 50 Å at room temperature. Two types of TEM specimens were prepared. One is for plan-view observation, which confirms the preservation of the 7×7 periodicity, and chemically thinned from the substrate side. The other is for cross-sectional observation along the <110> direction. During the preparation process, samples were not exposed to temperatures higher than 150 ℃ and Ar-ion thinned at the lowest energy of 3 kV. TEM was performed with AKASHI EM 002B operating at 200 kV in both plan-view and cross-sectional observations.

RESULT AND DISCUSSION

Diffraction Analyses

Figure 1 shows a TED pattern taken from the sample in the [111] plan-view observation. Bragg reflection spots from the buffer layer denoted by the hexagonal unit cell index (11) (which is the cubic index {220} of a bulk crystal), diffuse rings from the deposited a-Si layer, two-dimensional fundamental reflections denoted by (01) and (10) (1/3{224}) are seen. Furthermore, extra spots of fractional indices (h/7,k/7) which show the existence of the 7×7 superstructure are clearly observed. Since the superstructure spots were always accompanied by the diffuse rings, {11} and {10} spots, they are proved to come from the a-Si/ Si buffer layer interface.

Figure 2 (a) is an electron diffraction pattern taken from the area including the a-Si / Si buffer-layer interface in the <110> cross-section geometry. Bragg reflection spots and double diffraction spots of {200} are seen. Diffuse streaks running through the [111] direction, perpendicular to the interface, on respective bulk reflections are due to the drastic termination of the lattice periodicity at the interface. The interesting point to note in the pattern is that additional diffuse streaks (denoted by arrows), which arise from the interface, are visible among the diffraction spots. Figure 2 (b) illustrates the location of the observed diffuse streaks in Fig.2 (a). The (10) position corresponds to $1/3(\overline{2}\overline{2}4)$ periodicity of bulk crystal, which indicates one monolayer atomic array along the interface of the $[\overline{1}\overline{1}2]$ direction. Each diffuse streak is found to be located at $(h/7,k)$ fractional position and suggests the existence of 7 times periodicity of $1/3(\overline{2}\overline{2}4)$ at the interface.

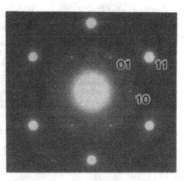

Figure 1   Transmission electron diffraction pattern of the plan-view observation taken from the area including the a-Si/Si interface.

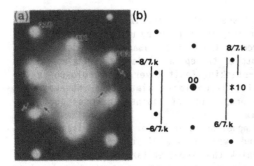

Figure 2 (a) Transmission electron diffraction pattern of the <110> cross-sectional observation taken from the area including the a-Si/Si interface. (b) Schematic illustration of (a).

HRTEM Image of Interface Superstructure

Figure 3 shows a typical cross-sectional HRTEM image of the a-Si / Si buffer layer interface region taken near the Scherzer focus. The specimen thickness of the observed region was thin enough to be before appearance of the half-spacing contrast [11]. The average terrace width at the interface was estimated to be about 500 Å. Since the specimen thickness (< 100 Å) is sufficiently less than the terrace width, the only one domain of the 7×7 superstructure is projected on the image.

The following distinct image variation can be noticed in the interface regions. First, as indicated by arrows, comparatively large bright spots are observed between the A and B layers at intervals of 23 Å. This periodicity corresponds to 7 times the $1/3(\overline{2}\overline{2}4)$ periodicity of 3.3 Å. Second, in the B layer, it is clearly observed that the elongated dark spots are gradually rotated relative to their orientations in the bulk crystal (below the C layer) within a period of 23 Å. Such contrast variation is also recognized in the A layer. The inset shows an optical diffractogram of the image. The fractional diffuse streaks observed in the electron diffraction pattern of

Fig. 2 (a) are reproduced in it. It is, therefore, confirmed that these diffuse streaks contribute to the image formation of the interface 7×7 superstructure which has the periodicity of 23 Å.

Figure 3 Cross-sectional HRTEM image of the a-Si / Si buffer layer interface showing the buried 7×7 superstructure. Inset is an optical diffractogram of the image.

Identification of Structure

Robinson et al. have analyzed the partial ordering extending into the amorphous region on the dimer-stacking fault layer by using X-ray diffraction [6]. Therefore, in order to reproduce the observed image, the model proposed by Robinson et al., which consists of two epitaxial layers having normal and reversed stacking on the dimer-stacking fault layer, was employed as the starting position for the multi-slice image simulation. However, the simulation images of the <110> projection of Robinson's model could not explain the details of our experimental images [12].

We, therefore, propose a new model shown in Fig. 4. This model consists of a dimer-stacking fault layer and two epitaxial layers having normal and reversed stacking regions in which the second epitaxial layer is extending over the dimer-chains and vacant corner sites. Thus the atomic tunnels along the <110> direction are formed between the atomic arrays (see small arrows) and the dimer-chains.

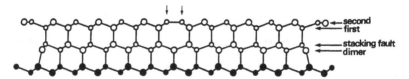

Figure 4 Model of the buried 7×7 superstructure in which the atomic arrays are indicated by small arrows.

By using this model, we carried out the image matching between the experimental and simulated images. In the simulation, atoms in respective epitaxial layers are assumed to have partial occupancies in order to vary the ordering gradually to the amorphous layer along the growth direction of [111]. Figure 5 (a) shows the part of the experimental image of Fig. 3 and the simulated image of the present model, where specimen thickness and an objective lens defocus value are assumed to be the experimental conditions of

75 Å and -320 Å, respectively. Figure 5 (b) also shows the experimental image
of the same area, taken with another defocus value of -710 Å and the
corresponding simulated image. Under such imaging conditions, it is noticed
that these two sets of images are complementary: The columns of the atom
pairs were elongated dark (bright) spots, while the tunnels bright (dark) in
Fig.5 (a) (Fig.5 (b)). Comparisons in both cases show good matching between
the experimental and simulated images. Particularly in Fig.5 (a), the large
bright spots as well as the rotation of the elongated dark spots representing
the atom pairs are clearly reproduced in the simulated image. Therefore, the
existence of the atomic arrays in the second epitaxial layer can be
identified, although their arrangements on (111) plane cannot be known from
our <110> projection.

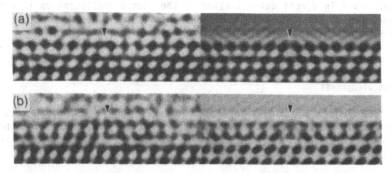

Figure 5 Comparisons between the experimental (left) and the
simulated (right) images where the atomic occupancy factors of first
and second epitaxial layers are assumed to be 0.9 and 0.63,
respectively. Defocus values are -320 Å (a) and -710 Å (b).

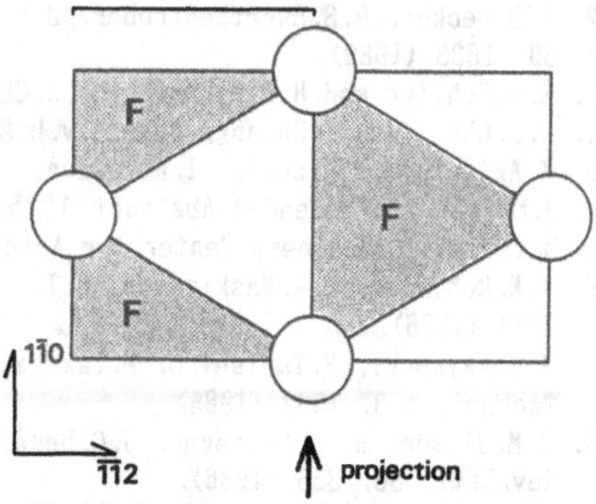

Further observations of thicker
regions of the specimen also showed
the presence of at least two more
epitaxial layers which had partial
atomic occupancies. However, it was
found from the image simulation
that these epitaxial layers did not
influence the image of the
underneath superstructure.

The origin of the contrast
variation of the atom pairs in the
B layer is qualitatively explained
as follows. Figure 6 illustrates
the present observation geometry of
a 7×7 unit cell. In the <110>
cross-section, both normal and

Figure 6 Observation geometry of a
unit cell of the 7×7 superstructure

reversed (labeled F) stacking regions are projected on the same position.
Considering within a half unit cell, it is noticed that normal stacking
regions are predominantly projected on the right side of half unit cell
image, whereas reversed stacking regions on the left side. Therefore, the

right side of the half unit cell exhibits dark spots elongated to the same orientation as the bulk crystal, while in the left side the elongated orientation is reversed.

CONCLUSION

We have demonstrate the direct imaging of the 7×7 superstructure buried at a-Si/Si(111) interface by the <110> cross-sectional HRTEM. Image matching technique between the experimental and computer simulated images allowed us to elucidate the interface structure on an atomic scale. The major point of this paper is direct determination of the atomic configuration in a few monolayer of the buried 7×7 superstructure. We could identified the structure at the only one atomic-column level.

Acknowledgement

The authors would like to thank T. Niino for her helpful assistance in experiments and K. Akimoto, A. Ogura, H. Ono, N. Aizaki and J. Matsui for their useful discussions. Thanks are also due to M. Ogawa, Y. Matsumoto, H. Watanabe and F. Saito for their encouragement.

References

1. G. Binnig, H. Rohrer, Ch. Gerber and E. Weibel, Phys. Rev. Lett. 50, 120(1983).
2. R. S. Becker, B. S. Swartzentruber, J. S. Vickers and T. Klitsner, Phys. Rev. B 39, 1633 (1989).
3. R. E. Schlier and H. E. Farnsworth, J. Chem. Phys. 30, 917 (1959).
4. D. J. Chadi and C. Chiang, Phys. Rev. B 23, 1843 (1981).
5. K. Akimoto, J. Mizuki, I. Hirosawa, T. Tatsumi, H. Hirayama, N. Aizaki and J. Matsui, in Extended Abstract 19th Conference Solid State Devices and Materials, (Business Center for Academic Societies, Tokyo, 1987) pp. 463.
6. I. K. Robinson, W. K. Waskiewicz, R. T. Tung and J. Bohr, Phys. Rev. Lett. 57, 2714 (1986).
7. K. Takayanagi, Y. Tanishiro, M. Takahashi and S. Takahashi, J. Vacuum Sci. Technol. A 3, 1502 (1985).
8. J. M. Gibson, H.-J. Gossmann, J. C. Bean, R. T. Tung and L. C. Feldman, Phys. Rev. Lett. 56, 355 (1986).
9. T. Tatsumi, N. Aizaki and H. Tsuya, Jpn. J. Appl. Phys. 24, L227 (1985).
10. A. Sakai, T. Tatsumi, T. Niino, H. Hirayama and K. Ishida, Appl. Phys. Lett. (to be published).
11. R. W. Glaisher, A. E. C. Spargo and D. J. Smith, Ultramicroscopy 27, 35 (1989).
12. A. Sakai, T. Tatsumi and K. Ishida, Surf. Sci. (to be published).

TOPOGRAPHY OF Si(111): CLEAN SURFACE PREPARATION AND SILICON MOLECULAR BEAM EPITAXY

R. T. TUNG, F. SCHREY AND D. J. EAGLESHAM
AT&T Bell Laboratories, 600 Mountain Avenue, Murray Hill, N.J. 07974

ABSTRACT

Line defects at the interfaces of epitaxial silicide layers grown at room temperature on Si(111) are found to correspond to steps on the original surface. This has enabled the examination, by transmission electron microscopy, of the topography of large areas of the Si surface after various treatments. Methods for removal of surface oxide and carbide are compared. Silicon molecular beam epitaxy (MBE) is shown to occur via step-flow mechanism at high temperatures, and through nucleation and growth of islands on terraces at low temperatures.

INTRODUCTION

Steps on clean semiconductor surfaces play a very important role in various physical and chemical processes. The use of double steps on miscut wafers may lead to a reduction of antiphase domain boundaries in heteroepitaxial systems.[1] A knowledge of the density of steps during epitaxial growth has led to improved interface sharpness.[2] Steps have also been employed in the growth of tilted superlattices with lateral dimensions smaller than the present capability of lithography.[3] Information regarding the topography of a growing surface under conditions identical to those employed in a conventional MBE chamber is still much in need.

The topography of surfaces has traditionally been inferred from diffraction studies.[4] [5] [6] [7] Recently, a flurry of in-situ microscopy studies[8] [9] [10] [11] [12] [13] has lead to examination of step arrangement in the real space. However, ultrahigh vacuum (UHV) microscopes are restricted in space around the specimen which severely limits the means of deposition and the vacuum condition. Recently, the growth of epitaxial $CoSi_2$[14] and $NiSi_2$[15] at room temperature was demonstrated. High quality silicide thin films have been grown on Si(111) by deposition of 2-3 monolayers (ML) of metal and co-deposition of disilicide at room temperature. Because of the absence of long range diffusion at room temperature, surface steps on the original Si(111) surface are preserved at the interface between silicide and silicon. The type B orientation of these silicide thin films requires a phase difference of 1/3[111] across a step. This exact correspondence of defects observed at the silicide interface, with transmission electron microscopy (TEM), to features on the original Si surface provides a convenient means of investigating large-area surface topography. Furthermore, since there is no constraint on sample preparation, growth under usual MBE conditions (e.g. growth rate ~ 1 Å/s, and $p < 1 \times 10^{-10}$ torr) may be studied. In this paper, this technique is used to study the surface topography from different oxide removal processes and the topography during Si homoepitaxy. Some of these results, as well as detailed experimental procedures, have already been presented.[16]

[17]

CLEAN SURFACE PREPARATION

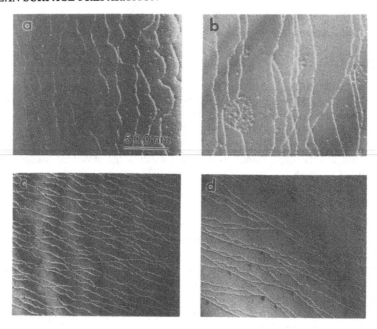

Fig. 1. Dark-field TEM images of CoSi$_2$ layers grown on a Si(111) surface. In (a) and (c), the oxide layer was removed by Si beam cleaning. For the surfaces shown in (b) and (d), the oxide layer was removed by direct heating to 900°C and 1000°C, respectively. Areas (a) and (b) were generated in one run on the same Si wafer by the use of a deposition shadow mask. Likewise, (c) and (d) are a different pair.

Si substrates were chemically cleaned and a protective oxide layer grown as the final step.[18] After thorough degassing of the substrates in UHV, the protective oxide layer was removed by either Si beam cleaning[19] or by heating to high temperature (> 900°C). Typical arrangements of surface steps after selected oxide removal procedures are shown in Fig. 1. Volitization of oxide upon Si beam impingement is uniform across the surface, resulting in a surface with an array of uniformly distributed steps.[14] The majority of these steps are of single height (3.14 Å). Direct heating of the Si substrate to 900°C also removes the oxide layer from the surface, as shown in Figs. 1(b) and 1(d). The impurity particles on the Si surfaces are carbide, as their density correlates with the amount of carbon observed by Auger electron spectroscopy (AES). On deliberately contaminated surfaces, positive identifications of the carbide particles have been made, as shown in Fig. 2. Without the Si beam, volitization initiates at isolated locations and proceeds laterally.[20] This process proceeds by surface diffusion and reaction of the Si and oxide at the peripheries of the oxide islands. With the same surface carbon concentration, carbide particles on directly heated Si surfaces are larger in size, but lower in density, than those on surfaces cleaned by Si beam. This is consistent with a scenario where the carbon impurity is contained in or on oxide islands as they decrease in size. When the size of the oxide islands reaches a point where carbon reaches a certain critical concentration, carbide formation becomes significant. Thus the locations of large carbide particles probably mark the positions where oxide islands finally leave the surface.

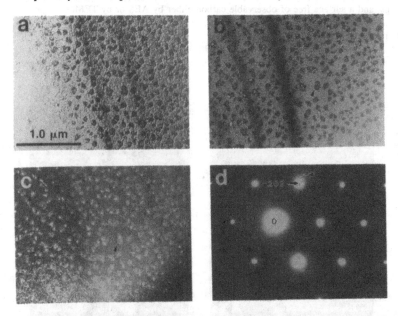

Fig. 2. TEM images and diffraction pattern of a CoSi$_2$ layer grown on a carbon-contaminated Si(111) surface which was etched in aqueous HF. (a) <2$\bar{2}$0> dark field, (b) dark field image using a type B related <111> diffracted beam, (c) image formed using a carbide related spot, (however, some intensity from a nearby <220> also entered the aperture) (d) diffraction pattern.

Carbon contamination on Si may be minimized by thorough chemical cleaning[18] and careful handling of the substrates. If the Si is not carefully handled, or if the Si is etched in aqueous HF prior to loading, high carbon content is found on the surface and a high density of carbide particles can be found, shown in Fig. 2. A surface only slightly contaminated with carbon may be remedied through the growth of a Si buffer layer. It is obvious that, prior to and in conjunction with the growth of a Si buffer layer, Si beam cleaning rather than direct heating should be used to remove the oxide layer. This is because the smaller the carbide particles, the easier it is to "bury" them with the growth of a buffer layer.[21] The presence of the impurity particles may significantly affect the epitaxial growth, through, for instance, pinning of the migrating steps. Step flow is expected to become of increasing importance as the growth temperature is raised. It is not surprising that we found much more uniform buffer layers to be grown at lower MBE temperatures (<640°C). After the growth of a >200Å thick buffer layer, the surface shows no carbon by AES analysis (< 0.2% ML). A brief anneal to >900°C restores the surface smoothness. The combination of Si beam cleaning, Si buffer growth at < 640°C, and a brief anneal seems to give the most satisfactory surface cleaning. A substrate prepared by these procedures always has evenly distributed, approximately parallel, single steps, and a surface free of observable carbon either by AES or by TEM.

Si HOMOEPITAXY

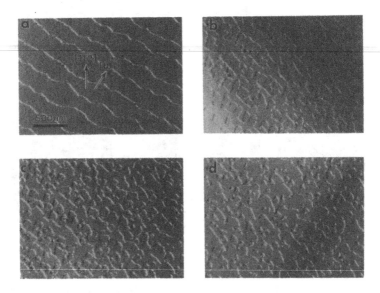

Fig. 3. Dark field TEM images of CoSi$_2$ layers grown on Si surfaces after Si MBE growth. (a) substrate surface, and surfaces after the growth of (b) 1 ML, (c) 2 ML, and (d) 3 ML, respectively, at 650°C, at a rate of 0.4 Å/s.

The surface topography during Si homoepitaxial growth is revealed by rapidly cooling to room temperature and by epitaxial growth of silicide. A deposition series is shown in Fig. 3. Similar studies suggest that, at a fixed deposition rate of ~ 1Å/sec, the growth proceeds via step-flow above 700°C, and via two dimensional nucleation and growth (2DNG) of Si islands on terraces at < 600°C.[16][17] The effect of deposition rates on the observed growth mode is demonstrated in Fig. 4. Migration of steps is seen to play a more dominant role at lower deposition rates. This is to be expected since an increase in deposition rate serves to increase the supersaturation.[22] The observed surface topographies are in agreement with the predicted trend of growth mode based on existing theories.[22] However, the theories seem to have neglected two important aspects of epitaxial growth – nucleation of islands and effects of step energetics. From micrographs in

Figs. 3 and 4, a clear preference for $<\overline{11}2>$ type steps over $<11\overline{2}>$ steps during growth is discerned. This apparent difference in step stabilities, first noted some years ago,[23] is known to lead to very different faceting behaviors on vicinal Si(111) surfaces.[7][17]

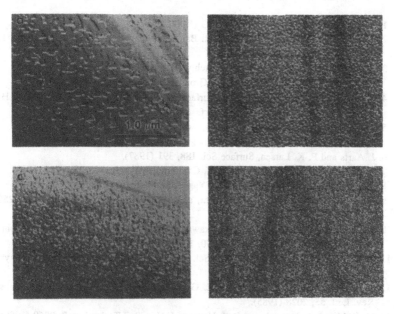

Fig. 4. Dark field TEM images after MBE growth of 2 ML Si at (a) 550°C and 0.02 Å/sec, (b) 550°C and 0.4 Å/sec, (c) 500°C and 0.02 Å/sec, and (d) 500°C and 0.4 Å/sec, respectively.

DISCUSSION

The successful application of the present technique to the study of Si surface topography is based on a faithful one-to-one correspondence of silicide interface dislocations to Si surface steps. Two conditions are vital to this relationship: limited diffusion at room temperature in the silicide growth process preserving the original surface steps at the epitaxial interface, and a change in the crystal symmetry at the (type B) interface leading to a phase difference across a step and, therefore, a dislocation. One notes that capping a surface with an amorphous layer or a regular heteroepitaxial layer, such as GeSi or type A silicide, also leads to interface steps. No dislocations are formed at these steps and therefore a lack of diffraction contrast prevents their observation by TEM.

In summary, we have shown that clean Si(111) surface topography may be imaged by TEM through epitaxial silicide formation. A difference is shown between topographies of directly heated and Si-beam cleaned surfaces. The observed overall surface topographies at different MBE temperatures and growth rates demonstrate the importance of considering nucleation in the existing theories.

REFERENCES

1. H. Kroemer, J. Cryst. Growth, **81**, 193 (1987).
2. D. Bimberg, D. Mars, J. N. Miller, R. Bauer and D. Oertl, J. Vac. Sci. Technol. B **4**, 1014 (1986).
3. P. M. Petroff, J. M. Gaines, M. Tsuchiya, R. Simes, L. Coldren, H. Kroemer, J. English and A. C. Gossard, J. Cryst. Growth **95**, 260 (1989).
4. K. D. Gronwald and M. Henzler, Surface Sci. **117**, 180 (1982); R. Altsinger, H. Busch, M. Horn, and M. Henzler, Surf. Sci. **200**, 235 (1988).
5. T. Sakamoto, N. J. Kawai, T. Nakagawa, K. Ohta, and T. Kojima, Appl. Phys. Lett. **47**, 617 (1985).
6. J. Aarts and P. K. Larsen, Surface Sci. **188**, 391 (1987).
7. R. J. Phaneuf, E. D. Williams, and N. C. Bartelt, Phys. Rev. B **38**, 1984 (1988).
8. Y. Tanishiro, K. Takayanagi, and K. Yagi, Ultramicroscopy **11**, 95 (1983).
9. Y. Ishikawa, N. Ikeda, M. Kenmochi, and T. Ichinokawa, Surface Sci. **159**, 256 (1985).
10. M. Ichikawa, T. Doi, and K. Hayakawa, Surface Sci. **159**, 133 (1985); M. Ichikawa and T. Doi, Appl. Phys. Lett. **50**, 1141 (1987).
11. W. Telieps and E. Bauer, Surface Sci. **162**, 163 (1985); Ber. Bunsenges. Phys. Chem. **90**, 197 (1986).
12. R. S. Becker, J. A. Golovchenko, E. G. McRae, and B. S. Swartzentruber, Phys. Rev. Lett. **55**, 2028 (1985).
13. U. Köhler, J. E. Demuth and R. J. Hamers, J. Vac. Sci. Technol. A **7**, 2860 (1989).
14. R. T. Tung and F. Schrey, Appl. Phys. Lett. **54**, 852 (1989).
15. R. T. Tung and F. Schrey, Appl. Phys. Lett. **55**, 256 (1989).
16. R. T. Tung and F. Schrey, Phys. Rev. Lett. **63**, 1277 (1989).
17. R. T. Tung, F. Schrey, and D. J. Eaglesham, 10th MBE Workshop, Raleigh, NC, Sept. 1989, to appear in J. Vac. Sci. and Technol..
18. A. Ishizaka and Y. Shiraki, J. Electrochem. Soc. **133**, 666 (1986).
19. M. Tabe, Jpn. J. Appl. Phys. **21**, 534 (1982).
20. R. Tromp, G. W. Rubloff, P. Balk, F. K. LeGoues, and E. J. van Loenen, Phys. Rev. Lett. **55**, 2332 (1985).
21. See, for instance, Y. Ota, Thin Solid Films, **106**, 1 (1983); and J. C. Bean, J. Cryst. Growth **70**, 444 (1984).
22. F. Allen and E. Kasper, in *Silicon-molecular beam epitaxy*, E. Kasper and J. C. Bean, eds., CRC press, Boca Raton, FL, 1988, Vol.1, p.65.
23. H. C. Abbink, R. M. Broudy, and G. P. McCarthy, J. Appl. Phys. **39**, 4673 (1968).

ELECTRONIC AND ATOMIC PROPERTIES OF
a-C:H/SEMICONDUCTOR INTERFACES

M. WITTMER*, D. UGOLINI** AND P. OELHAFEN**
*IBM T. J. Watson Research Center, Yorktown Heights, N.Y. 10598
**Institute of Physics, University of Basel, CH-4056 Basel, Switzerland

ABSTRACT

We have investigated the electronic and atomic properties of the interface between amorphous hydrogenated carbon (a-C:H) films and the semiconductor materials Si, Ge and GaAs with photoelectron spectroscopy, high resolution transmission electron microscopy and ion channeling technique. The different properties of the interfacial layers are summarized and compared to the adhesion quality of a-C:H films on these semiconductor materials.

1. THE IMPORTANCE OF a-C:H FILMS

Amorphous hydrogenated carbon (a-C:H) films find many applications in magnetic recording media [1,2], microelectronics [3-5] and optical components [6,7]. This is due to a large extent to their outstanding properties which gave them the attribute of 'diamond like carbon films'. Unfortunately, the adhesion of a-C:H films on many materials is poor. Since the adhesion is directly related to the electronic and atomic properties of the interface between film and substrate, a thorough knowledge of these properties is crucial for the improvement of the adhesion quality of the films. Towards this goal we have performed a detailed study of the interface properties of a-C:H films on three different semiconductor substrate materials, silicon, germanium and gallium arsenide. A detailed account of our investigation is presented in the literature [8-10]. The purpose of this paper is to summarize our results on Si and GaAs substrates and to compare them with the interface properties of a-C:H films on Ge substrates.

2. SAMPLE PREPARATION

All substrates were cleaned in UHV environment by sputtering with Ar^+ ions. The a-C:H films were deposited with either a Penning ion or Kaufman source operated with methane gas. The ion energy ranged between 100 and 500eV and the deposition rates were in the order of 6×10^{-4} to 5×10^{-2}nm/s. The electronic properties of the interface were investigated with XPS and UPS photoelectron spectroscopy (Leybold EA 10/100 and EA 11/100 spectrometers) and the atomic properties of the interface were studied with ion channeling and transmission electron microscopy. The details of the experimental procedures can be found in the literature [8-10].

3. INTERFACE PROPERTIES OF a-C:H FILMS ON Si, GaAs AND Ge

The electronic properties of the a-C:H/semiconductor interfaces are summarized in Table I which lists the bulk and interface core level energies measured with XPS. For comparison purposes we have added the results of the a-C:H/Au interface. It can be seen that the Au $4f_{7/2}$ core level does not shift at all whereas the C 1s core level shows a

TABLE I. Bulk and interface core level energies

Substrate	Core level	Binding energy (eV) bulk	interface	Shift (eV)
Si	C 1s	284.6	283.4	-1.2
	Si 2p	99.4	100.7	1.3
GaAs	C 1s	284.7	283.1	-1.6
	As 3d	41.0	42.6	1.6
	Ga 3d	19.0	19.7	0.7
	Ga 2p$_{3/2}$	1117.3	1117.7	0.4
Ge	C 1s	284.7	283.3	-1.4
	Ge 3d	29.3	30.6	1.3
Au	C 1s	284.6	284.2	0.4
	Au 4f$_{7/2}$	84.0	84.0	0.0

minor shift of 0.4eV. This indicates clearly that chemical bonds are not formed and that the a-C:H/Au interface is not reactive. This result is supported further by the absence of adhesion [11]. The a-C:H films deposited on Au can literally be blown off.

We now compare the results for the Au substrate with those for the Si substrate. From Table I we find that on Si both the Si 2p and C 1s core level energies measured in the interface region show a strong chemical shift with respect to their bulk values. This is in contrast to the Au substrate and signifies the formation of chemical bonds between Si and C. In fact, we have shown that the energy and the shape of the interface core levels are strikingly similar to those measured on bulk SiC [8]. From this we conclude that a SiC like compound forms at the interface between the Si substrate and the a-C:H overlayer. TEM analysis shows that the compound is amorphous and from ion channeling analysis we found that the interface region contains about 6×10^{16}Si/cm^2 [9]. It is therefore not surprising that we observed excellent adhesion of a-C:H films on Si substrates.

The situation on the GaAs substrate is different from that of Si. A glance at Table I reveals a strong shift of the As 3d and C 1s core levels in the interface region with respect to the bulk but only a minor shift of both the Ga 3d and 2p$_{3/2}$ core levels. We interpret the results such that chemical bonds are formed between As and C atoms but not be- tween Ga and C atoms. In fact, the existence of an As carbide compound (AsC$_3$) has been reported in the literature [12]. Therefore, we assume that a AsC$_3$ like compound forms at the interface between the GaAs substrate and the a-C:H overlayer. On the other hand, the interface core level energies of Ga are practically identical to those of metallic Ga [8]. This points to precipitation of metallic Ga at the a-C:H/GaAs interface. Again,

TEM analysis reveals that the interface layer is amorphous and ion channeling analysis shows that the amount of Ga and As atoms in the interfacial layer is about $2 \times 10^{16} at/cm^2$ [9]. We found that the adhesion of a-C:H films on GaAs is inferior to that on Si and attribute it to the disruption of Ga-As bonds and the precipitation of metallic Ga at the interface.

There is no Ge carbide compound known in the literature. Nevertheless, we observed good adhesion of a-C:H films on Ge substrates [10]. It is thus suspected that some kind of bonding must take place between Ge and C atoms. This is supported by the observed energy shifts of the Ge 3d and C 1s core levels at the interface compared to the bulk as shown in Table I. The amount of chemical shift is comparable to the a-C:H/Si interface. Therefore, we assume that a metastable compound must form at the a-C:H/Ge interface [10]. Recent results by Vedovotto et al. [13] and Sanders et al. [14] support our view.

Cross-sectional electron micrographs of the a-C:H/Ge(100) interface viewed in the [110] direction are shown in Fig. 1. The amorphous microstructure of the a-C:H overlayer is clearly discernible from the single crystal Ge substrate. The interface in Fig. 1a appears very smooth with a roughness of only 10 to 13Å. The enlarged view of the interface shown in Fig. 1b does not reveal a crystalline or polycrystalline second phase at the a-C:H/Ge interface. Thus, we conclude that the metastable interfacial compound is amorphous. Ion channeling analysis shows that the interfacial layer contains about $3 \times 10^{16} Ge/cm^2$ [10]. This is visible as a darker band in the micrograph of Fig. 1a because the atomic scattering amplitude of Ge is three times larger than that of carbon [15].

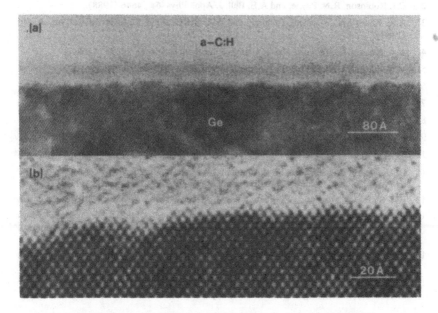

Fig.1. Cross-sectional electron micrographs of the a-C:H/Ge(100) interface viewed in the [110] direction showing a) a smooth interface and b) the absence of a microcrystalline second phase at the interface.

330

4. ADHESION QUALITY AND INTERFACE PROPERTIES

The quality of adhesion depends to a large degree on the properties of the interfacial layer that is formed between overlayer and substrate. The dominant adhesion mechanism of a-C:H films on Si, Ge and GaAs substrates is chemical bonding. It is responsible for the strong adhesion of a-C:H films observed on Si and Ge substrates. This is in contrast to Au substrates where the absence of chemical bonding results in a lack of adhesion. There is no doubt that the impingment of energetic $C_mH_n^+$ ions onto the substrate during deposition of the a-C:H films promotes carbide formation and, in case of Ge, is responsible for the formation of a metastable compound.

Atomic interdiffusion in the interfacial layer greatly supports the adhesion quality of a-C:H films on semiconductor substrates. In fact, interdiffusion and chemical bonding are interrelated because the formation of compounds necessitates the diffusion of atoms across the interface. There are factors, however, that can deteriorate adhesion. As we have seen on GaAs substrates the disruption of Ga-As bonds and the precipitation of metallic Ga are deleterious to the adhesion of the overlayer. In conclusion, we have observed that chemical bonding and atomic interdiffusion are necessary but not sufficient conditions for strong adhesion properties of a-C:H films on semiconductor substrates.

REFERENCES

1. H.-C. Tsai and D.B. Bogy, J. Vac. Sci. Technol. A *5*, 3285 (1987).
2. C.J. Robinson, R.N. Payne, and A.E. Bell, J. Appl. Phys. *64* , 4646 (1988).
3. A.A. Khan, J.A. Woollam, Y.Chung, and B.A. Banks, IEEE Electron Device Lett. *4*, 146 (1983).
4. A.A. Khan, J.A. Woollam, and Y. Chung, Solid-State Electron. *27*, 385 (1984).
5. M. Kakuchi, M. Hikita, and T. Tamamura, Appl. Phys. Lett. *48* , 835 (1986).
6. S. Craig and G.L. Harding, Thin Solid Films *97*, 345 (1982).
7. B. Dischler, A. Bubenzer, and P. Koidl, Appl. Phys. Lett. *42* , 636 (1983).
8. D. Ugolini, J. Eitle, P. Oelhafen, and M. Wittmer, Appl. Phys. A *48*, 549 (1989).
9. M. Wittmer, D. Ugolini, J. Eitle, and P. Oelhafen, Appl. Phys. A *48*, 559 (1989).
10. M. Wittmer, D. Ugolini, and P. Oelhafen, J. Electrochem. Soc. *137*, (in press).
11. D. Ugolini, P. Oelhafen, and M. Wittmer, in "Amorphous Hydrogenated Carbon Films", ed. by P. Koidl and P. Oelhafen, Les Editions de Physique, Les Ulis, France 1987, p.297.
12. F.A. Shunk, "Constitution of Binary Alloys", McGraw-Hill, New York 1985, p.49.
13. N. Vedovotto, J.M. Mackowski and P. Collardelle, Proc. Optical Interference Coatings Meeting, Tucson, April 12-16, 1988, Tech. Digest Series Vol. 6, Optical Society of America, Washington D.C., 1988.
14. P. Sander, M. Altebockwinkel, W. Storm, L. Wiedemann, and A. Benninghoven, J. Vac. Sci. Technol. B *7*, 517 (1989).
15. P. Hirsch, A. Howie, R.B. Nicholson, D.W. Pashley and M.J. Whelan, "Electron Microscopy of Thin Crystals", Krieger, New York 1977, p.503.

NON-EQUILIBRIUM MOLECULAR DYNAMICS SIMULATION OF THE RAPID SOLIDIFICATION OF METALS

CLIFF F. RICHARDSON AND PAULETTE CLANCY
School of Chemical Engineering, Cornell University, Ithaca, NY 14853, USA

ABSTRACT

The ultra-rapid melting and subsequent resolidification of Embedded Atom Method models of the fcc metals copper and gold are followed using a Non-Equilibrium Molecular Dynamics computer simulation method. Results for the resolidification of an exposed (100) face of copper at room temperature are in good agreement with recent experiments using a picosecond laser. At $T = 0.5\ T_m$, the morphology of the solid/liquid interface is shown to be similar to a Lennard-Jones model. The morphology of the crystal-vapor interface at 92% of T_m shows a significant disordering of the topmost layers. Difficulties with the EAM model for gold are observed. Comparison of the Baskes et al. and Oh and Johnson embedding functions are discussed.

SIMULATION METHOD

A crystal-vapor interfacial system of 4096 atoms and one of 8072 atoms was obtained in two stages. First, a bulk surfaceless system was equilibrated for 50,000 time steps at zero pressure and a desired substrate temperature (ie. an isothermal-isobaric simulation). This yields the density which is checked against the value obtained from previous work due to Foiles [1]. Second, a vapor containing ~ 10 atoms is placed above the solid. Four layers of atoms at the bottom of the solid are designated as the fixed lattice to emulate the presence of a deep substrate. Above this, four layers of atoms become the heat bath to maintain the substrate temperature. Above this, 24 layers of atoms (or 48 for the larger system), with 128 atoms per layer, form the rest of the crystal. This system is equilibrated for at least 100,000 time steps. Any surface disordering can be observed at this time.

The simulation of a rapid thermal process, such as is produced by laser processing, is made using a non-equilibrium Molecular Dynamics method recently developed in our group [2]. The energy fluence and pulse duration were matched to recent experimental data of MacDonald et al. [3] for the picosecond laser processing of copper and gold at room temperature (ie. 5 mJ cm^{-2} and 20 ps). The energy input to the system during this 20 ps "pulse" causes roughly half the system to melt. After the energy "pulse" is switched off, the system regrows quickly. The location of the transient solid/liquid interface is followed by an analysis of the properties of two-dimensional layers parallel to the original crystal/vapor interface. The layers are approximately one atomic diameter thick, ie. they encompass one lattice plane in the solid. The properties used in this analysis are density profiles, radial distribution functions, diffusion coefficients, order parameters and a computer graphical analysis.

Once the extent of the interfacial region is known (it typically extends over two or three "layers") the average interfacial properties are calculated. Of particular interest here are the position of the interface as a function of time (and hence its velocity) and the interface temperature as a function of time. The mechanisms involved in melting and resolidification are then deduced from the data produced.

The potential model used to describe the fcc noble metals, copper and gold is the Embedded Atom Method due to Baskes et al. [4]. It is a many-body, density-dependent potential fitted to properties of the solid, but not the liquid. Thus the correct melting point of the material is not guaranteed by use of this model.

RESULTS

(i) Results for (100) copper

The rapid melting and resolidification of an exposed (100) face of EAM copper was studied at three temperatures: $0.22\ T_m$, $0.53\ T_m$ and $0.92\ T_m$. The lowest temperature corresponds to room temperature and provided our first opportunity to test the quantitive capability of the non-equilibrium Molecular Dynamics simulation method. The results for the melt depth as a function of time is shown in Figure 1; this curve shows the expected qualitative features observed experimentally though, naturally, the length-scales of the simulation and experiment vary by orders of magnitude. The slope of this curve gives rise to the interface velocity. The maximum melt-in velocity is around $250\ ms^{-1}$, with a resolidification velocity, V_s, of $82\ ms^{-1}$. This compares well with the experimental result of "greater than $60\ ms^{-1}$" for V_s [3]. The undercooling produced at this temperature is around 100° (ie. ~7%). Once the location of the interfacial region has been made, the average temperature of that region can be obtained from the layerwise temperature/time data. Plotting smoothed values of the interface velocity against interface temperature gives the interface response function shown in Figure 2. The temperature for which the velocity is zero should correspond to the bulk melting point. Averaged over many such runs we estimate T_m to be 1380K (ie. within 2% of experiment); an unexpectedly good result which endorses the EAM model of copper. Chen et al. found a similarly good result for the melting point of EAM nickel [5]. Figure 2 shows no asymmetry around v=0, as expected for a system with a very small volume change on melting. A computer graphics analysis shows the initiation of melting due to the formation of a vacancy/interstitial pair. Rapid structure loss ensues followed by a loss of density. Resolidification on the (100) face proceeds in a continuous fashion without the nucleation of ledges, in a similar way to (100) Lennard-Jones models. We also found that while both system sizes (4096 and 8072 atoms) give the same resolidification velocity, the temperature reaches a constant undercooling only for the larger system. In the smaller system, it appears that the rapid heat conduction in metals requires a greater distancing between the solid/liquid interface and the heat bath.

The melting and resolidification of EAM copper was repeated at $0.52\ T_m$ in order to compare its behavior to a Lennard-Jones potential for which results already existed [2]. The simulation showed, somewhat to our surprise, that the extent of the solid/liquid interface of EAM copper was a little less diffuse than the equivalent Lennard-Jones interface at the same T/T_m value. However, as expected, the velocity of the regrowing interface was much faster for EAM copper than Lennard-Jones ($57\ ms^{-1}$ vs. $21\ ms^{-1}$).

A high temperature study was also made, at $0.92\ T_m$, in order to investigate the nature of the crystal/vapor interface and any manifestation of surface melting and to determine the interface velocity at small undercoolings.

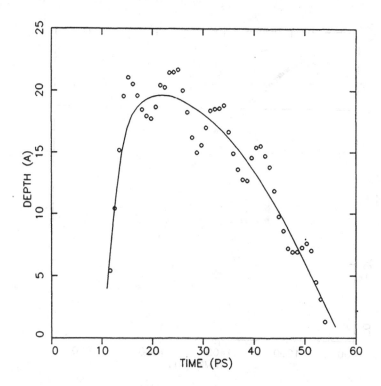

Figure 1. Melt depth versus time curve for the simulation of rapid solidification in (100) EAM copper at 300K using the larger system (8072 atoms). Circles are the raw simulation data which can be seen to oscillate strongly. The solid line is a best fit to data and is used in estimating the velocity.

The uppermost three layers of the crystal were found to be disordered, as shown by the order parameter, g(r) and the in-plane diffusion coefficients. This result of three disordered surface layers at $0.92\ T_m$ for an EAM model of copper is significantly more than the one layer of surface disorder found for a system of Lennard-Jones atoms at $0.92\ T_m$ [6]. At lower temperatures the crystal/vapor interface showed a simple termination of the bulk. At room temperature the inter-layer spacing at the surface was found to show a 6% contraction between the first two crystalline layers and a small contraction (0.5%) between layers 2 and 3. These results compare favorably with the results of Gustafsson [7] who used ion beam scattering to observe a 7.5% contraction and an 2.5% expansion for layers 1-2 and 2-3 respectively.

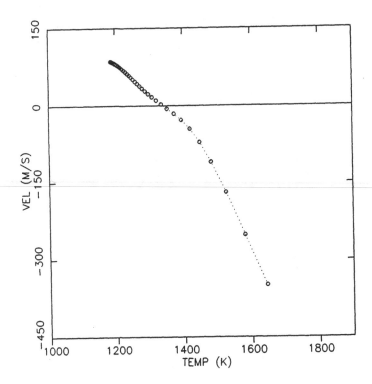

Figure 2. Smoothed values of the interface velocity as a function of interface temperature corresponding to the simulation results shown in Figure 1. Key as in Figure 1. The kink at around 1420K is probably due to difficulties in assigning the velocity close to the maximum in the melt depth curve.

A rapid solidification run at this temperature produced little undercooling and hence a relatively slow resolidification velocity of only 9 ms^{-1}. Since we are planning to study impurity segregation in EAM alloys, these results show that our simulations will have to be performed at temperatures very close to T_m to produce the slow interface velocities needed for the observation of segregation rather than trapping. We also studied the effect on the 300K (0.22 T_m) results of substituting the Foiles EAM model with the Oh and Johnson [9] EAM model. From a simulation viewpoint, there are no savings in terms of CPU time using the Oh and Johnson model. We found that the melting point of the Oh and Johnson model was ~ 150K too low. Thus the undercooling of the simulated system at 300K is less than in the experimental results and hence the resolidification velocity is less than we found for the Foiles model (V_s ~ 60 ms^{-1}). Thus, though this result for the resolidification of the Oh and Johnson model of EAM copper is numerically closer to experiment, it is not at comparable processing conditions.

(ii) Results for (100) EAM gold

MacDonald et al. [3] measured the resolidification velocity of gold at room temperature and found it to be over 100 ms^{-1}. However at 300K, using the same procedure as for copper, we found that the resolidification velocity was ~ 40 ms^{-1} (less than that for copper!). However, this simulation also suggested (from the temperature at which v=0) that the bulk melting point of EAM gold was ~ 965K, ie. ~ 350° below the experimental T_m value. This was confirmed by running an NPT bulk simulation at 1100K during which the atoms (originally placed on fcc lattice sites) melted to form a liquid. Since bulk surfaceless NPT simulations tend to superheat [8], the melting point of EAM gold is less than 1100K. We had previously verified that the solid properties of gold reproduced those of Foiles [1]. In view of the excellent reproduction of T_m for EAM nickel and copper, this result was disappointing. Since the melting point of EAM gold is ~ 965K, 300K is ~ 0.3 T_m. However the experimental results at 300K are closer to 0.2 T_m, this partially explains the low resolidification velocity found above. A rapid solidification NEMD simulation was then run at 0.2 T_m for a fairer comparison with experiment. The resulting resolidification velocity was found to be 47 ms^{-1}. Changing the form of the potential model, from the Foiles model to an Oh and Johnson model [9], did not change the melting point closer to the experimental value.

DISCUSSION

The results for the rapid solidification of EAM models of copper and gold are summarized in Table 1. While the EAM model for copper produces results surprisingly close to recent experimental data, the results for gold are less satisfactory, seemingly originating in deficiencies in the potential model. We are currently investigating a modified EAM model for gold.

TABLE I

Summary of results of NEMD simulation with a 20 ps "laser" pulse

Material	T_{subs}/K	T_m/K	V_s/ms^{-1}	Crystal/vapor surface disorder
(100) Cu	300	1364	82	No
(100) Cu	710	1332	62	No
(100) Cu	1240	1305	9	Yes
(100) Au	300	980	40	No
(100) Au	220	950	47	No

REFERENCES

1. S.M. Foiles, Phys. Rev. B, 32, 3409 (1985).
2. D.K. Chokappa, S.J. Cook and P. Clancy, Phys. Rev. B, 39, 10075 (1989).
3. C.A. MacDonald, A.M. Malvezzi and F. Spaepen, J. Appl. Phys., 65, 129 (1989).
4. M.S. Daw and M.I. Baskes, Phys. Rev. Lett., 50, 1285 (1983); idem, Phys. Rev. B., 29, 6443 (1984).
5. E.T. Chen, R.N. Barnett and U. Landman, Phys. Rev. B., 40, 924 (1989).
6. J.Q. Broughton and G.H. Gilmer, Acta. Metall., 31, 845 (1983). We corroborated this result to ensure that our designation of the extent of the disordered region was the same.
7. T. Gustafsson, unpublished work presented at a recent Solid State Physics seminar, Cornell University, November (1989).
8. D.K. Chokappa and P. Clancy, Molec. Phys., 61, 597, 617 (1987).
9. D.J. Oh and R.A. Johnson, J. Mat. Res., 3, 471 (1988).

PART VIII

Roughness and Interdiffusion

STUDY OF INTERFACIAL SHARPNESS AND GROWTH IN $(GaAs)_m(AlAs)_n$ SUPERLATTICES

D. Gammon, S. Prokes, D.S. Katzer[*], B.V. Shanabrook, M. Fatemi, W. Tseng[**], B. Wilkins and H. Dietrich
Naval Research Laboratory, Washington DC 20375-5000

ABSTRACT

X-ray diffraction, inelastic light scattering, photoluminescence and photoluminescence excitation spectroscopies have been employed to study the effects of growth conditions on the interfacial sharpness of GaAs/AlAs superlattices. In addition to describing which changes in the MBE growth procedures result in the sharpest interfaces, this study addresses the relative strengths and weaknesses of the experimental probes employed in the examination of interfacial irregularities.

INTRODUCTION

Advances in growth technologies allow artificially structured materials to be grown where individual layer thicknesses approach a monolayer. Many novel changes in material properties and the application of these materials to opto-electronic and electronic devices require that the interfaces between the epitaxial layers be near perfect. Deviations from perfection arise because of intermixing of the constituents of one layer into another, from growth conditions that do not allow two dimensional island growth or from the formation of misfit dislocations. Although in-situ characterization tools such as reflection high energy electron diffraction and AUGER spectroscopy provide some information regarding these imperfections, ex-situ characterizations tools, such as x-ray diffraction [1-5], photoluminescence (PL)][6-12], photoluminescence excitation spectroscopy (PLE)[13] and inelastic light scattering [14-18] have proven to be very valuable. We have employed these ex-situ characterization tools to the study of GaAs/AlAs superlattices. The combination of all of the techniques allows detailed information to be gained about the impact of MBE growth parameters on the perfection of the interfaces. Furthermore, because each of the techniques has been employed on the same samples, the comparison between the results from the different techniques allows the relative strengths and weaknesses of each probe to the examination of interfacial perfection to be determined.

GROWTH CHARACTERISTICS

The GaAs/AlAs superlattices were grown with a VG V80H MBE machine at substrate temperatures between 520-640C with a growth rate of ~1μm/hr. The As to Ga flux ratio was lowered to a value slightly above Ga rich growth conditions. Some samples were grown with interruptions of 90 seconds between the growth of alternate layers. During these interruptions only the shutters to the cation ovens were closed. All of the samples discussed in this paper had periods of 37 monolayers and the total superlattice thickness was ~3000Å.

EX-SITU CHARACTERIZATION

The x-ray diffraction measurements were performed on a powder diffractometer using CuKα radiation. The optical measurements were performed with a double monochromator with photon counting detection. A dye laser with Pyridine 2 dye pumped by an argon ion laser was used for the PLE measurements. The inelastic light scattering measurements were performed with either an argon ion laser or dye laser.

X-ray Diffraction

Shown in Figure 1(a) is an x-ray diffraction scan of a GaAs/AlAs superlattice grown at 580 C with 90 second interruptions at the interfaces. The large number of satellites observed in this sample proves that the interfaces are fairly abrupt. An analysis of the separation between satellites indicates that the period of the super-lattice is 37 monolayers. Shown in Figure 1(b) is a kinematic simulation [5]of the x-ray diffraction scan. This simulation was performed with an interfacial broadening of 2 monolayers, with individual layer thicknesses of 17 monolayers of GaAs and 20 monolayers of AlAs. If the period employed in the simulation was kept constant and the individual layer thicknesses were varied, the relative intensities of the even and odd satellites were not in as good agreement with the experimental data. Furthermore, if the interfaces were assumed to be perfectly sharp or broadened by 3 monolayers the intensities of the high order satellites were too large or small, respectively. For this sample, the comparison between the x-ray diffraction measurement and the simulation indicates that the interfacial broadening is between one and two monolayers. Because the simulation assumes that the interfaces are translationally invariant perpendicular to the growth direction, we obtain no information regarding the size or nature of islands at the interface.

We have also examined samples grown without growth interruptions at substrate temperatures between 520-640 C. In all cases, the intensities of the high order satellites were smaller for samples grown at the higher temperatures. This observation implies that the lower growth temperatures result in more abrupt interfaces. The samples grown at 520 C exhibited slightly smaller intensities for the high order satellites than those exhibited by the interrupted growth sample. Specifically, comparisons with the kinematic model indicated that the interfaces grown at 640 C and 520 C were broadened by 3-4 and 2-3 monolayers, respectively.

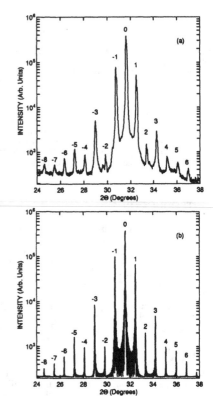

Figure 1 (a) X-ray diffraction scan of a $(GaAs)_{17}(AlAs)_{20}$ superlattice. (b) Simulation of a $(GaAs)_{17}(AlAs)_{20}$ superlattice. The numbers in the figure label the satellites.

Photoluminescence and Photoluminescence Excitation Spectroscopy

PLE spectroscopy has been particularly useful in characterizing the band structure of quantum wells because it's spectral form resembles that of an absorption spectrum. The PLE measurements that we report here were performed by observing the efficiency with which low energy impurity related luminescence is excited by above band gap excitation. PL and PLE have proven to be valuable in determining the size of islands that exist at the interfaces of GaAs/AlAs quantum wells [8,9,19]. This information is obtained because the size of the exciton Bohr orbit transverse to the growth direction is ~200 Å. If the transverse dimension of the interfacial fluctuations is much smaller or much larger than the Bohr orbit, the observed exciton spectral widths are very narrow and the interfaces have been characterized by the terminology pseudo-smooth and truly smooth, respectively [11,12,19]. In the case where the dimension of the Bohr orbit is comparable to that of the interface fluctuation, the exciton spectral line widths are large and the interfaces are classified as rough.

Shown in Figure 2(a) is the PL and PLE spectra obtained from superlattices prepared at 520 C with normal MBE growth. Similar spectra were obtained from the samples grown at the higher temperatures. The PL energy of first heavy hole exciton (1H) is shifted by ~ 10 meV from the peak position of the first heavy hole exciton (1H) observed in the PLE. In addition, the line widths observed in the PL and PLE are wide (10-15 meV). These observations imply at least one of the two interfaces of the quantum well is rough. The separation between the peaks of the PL and PLE occurs because the excitons recombine after thermalizing to the low energy regions of the quantum well (i.e. regions of larger average quantum well width). The exciton spectral line widths are broad because of inhomogeneous broadening arising from rough interfaces. In contrast, and as shown in Figure 2(b), the PL and PLE spectra of a sample grown at 580 C with 90 seconds growth interruption between interfaces are characterized by line widths that are significantly narrower (< 5meV). Furthermore, these spectra exhibit a multiplicity of peaks near the energy of the heavy hole (1H(a)-1H(c)) and light hole (1L(b)-1L(c)) excitons. Calculations based on the effective mass approximation that employ the accepted values of the carrier masses and band offsets of the GaAs/AlAs material system [20] indicate that the energy separation of these features is that characteristic of quantum wells that differ in width by a monolayer. These observations imply that one of the interfaces of the quantum well is truly smooth and that the other must be either pseudo-smooth or truly smooth. Furthermore, one notices that there is a correspondence between the peak energies observed in the PL and PLE spectra of Figure 2(b). The regions of the quantum well with width that give rise to the 1H(a) emission dominate the PL spectrum. In contrast, the PLE spectrum indicates that well widths that result in excitonic spectral energies of 1H(b) are dominant. Because the PLE spectrum resembles the absorption spectrum, we conclude that a large percentage of the quantum wells of the superlattice are characterized by well widths that result in the 1H(b) absorption. The PL is dominated by the 1H(a) emission because of intrawell thermalization to the low energy regions of the quantum well.[13]

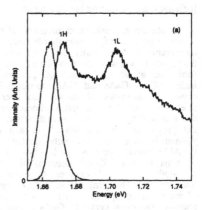

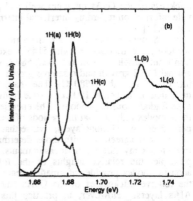

Figure 2 PL and PLE spectra from samples grown (a) at 520 C with conventional MBE and (b) at 580 C with growth interruptions. 1H and 1L label the heavy and light hole excitons, respectively.

The comparison between the spectra shown in Figure 2(a) and 2(b) indicates the value that PL and PLE spectroscopies have at characterizing interfacial structure that was not obtained with x-ray diffraction measurements. However, information regarding the thicknesses of the AlAs and GaAs layers of the quantum well can not be determined with the accuracy of the x-ray measurements. Specifically, the energies of the excitons are not strongly affected by variations in the 20 monolayer barrier widths of our quantum wells. In addition, although the absolute energies of the excitons are strongly affected by changes in the quantum well widths, the energies of the excitons obtained from one band effective mass models should not be expected to accurately characterize the width of the quantum well [16]. This inaccuracy occurs because of nonparabolicities of the electron dispersion curve of bulk GaAs and because of the simplistic boundary conditions normally employed for matching the envelope functions of the electron and hole states at the GaAs/AlAs interface.[21,22] However, after a calibration has been made between the structural parameters determined by x-ray diffraction and the corresponding energies obtained from PL and PLE, these optical measurements could be useful for determining the width of the quantum wells with similar interfacial structure.

Inelastic Light Scattering Measurements

Raman scattering measurements of the vibrational properties of superlattices have been shown to provide information that is complimentary to both luminescence and x-ray diffraction measurements. Specifically, the energies of the folded acoustic phonon modes combined with the speeds of sound in the two layers gives information about the period of the super-lattice.[18] In GaAs and AlAs, the dispersion curves for the optical phonons do not overlap in energy. Therefore, an optical phonon mode that is travelling along the growth direction can not propagate into the next layer and the phonons become confined to either the GaAs or the AlAs layers. The envelope function for the phonons is modeled by assuming that the amplitude of the oscillation drops to zero at the interface.[18] The combination of the energies of the confined optical modes with the bulk phonon dispersion curve of GaAs reveal the thickness of the GaAs quantum well and provide information regarding interfacial intermixing.[16,17]

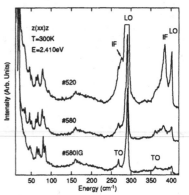

Figure 3 Raman scattering spectra from superlattices grown at 520 C (#520), 580 C (#580) with normal MBE techniques and a sample grown at 580 C with interruptions (#580IG). The energies of the interface phonons and the TO phonons are labeled by IF and TO, respectively.

Shown in Figure 3 are inelastic light scattering spectra obtained with 5145 Å excitation from samples grown at 520 C, 580 C with normal MBE growth and a sample grown with interrupted growth at 580 C. The sharp features that occur below 100 cm^{-1} arise from the zone folded acoustic phonons. The energies of these modes indicate that the periods of the superlattices are ~37 monolayers, a value that is in excellent agreement with those determined from the x-ray diffraction measurements. In principle, the relative heights of the folded acoustic modes provide information regarding the relative thicknesses of the GaAs and the AlAs layers. However, in practice this information can be difficult to obtain because of resonance effects that modulate the relative intensities of the folded acoustic phonon modes observed in the Raman scattering measurement.[18] In the regions from 250-300 cm^{-1} and from 350-405 cm^{-1} are the confined optical phonon modes of GaAs and AlAs, respectively. The confined LO optical modes will be discussed later. The transverse optic (TO) and interface (IF) mode are forbidden in this back-scattering geometry. The appearance of these modes in the spectra indicate that there is disorder, presumably interfacial, in the superlattice structure. It is interesting to note that the intensities of the modes are the smallest in the sample grown with interruptions. This sample also exhibited PL and PLE spectra

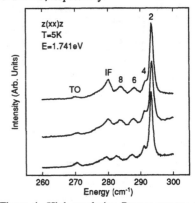

Figure 4 High resolution Raman spectra of the confined GaAs optical phonons of superlattices grown at 520 C (top), 580 C (middle) with normal MBE techniques and at 580C with interruptions.

characteristic of truly smooth interfaces and the x-ray diffraction indicated that the interfaces were between 1-2 monolayers broad. Although current Raman scattering studies have not been able to relate the intensities of the interface phonons quantitatively to types of disorder at the interface, it is an ongoing research topic in inelastic light scattering studies[23].

Shown in Figure 4 is a high resolution scan of the confined optical phonon modes of the GaAs layers in the superlattices. As noted previously, the forbidden interface modes become smaller as the perfection of the interfaces becomes greater. The peaks of the m = 2,4,6,8 confined optical phonon modes are denoted in the Figure. If the energies of the confined LO phonon mode of index m, E_m, follow the relationship,

$$E_m = E_{LO} - Cm^2/(n+1)^2,$$

where E_{LO} is the k=0 energy of the LO phonon in the bulk and n is the number of monolayers of GaAs, then the composition profile of the quantum well is rectangular.[17]   The dispersion curve for GaAs indicates that $C = 52$ cm^{-1}. For the sample with growth interruptions this relationship is obeyed and the well width is 16-17 monolayers, a value that is in good agreement with the diffraction measurements. For the samples grown with normal MBE techniques (i.e. no interruptions) this relationship is only approximately obeyed. Therefore, the composition profiles in these samples are not rectangular and interfacial broadening is more significant in these samples.[17]

Employing resonant Raman scattering techniques, where resonances with the 1H(a)-1H(c) excitonic transitions of Figure 2 are employed, we have been able to show that Raman scattering measurements are sensitive to interfacial fluctuations typical of truly smooth interfaces.   The details of these observations as well as additional information regarding the samples grown without interruptions will be discussed elsewhere in the proceedings of this conference.[24]

CONCLUSIONS

We have shown that inelastic light scattering and x-ray diffraction measurements provide information regarding the period and relative layer thicknesses of GaAs/AlAs superlattices. PL, PLE and inelastic light scattering measurements provide additional information regarding the nature of the island structure at the interface between the two materials.

ACKNOWLEDGEMENTS

We wish to acknowledge the partial support of the Office of Naval Research.

REFERENCES

*   ONT Postdoctoral Fellow
**   Current address: National Institute for Standards and Technology, Gathersburg, MD 20899

[1]   Armin Segmuller, P. Krishna and L. Esaki, J. Appl. Crys. 10, 1 (1977).
[2]   R.M. Fleming, D.B. McWhan, A.C. Gossard, W. Weigman and R.A. Logan, J. Appl. Phys. 51, 357 (1980).
[3]   V.S. Speriosu and T. Vreeland, Jr., J. Appl. Phys. 56, 1591 (1984).
[4]   S. Koshiba, S. Nanao, O. Tsuda, Y. Watanabe, Y. Sakurai, H. Sakaki, H. Kawata and M. Ando, J. Crystal Growth, 95, 51 (1989).
[5]   J.M. Vandenberg, R.A. Hamm, M.B. Panish and H. Temkin, J. Appl. Phys. 62, 1278 (1987).
[6]   C. Weisbuch, R.C. Miller, R. Dingle, A.C. Gossard and W. Weigmann, Solid State Commun. 38, 709 (1981).
[7]   L. Goldstein, Y. Horikoshi, S. Tarucha and H. Okamoto, Jpn. J. Appl. Phys. 22, 1489 (1983).
[8]   T. Hayakawa, T. Suyama, K. Takahashi, M. Kondo, S. Yamamota, S. Yano and T. Hijikata, Appl. Phys. Lett. 47 952 (1985).
[9]   B. Deveaud, J.Y. Emery, A. Chomette, B. Lambert and M. Baudet, Appl. Phys. Lett. 45, 1078 (1984).
[10]   D.C. Reynolds, K.K. Bajaj, C.W. Litton, P.W. Yu, Jasprit Singh, W.T. Masselink, R. Fisher and H. Morkoc, Appl. Phys. Lett. 46, 51 (1985).

344

[11] M. Tanaka, H. Sakaki and Y. Yoshino, Jpn. J. Appl. Phys. 25, L155 (1986).

[12] H. Sakaki, M. Tanaka and Y. Yoshino, Jpn. J. Appl. Phys. 24, L417 (1985).

[13] R.C. Miller, C.W. Tu, S.K. Sputz and R.F. Kopf, Appl. Phys. Lett. 49, 1245 (1986).

[14] C. Colvard, R. Merlin, M.V. Klein and A.C. Gossard, Phys. Rev Lett. 43, 298 (1980).

[15] A.K. Sood, J. Menendez, M. Cardona and K. Ploog, Phys. Rev. Lett. 54, 2115 (1985).

[16] G. Fasol, M. Tanaka, H. Sakaki and Y. Horikoshi, Phys. Rev. B38, 6056 (1988).

[17] Bernard Jusserand, Francois Alexandre, Daniel Paquet and Guy Le Roux, Appl. Phys. Lett. 47, 310 (1985).

[18] B. Jusserand and M. Cardona, in *Light Scattering in Solids V*, edited by M. Cardona and G. Guntheridt (Springer, Berlin, 1989), p. 49.

[19] Jasprit Singh, K.K. Bajaj and S. Chaudhuri, Appl. Phys. Lett. 44 805 (1984)

[20] R.C. Miller, D.A. Kleinman and A.C. Gossard, Phys. Rev. B29, 7085 (1984).

[21] S.R. White and L.J. Sham, Phys. Rev. Lett. 47, 879 (1981).

[22] L.J. Sham, Superlattices and Microstructures 5, 335 (1989).

[23] D. Gammon, L. Shi, R. Merlin, G. Ambrazevicius, K. Ploog and H. Morkoc, Superlattices and Microstructures 4, 405 (1988).

[24] D. Gammon, B.V. Shanabrook, S. Prokes, D.S. Katzer, W. Tseng, B. Wilkins and H. Dietrich, this conference.

A WEAK BEAM IMAGING TECHNIQUE FOR THE CHARACTERIZATION OF INTERFACIAL ROUGHNESS IN (InGa)As/GaAs STRAINED LAYER STRUCTURES

J. Y. Yao, T. G. Andersson, and G. L. Dunlop*.
Department of Physics, Chalmers University of Technology, S-412 96 Göteborg, SWEDEN
* Department of Mining and Metallurgical Engineering, University of Queensland, St. Lucia 4067 Queensland, AUSTRALIA

ABSTRACT

A transmission electron microscope weak beam imaging technique has been developed for the characterization of interfacial roughness in lattice strained (InGa)As/GaAs multiple layered structures. In this technique, the heterointerfaces of (100) type strained layers are imaged in an inclined projection with a g_{311} diffracted reflection at off-Bragg conditions which gives an enhanced contrast from variations in strained layer thickness. A calculation based on the kinematic theory of contrast was made in order to gain a better understanding of the contrast. The calculation suggests that the observed contrast is due to monolayer scale variations in thickness of the strained layers.

INTRODUCTION

The structural quality of lattice strained semiconductor heterostructures is of importance for their electronic and optoelectronic properties [1, 2]. Heterostructures of high structural quality are of considerable interest for application in high speed and optoelectronic devices [3, 4]. Recently a transmission electron microscope (TEM) weak beam imaging technique has been developed in our laboratory for the characterization of interfacial microstructures in lattice strained (InGa)As/GaAs multiple layers [5]. This technique enables the visualization of detailed interfacial microstructures, e. g., interfacial roughness, and is an aid to the fabrication of heterostructures of high structural quality. In practice, a conventional (110) type cross-section TEM specimen is tilted so that the Ewald sphere cuts between the split reciprocal lattice points, g_{311}, of unstrained GaAs and strained (InGa)As. Compared to the two beam condition for g_{311}, the contribution to the intensity of the g_{311} diffracted beam from GaAs is decreased since its reciprocal lattice point is off the Bragg condition. By contrast the contribution from the strained layer is increased since its reciprocal lattice point lies closer to the Ewald sphere than it would be in the GaAs g_{311} two beam condition. Thus in this imaging condition, the contrast is sensitive to the detailed microstructure of the strained (InGa)As layer.

In this article, a simplified theoretical calculation for the nature of the contrast in the weak beam imaging of strained single layer structures and the results from experimental observation will be presented. Contributions to the contrast from superlattice reflections of multiple layered structures, as observed for (AlGa)As/GaAs multiple layers [6], have not been included in this calculation.

THEORETICAL CALCULATION

The interpretation of contrast is based upon a simplified theoretical calculation, in which the kinematical theory of electron diffraction contrast [7] has been used. This theory has been found to give good qualitative agreement with experiments where images were formed by reflections with large diffraction deviations. It should thus be applicable to the present problem and should give a qualitative description of the nature of the contrast. An alternative approach could be to use the dynamical theory of contrast [8] for a more exact prediction of the contrast.

Figure 1 shows a schematic drawing of an $In_xGa_{1-x}As/GaAs$ strained single layer structure of (100) type for the purposes of this calculation. The geometry of the TEM specimen is of (011) type, i. e., the foil normal is parallel to the [011] direction, and a column, for calculation using the column approximation, is along the z-axis as shown Fig. 1(a). Figure 1(b) shows a construction of the Ewald sphere at the $3\bar{1}\bar{1}$ reciprocal lattice point. Assume that (i) the

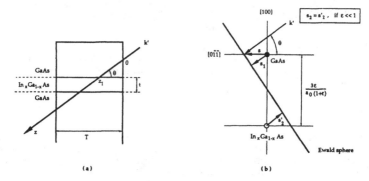

(a)   **(b)**

Fig. 1. Schematic diagram for the calculation of contrast associated with the g_{311} weak beam imaging. (a) Geometry of an $In_xGa_{1-x}As/GaAs$ strained single layer structure. The strained layer and the specimen thicknesses are t and T respectively. (b) Construction of the Ewald sphere at $g_{3\bar{1}\bar{1}}$. The vector s is the diffraction deviation of the specimen foil and s_1 and s'_2 ($s'_2 = s_2$, if $\varepsilon \ll 1$) are the respective distances from the reciprocal lattice points of the unstrained GaAs and the strained $In_xGa_{1-x}As$ to the Ewald sphere.

values of the extinction distance (ξ_g) are the same for all of the layers and (ii) all the layers are sufficiently thick so that the summation of contributions to the scattered wave from each atomic plane can be approximated by an integral (this is an approximation for the convenience of calculation). The strain in the $In_xGa_{1-x}As$ layer is a lattice displacement with respect to the GaAs lattice. The intensity of the diffracted $g_{3\bar{1}\bar{1}}$ beam can be determined according to the kinematic theory and expressed as

$$I = \left(\frac{1}{\xi_g}\right)^2 \{ I_b + I_a \cos[2\pi s_1 z_1 + \alpha] \} \tag{1}$$

where

$$\alpha = \pi s_1 \left(\frac{t}{\sin\theta} - \frac{T}{\cos\theta} \right) \tag{2}$$

$$I_b = \left(\frac{1}{s_1}\right)^2 \sin^2\pi\left(\frac{s_1}{\cos\theta}T + \frac{s_2-s_1}{\sin\theta}t\right) + \left(\frac{1}{s_1} - \frac{1}{s_2}\right)^2 \sin^2\pi\left(\frac{s_2}{\sin\theta}t\right) \tag{3}$$

and

$$I_a = \frac{2}{s_1}\left(\frac{1}{s_2} - \frac{1}{s_1}\right) \sin\pi\left(\frac{s_2}{\sin\theta}t\right) \cdot \sin\pi\left(\frac{s_1}{\cos\theta}T + \frac{s_2-s_1}{\sin\theta}t\right) \tag{4}$$

The intensity distribution Eq. (1) is therefore in the form of a cosine wave, having a depth periodicity of s_1^{-1} with a phase shift α, Eq. (2). The cosine wave has an amplitude I_a, Eq. (4), superimposed on a background I_b, Eq. (3). The diffraction deviations s_1 and s_2 in the above expressions as well as s, the diffraction deviation of the specimen, are illustrated in Fig. 1(b) (where $s_2 \approx s_2'$ provided that ε, the layer strain with respect to the unstrained GaAs due to the Poisson's effect, is much less than 1). The periodicity of the fringes on the micrograph can be determined according to the geometry shown in Fig. 1(a) and expressed as

$$p = \frac{\sin\theta \cdot \cos\theta}{s_1} \tag{5}$$

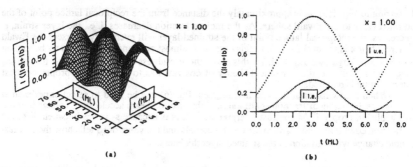

Fig. 2. Maximum intensities of the strained layer fringes, for an InAs/GaAs strained single layer structure at a g(3g) weak beam condition, (a) dependencies of the intensities upon the thicknesses of the strained layer t and the specimen foil T; and (b) the dependence of the upper and lower intensity limits on the thickness of the strained layer.

Table I. Physical parameters used for the calculation of $I_b + |I_a|$ (see Fig. 2).

λ (wave length of 200 keV electrons) [Å]	0.0251
a_0(GaAs) [Å]	5.6535
a_0(InAs) [Å]	6.0585
ϵ (relative to unstrained GaAs due to the Poisson's effect) [%]	14.3
s_1 [Å^{-1}]	8.60×10^{-3}
s_2 [Å^{-1}]	-2.38×10^{-2}
θ [°]	25.9

which is determined by the parameters of the unstrained GaAs layer only. The phase shift α is not expected to be sensitive to the variation of the strained layer thickness, t, since t is only a few monolayers. Unlike the periodicity, the amplitude I_a and the background I_b depend not only on the parameters of the unstrained GaAs layer, i. e., s_1, T and θ, but also on parameters of the strained layer, i. e., t and s_2. Since variation in strained layer thickness can be described by the variation of t and the local stress field induced by the structural imperfections can be approximated by a variation of s_2, the intensity I is sensitive to the microstructure of the strained layer. The resulting fringes are here called "strained layer fringes" since their intensities depend on the parameters of the strained layer.

The maximum intensity of the fringes is given by $I_b + |I_a|$, which is a periodic function of both T and t under a given diffraction condition. This is shown in Fig. 2 for an InAs/GaAs (x = 1.00) strained single layer structure at a g(3g) diffraction condition. The physical data used for the calculation of this dependence are listed in Table I. As shown in Fig. 2(a), the normalized intensity varies with both T and t, and it is very sensitive to variations of t on a monolayer scale. For t = 0, i. e., no strained layer, the fringe intensity for different values of T is much lower than for t = 3 ML (monolayer). Figure 2(b) shows the range of the dependence of the maximum intensity, $I_b + |I_a|$, on t for all possible values of T. Thus for a given value of t the maximum intensity variation is limited by the upper envelop, $I_{u.e.}$, and the lower envelop, $I_{l.e.}$. As can be seen in Fig. 2(b), for t less than 3 ML the maximum intensity increases with increasing t, while the opposite is true for t larger than 3 ML but less than 6 ML. The value of t corresponding to the maximum of $I_{u.e.}$ and $I_{l.e.}$ is therefore of importance for the interpretation of fringe contrast. This parameter, denoted by t_M, can be determined from Eqs. (3) and (4) and expressed as

$$t_M = \frac{\sin\theta}{2 |s_2|} \qquad (6)$$

Referring to Fig. 1(b), s_2 is approximately the distance from the reciprocal lattice point of the strained layer to the Ewald sphere. As the indium fraction decreases, i. e., the layer strain, ε, decreases, the reciprocal lattice point of the strained layer will eventually intersect the Ewald sphere when x is about 0.2 ~ 0.3. In this case, s_2 is almost zero and this leads to t_M becoming dramatically large [see Eq. (6)]. At this stage the usefulness of the kinematical theory has reached its limit and the dynamical theory of contrast has to be used if a proper contrast interpretation is required.

When a new material is to be investigated, Eq. (6) can be used to select experimental parameters, including diffraction conditions and the electron beam accelerating voltage in order to establish a proper TEM contrast interpretation. This is because s_2 varies between different diffraction conditions and between different materials and the value of t_M tells how the contrast should change with variation in the strained layer thickness.

EXPERIMENTAL OBSERVATION

Samples used for this study were grown by molecular beam epitaxy on GaAs (100) substrates in a Varian MBE-360 system [5]. The thin $In_xGa_{1-x}As$ layers were separated by thick GaAs layers (100 ~ 200 nm) and the thickness of each successive $In_xGa_{1-x}As$ layer was increased by nominally one monolayer (2.82 Å for GaAs). The strained layer thickness, t, and the indium fraction, x, were predetermined by means of the beam flux calibration. Cross-section thin specimens of (110) type were prepared from these MBE samples for investigation by TEM using a one-side non-rotation ion beam thinning technique [9]. The resulting thin foils were investigated in a Jeol 2000FX instrument operated at 200 kV.

Figure 3 is a g_{311} weak beam image with its corresponding diffraction pattern, in which the diffracted beam used to form the image is arrowed. The image shows three strained layers in an $In_{0.8}Ga_{0.2}As/GaAs$ multiple layered structure. The thicknesses of the individual layers are given in monolayer, ML. The 1-ML structure has quite uniform strained layer fringes and their intensities are the lowest of the three layers. The 2-ML structure has a domain-like pattern in the intensities of the fringes and on average the intensity is higher than the 1-ML structure. Finally, the 3-ML structure has bright disk-like contrast (arrowed and denoted by 3D) in the fringes. The scattered bright disks are 10 ~ 20 nm in size. Except for the bright disks the intensity of the fringes in the 3-ML structure is similar to that in the bright regions of the 2-ML structure.

From Eq. (6), it can be determined that t_M is approximately 4.5 ML for this structure under the diffraction conditions shown in Fig. 3. The intensity of the fringes should, therefore, increase with increasing layer thickness for these three layers. The intensities of the fringes in the three layer structure is in general agreement with the above theoretical predictions, but there are

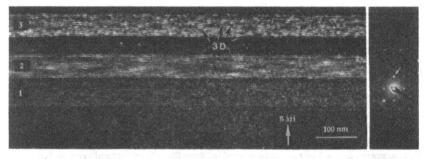

Fig. 3. A g_{311} weak beam image of a set of $In_{0.8}Ga_{0.2}As/GaAs$ layers taken at a diffraction condition close to g(3g). The domain-like contrast from the 2-ML structure and the disk-like contrast in the 3-ML structure (as arrowed and denoted by 3D) reveal interfacial morphologies corresponding to variations of the strained layer thickness on a monolayer scale.

many details, e. g., the domain-like and disk-like contrast, that are probably related to the detailed interfacial microstructures. The theoretical model suggests the following:

(i) Although the intensity of the fringes from the 1-ML structure is the lowest among the three layers, it is relatively more uniform than the other two. The thickness t of this strained layer is thus the most uniform one.

(ii) the domain-like contrast for the 2-ML structure is consistent with a difference in thickness between bright and dark regions. It can be noted that the dark areas are of approximately the same intensity as that for the 1-ML structure. The real thickness of this layer probably varies from 1 ML, in the regions of darker contrast, to 2 ML or more in the bright domains. The domain-like pattern is thus consistent with the presence of an interfacial roughness resulting from a 2-dimensional growth mode, i. e., growth by the addition of atoms to ledges that propagate across the growth surface. The non-homogeneous nature of the intensity in the regions of bright contrast suggests that parts of this layer may be more than 2 ML thick.

(iii) The scattered bright disk-like contrast shows randomly distributed areas that are thicker than their surroundings. The disk-like contrast thus suggests an interfacial roughness arising from a 3-dimensional mode of growth.

As can be seen in Fig. 2(a), variation of the specimen thickness can, of course, change the intensity of the fringes. To verify the dependence of the domain-like contrast for the 2-ML structure an investigation was carried out in which two g_{311} weak beam images of the same strained layer were taken from two different angles as schematically shown in Fig. 4(a). Interfacial microstructures in the strained layer, represented by "A", give rise to contrast in the

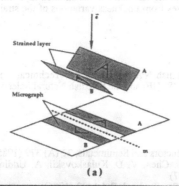

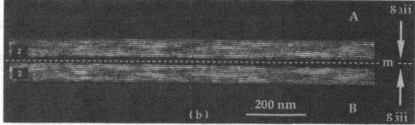

Fig. 4. (a) Schematic diagram of the experimental arrangement to verify that domain-like contrast is due to the interfacial roughness. (b) A pair of g_{311} weak beam images of a 2-ML-thick $In_{0.8}Ga_{0.2}As/GaAs$ layer structure obtained in the way shown in (a). The mirror symmetry about the line "m" of the domain-like contrast in the strained layer fringes in A and B can be recognized.

two micrographs for the two different incident angles of the electron beam. Contrast in the two micrographs that arises from interfacial microstructure should have a mirror symmetry about the line "m" as shown. The contrast due to the thickness variation of the specimen foil, however, is not expected to have such a symmetry. Figure 4(b) shows a pair of such g_{311} weak beam images in which a mirror symmetry for the domain-like contrast about "m" can be recognized. This investigation, therefore, proves that the domain-like contrast for the 2-ML structure arises mainly from the interfacial morphologies of the strained layers.

It can be determined from Eq. (5) that the periodicity of the strained layer fringes are 4.58 nm and 3.09 nm for $g(3g)$ and $g(4g)$ respectively. The periodicity of the fringes shown in Fig. 3 lies between these two values and is 4.0 ± 0.1 nm for all three layers. The Kikuchi line of the 3g diffraction spot, as marked in Fig. 3, shows that the diffraction condition is between $g(3g)$ and $g(4g)$. The observed periodicity is, therefore, in qualitative agreement with the theoretical calculation.

Since the contrast difference between two adjacent fringes is low, the spatial resolution of contrast in the plane of the interface is limited by the spacing of the fringes. This limit can thus be defined as the spatial resolution in the interface plane along the direction perpendicular to the fringes. For the strained layers shown in Fig. 3, the periodicity of the fringes is 4.0 ± 0.1 nm. The resolution is thus about 9 nm in the plane of the interface.

SUMMARY

To summarize, we have presented a TEM weak beam imaging technique, in which the g_{311} reflection is used to form the image, for the characterization of the interfacial roughness in lattice strained (InGa)As/GaAs multiple layers. It has been shown by both calculations and experiments that strained layer fringes can be observed by this weak beam imaging. The intensities of these fringes are highly sensitive to the variations in the strained layer thickness, and, under certain conditions, contrast that arises from thickness variations of the strained layer on a monolayer scale can be obtained.

ACKNOWLEDGEMENT

Support from the Swedish National Board for Technical Development (STU), STU's Technical Research Council (STUF) and the Swedish Natural Science Research Council (NFR) is gratefully acknowledged.

REFERENCES

1.   P. M. Petroff, Semiconductors and Semimetals, 22 (A) 379 (1985)
2   T. G.Andersson , Z. G. Chen, V. D. Kulaskovskii, A. Uddin, and J. T. Vallin, Appl. Phys. Lett. 51 752 (1987)
3   G. C. Osbourn, J. Vac. Sci. Technol. B 4 1423 (1986)
4   T. P. E. Broekaert, W. Lee, and C. G. Fonstad, Appl. Phys. Lett. 53 1545 (1988)
5   J. Y. Yao, T. G. Andersson, and G. L. Dunlop, in Chemistry and Defects in Semiconductor Heterostructures, Mat. Res. Soc. Symp. Proc. 148 p. 303-308 (1989)
6   R. Vincent, D. Cherns, S. J. Bailey and H. Morkoç, Phil. Mag. Lett. 56 1 (1987)
7   P. B. Hirsch, A. Howie, R. B. Nicholson, D. W. Pashley, and M. J.Wheloan, ELECTRON MICROSCOPY OF THIN CRYSTALS 4th ed.(BUTTERWORTHS, London, 1971) p. 156-194
8   See ref. 7 p. 208-246
9.   J. Y. Yao, and G. L. Dunlop. Inst. Phys. Conf. Ser. No. 93 2 93 (1988)

INTERFACIAL STABILITY AND INTERDIFFUSION EXAMINED
AT THE ATOMIC LEVEL

Y. KIM, A. OURMAZD, R.J. MALIK[*] AND J.A. RENTSCHLER

AT&T Bell Laboratories, Holmdel, NJ 07733

*AT&T Bell Laboratories, Murray Hill, NJ 07974

ABSTRACT

Using chemical lattice imaging in combination with vector pattern recognition, we obtain quantitative profiles of the chemical change across single interfaces with atomic plane resolution. We thus study interdiffusiuon across single GaAs/AlGaAs interfaces as a function of temperature, depth of interface beneath the surface, and doping. Since our technique is sensitive to interdiffusion coefficients as small as 10^{-20} cm^2/s, we can study atomic level changes at a single interface at the low temperatures used for many device processing steps (~700C). Our results show interdiffusion, and hence the layer stability depend not only on temperature and doping, but also on the distance of the interface from the surface. The implications of these results for the stability of multilayered structures are discussed.

INTRODUCTION

Modern crystal growth produces interfaces across which the composition changes precipitiously. Such systems are far from equilibrium and can relax by the introduction of extended defects, or by chemical interdiffusion. Here we investigate the chemical relaxation of interfaces, caused by interdiffusion between lattice-matched semiconductors. Using chemical lattice imaging combined with digital pattern recognition [1,2], we measure diffusion coefficients as small as 10^{-20} cm^2/s at single interfaces.

Our earlier work [3,4] on the HgCdTe/CdTe system has shown that interdiffusion depends stronly on the depth of the interface beneath the surface. Here we extend our work to the GaAs/AlGaAs system, measuring the interdiffusion coefficient as a function of temperature, doping and interface depth. Our results establish that (a) the depth-dependence of the interdiffusion coefficient is a general effect, (b) C-doping significantly enhances the interdiffusion coefficients from less than 10^{-21} cm^2/s to ~10^{-19} cm^2/s at 700C, and (c) that a stack of chemical interfaces can be used to investigate at the microscopic level the injection of native point defects from the surface.

EXPERIMENTAL

20 periods, each consisting of 50Å C-doped GaAs/50Å undoped GaAs/50Å Al$_{0.4}$Ga$_{0.6}$As were grown on GaAs (100) by Molecular Beam Epitaxy(MBE) at 600C. The carbon doping level was 3.7×10^{19} cm^{-3}. Bulk samples were annealed in the temperature range 650C to 750C in an evacuated ampule under As poor conditions, to induce enhanced intermixing in the p-type layers [5]. The annealing time at each temperature was chosen to produce the same small amount of carbon diffusion. TEM samples were prepared chemically, and chemical lattice images in the <100> zone axis were obtained at an accelerating voltage of 400KV with a JEOL 4000-EX high resolution transmission electron microscope. A digital pattern recognition method was used to obtain composition profiles at each interface before and after annealing [3,4].   •

Figure 1. Chemical lattice image of a GaAs/AlGaAs interface at a depth of 2500Åfrom the surface, annealed at 650C for 2.5hrs.

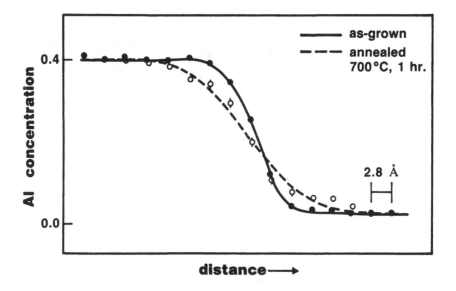

Figure 2. Composition profiles of a C:GaAs/AlGaAs interface at a depth of 300Å, as-grown (solid line), and after 700C, one hour anneal (dotted line). One standard deviation error bars are shown.

RESULTS AND DISCUSSIONS

Figure 1 is a chemical lattice image of a GaAs/AlGaAs interface. The analysis of such chemical images by the procedure described in Ref. 3 yields quantitative composition profiles across single interfaces. The composition profiles shown in Figure 2 refer to a single interface at a depth of 300Å, before and after annealing at 700C for one hr. Starting with the initial composition profile for a given interface, we solve the linear diffusion equation to fit the profile after the anneal, using the interdiffusion coefficient D as the adjustable parameter. This is carried out by a Marquardt procedure, which also yields the uncertainty in the deduced value of the fitting parameter D [6]. In this way, we measure D as a function of annealing temperature and interface depth.

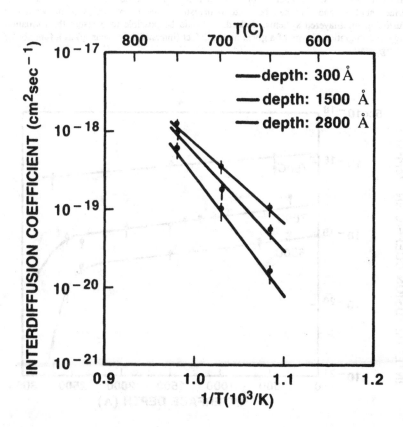

Figure 3. Arrhenius plot of the interdiffusion coefficient at C:GaAs/AlGaAs interfaces at three different depth.

Figure 3 is an Arrhenius plot of D vs $1/kT$ for C-doped interfaces at three different depths beneath the surface. Remarkably, the magnitude of the interdiffusion coefficient, as well as the activation energy for intermixing change strongly with depth. Since this behavior is also observed in the HgCdTe/CdTe system [3], we conclude that the depth-dendence of the interdiffusion coefficient is a general effect. This depth-dependence is more clearly displayed in Figure 4, where ln D is plotted as a function of the interface depth. At the lower temperatures (700C and particularly at 650C), ln D initially decreases linearly with increasing distance from the surface, but drops exponentially beyond a certain critical depth.

Our analysis, to be reported elsewhere, shows this effect to be related to the injection of point defects from the sample surface. In particular, interdiffusion in these systems is assisted by the presence of native point defects (interstitials and vacancies), whose concentration is negligible in our as-grown samples. For interdiffusion to occur, such native defects must be injected from the sample surface during the anneal. The interdiffusion coefficient is a sensitive function of the concentration of these defects at the particular interface studied, and thus can be used to investigate the microscopics of native defect diffusion in multilayered systems. Indeed, it should be possible to measure the formation energy and migration energy of a given native defect (interstitial or vacancy) as a function of its charge state.

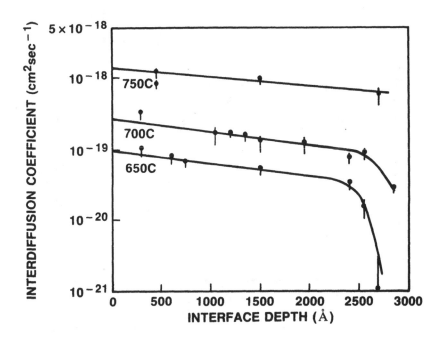

Figure 4. Plot of the ln D vs interface depth (z) at three different temperatures, for C:GaAs/AlGaAs interfaces.

Returning to interdiffusion, two important points emerge. First that the interdiffusion coefficient varies strongly with depth. Thus a measurement of this parameter is meaningful only if it refers to a single interface at a known depth. Second, it follows that the interface stability is also depth-dependent. Thus the layer depth must be regarded as an important design parameter in the fabrication of modern devices. This effect assumes additional importance when interdiffusion is also concentration dependent, leading to strong intermixing at very low temperatures [4].

CONCLUSIONS

The combination of chemical lattice imaging and vector pattern recognition allows the sensitive and quantitative measurement of interdiffusion at single interfaces, thus directly revealing the interdiffusion coefficient to be strongly depth-dependent. This effect stems from the need for the presence of native point defects for low temperature interdiffusion. Since the concentration of such defects in our as-grown sample is negligible, they must be injected from the sample surface during the anneal. Thus a stack of interfaces may be used to track the injection of native point defects, and measure their formation and migration energies. This represents a unique opportunity to measure parameters of fundamental and technological importance in compound semiconductors.

ACKNOWLEDGMENT

We acknowledge valuable discussions with A. Bourret.

REFERENCES

1. A. Ourmazd, W.T. Tsang, J.A. Rentschler, and D.W. Taylor, Appl. Phys. Lett. $\underline{50}$ 1417 (1987).

2. A. Ourmazd, D.W. Taylor, J. Cunningham and C.W. Tu, Phys. Rev. Lett. $\underline{62}$ 933 (1989).

3. Y. Kim, A. Ourmazd, M. Bode and R.D. Feldman, Phys. Rev. Lett. $\underline{63}$ 636 (1989).

4. Y. Kim, A. Ourmazd, R.D. Feldman, J.A. Rentshler, D.W. Taylor and R.F.Austin, MRS Sypm. Proc. Vol. $\underline{144}$ 163 (1988, Boston) Edited by D.K. Sadana, et al.

5. D.G. Deppe and N. Holonyak, Jr., J. Appl. Phys. $\underline{64}$ R93 (1988).

6. W.H. Press, B.P. Flannery, S.A. Teukolsky and W.T. Vetterling, Numerical Recipes (Cambridge University Press, Cambridge, 1986), chapter 14.

A STUDY OF INTERFACES IN GaAs/AlAs SUPERLATTICES
USING PHONONS

D. Gammon, B.V.Shanabrook, S. Prokes, D.S. Katzer*, W. Tseng**, B. Wilkins
and H. Dietrich
Naval Research Laboratory, Washington DC 20375-5000

ABSTRACT

Vibrational Raman scattering has been used in conjunction with luminescence, luminescence excitation spectroscopy, and x-ray diffraction to study the interfaces of GaAs/AlAs superlattices grown by MBE with and without growth interruptions at the interfaces. The confined LO phonon spectra clearly indicate that in
the sample grown with growth interrupts at least one of the interfaces in each GaAs
layer is truly smooth. In addition it is shown that the intensity of the interface
phonons probed via Raman scattering is sensitive to the state of the interface.

INTRODUCTION

It has recently become possible to grow GaAs/AlAs superlattices with very
high quality interfaces[1]. Novel MBE growth procedures such as the growth interruption technique enhance the growth of two dimensional islands at the interface
with resulting island sizes which have been estimated to be as large as 2000Å[2].
In contrast, samples grown with conventional MBE are thought to exhibit interfaces that are characterized by islands whose lateral dimension is ~ 10Å and only
in the best samples approach interlayer roughness of 1 monolayer[2]. These
samples have typically been characterized by x-ray diffraction [3], luminescence
[4], and Raman scattering[5] in terms of one dimensional models in which the interfaces are slightly interdiffused but laterally homogeneous. Because these new
MBE growth techniques create structures in which the interfaces are clearly not
homogeneous laterally, characterization techniques are now required to be sensitive to the spatial extent of the islands at the interface. Clearly a one dimensional
model with roughness represented only as interdiffusion is limited in characterizing this class of structures in which continued improvements in interface quality
will consist of increasing island sizes. In addition, our understanding of probes of
interfacial irregularities need to be improved so that the differences between the
two interfaces on each side of the GaAs layer can be better quantified[2].

The most convincing evidence for the existence of islands is provided by
luminescence[6] and luminescence excitation spectroscopies. In luminescence
measurements the probe of interfacial irregularities is the exciton. As the island
sizes become larger, the exciton line in the spectra splits into multiplets with energy splittings corresponding to those characteristic of quantum wells with widths
differing by a monolayer. This implies that at least one of the interfaces is perfect
over lateral distances larger than an exciton diameter (200Å in GaAs). For well
and barrier widths on the order of 50Å the exciton is direct (in real space and k
space) and is confined to a single well. The exciton energy is thus easily calculated
and sensitive to intra-well fluctuations in layer thickness. For thinner layers, the
exciton changes character and may not be as useful. In the smaller period superlattices, where excitons have not been as valuable in determining the type of interfacial disorder, a possible alternative is the optical phonon. Optical phonons in

GaAs/AlAs superlattices are confined to individual layers [5] and have energies which are sensitive to layer thicknesses in a way analogous to electrons, and in fact, show clearly well width fluctuations due to large island growth. In contrast to electrons, however, phonons are confined even in very thin layers. In addition, there exist phonons known as interface phonons which are localized at the interfaces and thus should provide microscopic information about the interface[5]. We present a Raman scattering study of the interfaces in samples grown by the growth interruption technique as well as conventional MBE. The samples we have chosen to study have layer thicknesses suitable for characterization with luminescence, luminescence excitation spectroscopy and x-ray diffraction. We correlate our Raman scattering results with those of the other techniques. Additional details from the other measurements will be presented elsewhere in these proceedings[7].

GROWTH AND CHARACTERIZATION

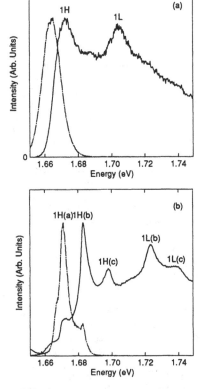

Fig. 1 PL and PLE from samples (a) #520 and (b) #580IG. 1H and 1L label the heavy and light hole excitons, respectively.

Three samples grown by MBE with 3000Å thick GaAs/AlAs superlattices and .5μm buffer layers on (001) GaAs substrates were studied. Two samples were grown by conventional MBE with substrate temperatures T=580C (sample #580) and T=520C (#520), and one sample was grown at T=580C in exactly the same way except that growth was interrupted (by shuttering the Ga and Al sources) at each interface for 90 seconds (#580IG). The intensities and splittings of the satellites in the x-ray diffraction measurements from the (200) planes can be fit by assuming 17 and 20 monolayers for the GaAs and AlAs layers, respectively, and by assuming a periodic superlattice with roughness at each interface of 2 monolayers for #580 and #520 and 1 monolayer for #580IG[7]. The Raman scattering measurements were done in the backscattering geometry with light polarized along the crystal axes: x=[100] and y=[010].

The luminescence and luminescence excitation spectra for two of the samples are shown in Figure 1. The spectra from the other sample (#580) is very similar to #520. Broad excitonic transitions appear in both the luminescence and excitation spectra for the two samples grown without growth interrupts (#520 and #580). This implies that the interfaces are characterized by islands with dimensions comparable in size to the Bohr orbit of the exciton[7]. The other sample (580IG), however, shows narrow lines which are split by energies which correspond to one monolayer fluctuations in the width of the quantum well. As mentioned above this indicates that at least one of the interfaces has large regions which are truly smooth. From previous studies we believe that under these growth conditions the interface where AlAs is grown on GaAs (top interface) is truly smooth over distances much larger than the exciton diameter, but that the other interface (bottom) is psuedosmooth (i.e. the islands are much smaller in size than the dimensions of the exciton). A recent study has shown that the bottom interface requires a higher growth temperature for the interface to become truly smooth[2].

RAMAN SCATTERING

There are three types of excitations observed in a Raman scattering measurement: folded acoustic phonons, confined optical phonons (LO and TO), and interface phonons (IF)[5]. Because of the superlattice periodicity, the acoustic phonon branches are folded over into a new Brillioun zone and give rise to a series of doublets in the low frequency Raman spectrum. The period of the superlattice can be found simply from the energies of the doublets and knowing the speeds of sound in GaAs and AlAs. In principle the relative intensities of the peaks are related through photoelastic theory to the sharpness of the interfaces and to the relative layer thicknesses of the materials in the superlattice. For our samples a resonance in the scattering is causing an additional modulation of the intensities of the folded acoustic phonon modes which makes such an analysis difficult.

Confined optical and interface phonons are derived from the bulk optical phonons of the constituent materials. The interface phonons are vibrations which decay exponentially away from the interfaces. Their energies are well described by a dielectric continuum model[5]. These phonons exist only for nonzero wavevectors parallel to the interfaces, and thus, in a backscattering geometry some type of symmetry-breaking disorder is required for these modes to be excited[8]. The precise origin of this disorder is still being studied, however our data shows that there is a clear correlation of the intensity of the IF with interfacial roughness. The other type of vibrations, confined optical phonons, are derived from the bulk phonons by requiring that the envelope function of the vibrational displacement goes to zero at the interfaces. This requirement is a result of the fact that the dispersion curves of the optical phonons of GaAs and AlAs do not overlap, and thus the vibrations are confined to one layer or the other. The energies of the confined modes are found in a similar way to those of confined electrons. For perfect interfaces the energies of the confined LO modes are found from an infinite square well model and given by,

$$E = E_{LO} - Cm^2/(n+1)^2, \qquad (1)$$

where C is taken from the bulk dispersion relation for the GaAs or AlAs phonons ($C=52$ cm^{-1} for GaAs), n is the number of monolayers in the layer, m is the order of the confined phonon and E_{LO} is the k=0 energy of the bulk LO phonon. The (n + 1) in the denominator comes from the boundary condition for the discrete lattice[5]. Equation 1 will not be obeyed if the interfaces are not abrupt. Specifically, for cases where there is severe interdiffusion at the interface, the quantum well will be better modelled as parabolic and the splittings between the energies of the confined phonons will go linearly with m. Thus, whether the energies go quadratically or linearly with m provides information on the abruptness of the interfaces. Although this is somewhat limited information and useful only for fairly rough interfaces where the thickness of the interfaces is a large fraction of the layer thickness, it has been used in many studies of interfaces in conventional MBE grown samples[9,10,11]. As shown below, it is also possible to measure well width fluctuations through the phonon spectrum when the interfaces become truly smooth. Therefore, our measurements indicate that Raman scattering is useful in studying higher quality interfaces[12].

Figure 2 displays the room temperature phonon spectra measured with the 5145Å laser line. In the spectra from these samples there are at least five orders of doublets at low energies due to folded longitudinal acoustic phonon scattering. The center energies of these doublets occur at multiples of $E=16$ cm^{-1}. A period of 37 monolayers is calculated for all three samples which is in excellent agreement with x-ray diffraction. There are no major differences in the folded phonon scattering for the three samples. The broad feature occurring around $E=160$ cm^{-1} is due to second order phonon scattering and will not be discussed. The n=2 LO phonon peak from the GaAs and AlAs occurs at 291 cm^{-1} and 401 cm^{-1}, respectively. Weaker lines due to the corresponding TO phonons, which are normally forbidden in backscattering from the (100) surface, also appear. In addition, in #580 and #520 there are broader lines be-

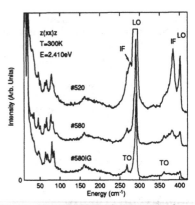

Fig. 2 Raman scattering spectra from samples #520, #580, and #580IG. The energies of the interface phonons and the TO phonons are labeled by IF and TO, respectively.

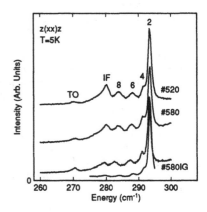

Fig. 3 High resolution Raman spectra of the GaAs optical phonons. The top three traces are of the three samples, #520, #580 and #580IG, at $E_L=1.741$eV. The bottom trace is #580IG at $E_L=1.647$eV.

tween the TO and the LO peaks labeled IF which correspond to interface phonons. These features are much weaker or absent in the sample with growth interrupts. The IF and LO scattering is larger by a factor of five in #520. This is possibly due to an enhanced defect-induced Fröhlich scattering mechanism but it is not yet well understood. Selection rules are such that the m=odd modes are not excited in this scattering geometry[5].

High resolution Raman scattering spectra in the energy region around the GaAs optical phonons is shown in Figure 3. The top three spectra in Figure 3 were taken at T=5K with 1.741 eV radiation. The confined phonons for n=2,4,6 and 8 are clearly seen in all three samples. The interface phonon scattering is present in all three samples, but it is again considerably weaker in the sample grown with interrupts (580IG). The energies of the confined phonons are constant for laser energies in this energy region. As the laser energy is reduced the luminescence becomes too strong and the phonons above m=2 cannot be resolved. However when the laser energy is tuned below E_L=1.68 eV for #580IG the phonon spectrum is again resolved but the confined phonons occur at different energies. A comparison between the phonon spectra from sample #580IG with the two different laser energies is shown at the bottom of Figure 3. This change in the phonon spectrum can be understood in terms of mov-

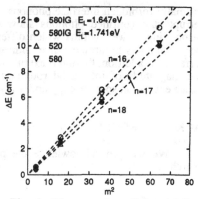

Fig. 4 The energies of the confined phonons shown in Figure 3 relative to the bulk LO phonon energy are plotted versus the mode number squared, m^2. The dashed lines are the results of calculations using Eqn. 1.

ing the laser energy out of resonance with the exciton in one well and into resonance with the exciton in a wider well. The relationship between the two laser energies and the luminescence excitation spectrum can be seen in Figure 1. The strong luminescence intensity does not allow us to measure the resonance profile, but an analysis of the phonon energies as discussed below supports this interpretation.

In Figure 4 the energy shifts of the confined LO phonons are plotted as a function of m squared. For a square well the energy shifts will lie on a straight line. The calculated energy shifts for several GaAs layer thicknesses are also shown. The energy shifts for 580IG clearly lies on a straight line for both laser energies, whereas the energy shifts for the samples grown without interrupts deviates slightly. In addition, the layer thicknesses calculated from the energy shifts for the two different laser energies differ by exactly one monolayer. This is clear evidence that there are large islands in at least one interface. If both interfaces were truly smooth we would expect the data to lie right on one of the calculated lines for an integer number of monolayers. This is in fact what happens, but because the calculation for the phonon energies is somewhat uncertain, we can not be sure whether or not this agreement in fortuitous. In part, this uncertainty is due to the errors in the value of the C coefficient in Eq. 1 and partly due to some uncertainty in the boundary conditions at the interfaces which are used in the calculation. Thus we do not know from our measurements whether one or both of the interfaces are truly smooth.

SUMMARY

In summary, we have studied samples grown with and without growth interrupts at each interface with Raman scattering. The Raman scattering probes the interface through the energies of confined LO phonons, and through the intensities of normally forbidden interface phonons. It is found that in a sample grown with interrupts the confined phonons show clearly two distinct sets of energies corresponding to well widths which differ by exactly one monolayer, and thus in a way analogous to luminescence, indicate that there is at least one truly smooth interface. In addition, the interface phonon scattering is shown to be correlated with the magnitude of interfacial roughness as measured by luminescence, luminescence excitation spectroscopy and x-ray diffraction[7].

ACKNOWLEDGEMENTS

We wish to acknowledge the partial support of the Office of Naval Research.

REFERENCES

*   ONT Postdoctoral Fellow
** Current address: National Institute for Standards and Technology, Gathersburg, MD 20899

[1]   H. Sakaki, M. Tanaka and J. Yoshino, Jpn. J. Appl. Phys. 24, L417 (1985).
[2]   M. Tanaka and H. Sakaki, Superlattices and Microstructures 4, 237(1988).
[3]   R.M. Fleming, D.B. McWhan, A.C. Gossard, W. Weigman and R.A. Logan, J. Appl. Phys. 51, 357 (1980).
[4]   C. Weisbuch, R.C. Miller, R. Dingle, A.C. Gossard and W. Weigmann, Solid State Commun. 38, 709 (1981).
[5]   B. Jusserand and M. Cardona, in Light Scattering in Solids V, edited by M. Cardona and G. Guntheridt (Springer, Berlin, 1989), p. 49.
[6]   C.T. Foxon, J. Crystal Growth 95, 11 (1989).
[7]   D. Gammon, S. Prokes, D.S. Katzer, B.V. Shanabrook, M. Fatemi, W. Tseng, B. Wilkins and H. Dietrich, this conference.
[8]   D. Gammon, L. Shi, R. Merlin, G. Ambrazevicius, K. Ploog, and H. Morkoc, Superlattices and Microstructures 4, 405 (1988).
[9]   B. Jusserand, F. Alexandre, D. Paquet, and G. LeRoux, Appl. Phys. Lett. 47, 301 (1985).
[10] D. Levi, S.L. Zhang, M.V. Klein, J. Klem, and H. Morkoc, Phys. Rev. B 36, 8032 (1987).
[11] G.W. Wicks, J.T. Bradshaw, and D.C. Radulescu, Appl. Phys. Lett. 52, 570 (1988).
[12] G. Fasol, M. Tanaka, H. Sakaki and Y. Horikoshi, Phys. Rev. B38, 6056 (1988).

PART IX

Grain Boundaries and Nanophase Materials

STRUCTURE OF A NEAR-COINCIDENCE Σ9 TILT GRAIN BOUNDARY IN ALUMINUM

M. J. Mills, G. J. Thomas, M. S. Daw, and F. Cosandey*
Sandia National Laboratories, Livermore, CA 94551-0969
*Department of Mech. and Mat. Sci., Rutgers University, Piscataway, NJ 08854.

ABSTRACT

A systematic study of the structure of tilt grain boundaries in aluminum has been initiated. High resolution transmission electron microscopy is being used to examine the interface structure of several bicrystals with <110> tilt axes. In this paper, we report the structure determination of a grain boundary close to the Σ9 (2$\bar{2}$1) symmetric orientation. The grain boundary plane, which appears wavy at lower magnification, is actually composed of atomically flat microfacets. Two distinct, symmetric structures with (2$\bar{2}$1) boundary planes have been identified within individual microfacets. These observations have been compared with structures calculated using the Embedded Atom Method. The semi-quantitative comparison between the observed and predicted grain boundary structures is accomplished using multislice image simulations based on the calculated structures. The results of these comparisons and the evaluation of the relative energies of the microfacets are discussed.

INTRODUCTION

Grain boundaries play a key role in determining the macroscopic behavior of polycrystalline materials. A growing body of evidence indicates that the physical properties of grain boundaries depend upon their structure. For example, Watanabe [1] has demonstrated experimentally that high-coincidence boundaries tend to exhibit greater resistance to grain boundary sliding and intergranular failure at elevated temperatures in pure metals. The transmission of slip from one grain to the next is also postulated to depend strongly on grain boundary structure [2]. The segregation of impurities to grain boundaries, and diffusion within them, is also likely to be strongly influenced by its detailed atomic structure.

Extensive theoretical work has been directed toward computing the structure, energetics and properties of a variety of grain boundaries. Recent advances in the resolution of medium-voltage transmission electron microscopes have made possible the experimental examination of a limited number of grain boundaries in close-packed metals–specifically pure tilt boundaries in which the tilt axis is along a low index direction. Therefore, the direct comparison between theoretical and experimental grain boundary structures in metals is now possible for certain special geometries.

In this study, a nearly pure tilt grain boundary in a bicrystal of aluminum has been examined. Study of the pure metal case is important in its own right, and also as a prelude to consideration of impurity effects. The controlled bicrystal geometry makes possible a detailed determination of the atomic structure of the boundary and also approaches the theoretical ideal of an infinitely long, periodic boundary. Thus, these specimens are ideally suited for attempts to quantitatively match theoretical and experimental grain boundary structures. The observed structures are compared with image simulations based on atomistic calculations using the Embedded Atom Method (EAM) [3].

EXPERIMENTAL AND COMPUTATIONAL PROCEDURES

The aluminum bicrystal used for these observations was provided by C. Goux and M. Biscondi of the Ecole National Superiere des Mines, St. Etiennes, France. A bicrystal very close to the symmetric Σ9 (2$\bar{2}$1) coincidence site lattice (CSL) orientation was prepared using a horizontal zone melting technique, with a [110] growth direction common to both crystals. Thin foil specimens perpendicular to the growth direction were obtained by electrical discharge machining and subsequent electropolishing in a 70% methanol/30% nitric acid solution. HRTEM was performed using a JEOL 4000EX operating at 400kV. The important microscope parameters are: spherical aberration coefficient = 1.0 nm, spread of defocus = 10 nm and semi-angular beam divergence = 0.7 mrad.

The observed boundary structures have been compared with relaxed atomic configurations calculated with EAM. A detailed description of this method and its application to grain boundary calculations can be found elsewhere [4]. The structures presented were calculated using a molecular statics approach with the local volume potential for aluminum developed by Voter, et al. [5]. A Σ9 (2$\bar{2}$1) boundary was constructed with a 3.5 nm thick slab of atoms on either side of the boundary. The relaxations were performed with free surfaces parallel to the (2$\bar{2}$1) boundary plane and periodic conditions in the other orthogonal directions. In the search for the lowest energy configurations, the starting structures were varied by applying different rigid-body displacements of one crystal relative to the other, and by removing atom layers at the boundary. Simulated images based on these calculated structures have been generated by the multislice method using the Electron Microscopy Software package of Stadelmann [6]. Supercells for the image simulations had projected dimensions 3.4 nm x 0.86 nm and were sampled at 512 x 128 points.

RESULTS AND DISCUSSION

In the CSL construction, the symmetric Σ9 tilt boundary has a (2$\bar{2}$1) boundary plane and a 38.9° disorientation between crystals. Lower magnification views of two portions of the bicrystal boundary are shown in Figures 1a and 1b. In fact, the tilt disorientation measured from these HRTEM micrographs is about 39.5° and a small twist component of approximately 0.3°

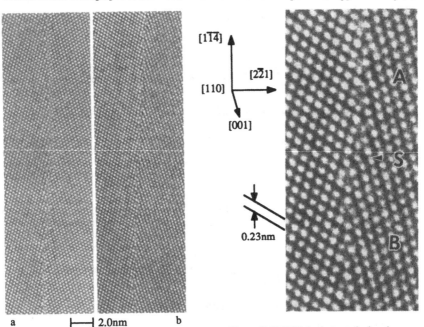

a   ⊢——⊣ 2.0nm   b

Figure 1: Overall boundary structure in which the boundary plane is (a) close to the symmetric (2$\bar{2}$1) and (b) about 7° from the (2$\bar{2}$1). In all figures indicated directions refer to left-hand crystal.

Figure 2: HRTEM micrograph showing a portion of the boundary with two microfacets (A and B) and a grain boundary step (S). White atomic column image obtained using defocus of -70nm and a crystal thickness of about 8nm.

about the [2$\bar{2}$1] has been determined from Kikuchi pattern analysis. In addition, the boundary plane can vary considerably. For example, in Figure 1a the overall boundary plane is aligned almost exactly along (2$\bar{2}$1), while in Figure 1b the overall boundary plane deviates by about 7° from the symmetric (2$\bar{2}$1) orientation. Although these boundaries appear to be wavy and lacking periodicity along their length, closer inspection reveals that they are actually composed of short, atomically flat boundary segments. This microfacetting of the boundary is best observed by inclining the micrograph and sighting along the boundary plane. In a magnified view of part of a boundary shown in Figure 2, these microfacets (indicated by A and B) are seen to lie precisely in a (2$\bar{2}$1) symmetric orientation and are separated by steps in the boundary plane (indicated by S).

As can be seen in Figure 2, the microfacets A and B exhibit two distinctly different structure types (see also the detailed images of Figures 4 and 6). Microfacet A has glide-plane symmetry along the (2$\bar{2}$1) boundary plane, with a relative translation of one crystal with respect to the other in the [1$\bar{1}$4]. The magnitude of this translation is about 0.05 nm. In contrast, microfacet B exhibits a mirror-plane symmetry with no relative translation in the boundary plane. The type A and type B microfacet structures are found to alternate periodically along most of the boundary regions which have been observed. The appearance of both structures for essentially the same imaging conditions (objective lens defocus and specimen thickness) clearly supports the concept of structural multiplicity which has been proposed based on theoretical studies [7]. It should be noted that regions with different translational states have been observed using diffraction contrast imaging in a Σ9 boundary in an FCC Cu-Si alloy [8].

The apparent multiplicity in the structure of the symmetric microfacets has been investigated using EAM calculations. Four unique, stable structures have been found after testing numerous starting configurations. The minimized structures are shown in Figures 3 and 5 and the calculated boundary energies for each are listed in Table 1. The filled and open symbols indicate atoms in adjacent (220) layers. The corresponding image simulations for these structures are also shown. The projected potential for each structure is superimposed on the simulation so that the relationship between image intensities and atomic column positions can be seen readily.

The first two structures shown in Figures 3a and 3c both have glide-plane symmetry in the [110] projection, with a relative translation of 0.06 nm in the [1$\bar{1}$4]. These structures differ in that there is a half-period (a/4[110]) shift of one crystal with respect to the other along the tilt axis in Figure 3a, while no such shift is present for Figure 3b. This shift is apparent by comparing the matching of the [001] rows across both boundaries. However, in the [110] projection these two structures are indistinguishable, as demonstrated by the image simulations of Figures 3c and 3d, and they have very similar boundary energies.

Both simulations provide an excellent qualitative match with the experimental image for the type A microfacet which is shown in detail in Figure 4. Comparison of the projected potential positions with the intensities in the simulated images demonstrates that there should be a very close correspondence between image intensities and atomic column positions. Therefore, the same projected potential has been superimposed on the experimental image in Figure 4, and good agreement is also obtained between the calculated structure and the image intensities. In order to determine whether a shift is actually present along the tilt axis, a diffraction contrast technique such as the α-fringe method [9] will be necessary.

The calculated structure shown in Figure 5a displays mirror-plane symmetry with no relative translation in the [1$\bar{1}$4]. However, inspection of the simulated images for this structure shown in Figure 5c reflects a local grain boundary structure which is clearly different from the experimental image for the type B microfacet shown in Figure 6. Notice that there are two image intensities at arrowed position in Figure 5c, and only one in the experimental image. It should be noted that this structure has a much higher energy than the other structures considered in this study.

The three structures discussed thus far (Figures 3a-b and 5a) are very similar to those previously calculated for the Σ9 (221) 38.9° by Pond, et al. [10] using a pair potential for aluminum. While structures consistent with the type A microfacet have been calculated, a match to the type B microfacet could not be found while maintaining the same periodicities as in the previous calculations. However, if reconstruction is allowed along the tilt axis, the structure shown in Figure 5b results. The periodicity along the tilt axis is now a[110], with the only difference between half-periods being a slight shift in the position of one atom in alternate (110)

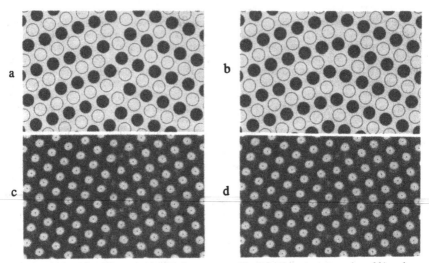

Figure 3: Atom plots of the two calculated structures with glide-plane symmetry (a and b) and the corresponding image simulations (c and d) for comparison with Figure 4 (defocus =-70, crystal thickness = 7.5nm). The atom positions are superimposed as black symbols on the simulations.

[1$\bar{1}$4]

[2$\bar{2}$1]

[110]

[001]

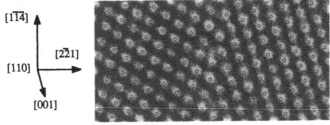

Figure 4: HRTEM micrograph showing the detailed structure of the type A (glide-plane symmetric) structure. White atomic column image obtained using a defocus of -70nm and for a crystal thickness of about 8nm. Superimposed as open circles are the projected atom positions for both calculated structures of Figure 3.

Table 1: The symmetry and boundary energies for the stable structures calculated EAM. Structure types are distinguished as having glide-plane (GP) and mirror-plane (MP) symmetry.

Structure Shown in Figure:	Structure Type	Boundary Energy (mJ/m^2)
3a	GP	281
3b	GP	300
5a	MP	435
5b	MP	307

layers. This "zig-zagged" atomic column is indicated by the arrow in Figure 5b. The structure has been tested with respect to its stability and it is found to persist even when small shear displacements between the two crystals are imposed prior to minimization. This demonstrates that the structure indeed corresponds to a local energy minimum, and is not simply metastable and dependent on precise symmetry conditions.

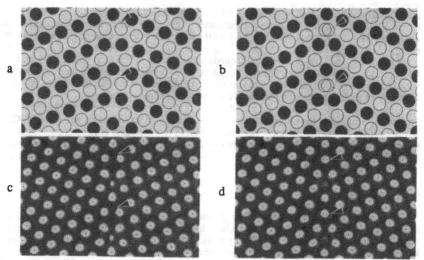

Figure 5: Atom plots of the two calculated structures with mirror-plane symmetry (a and b) and the corresponding image simulations (c and d) for comparison with Figure 6 (defocus =-70, crystal thickness = 7.5nm). The atom positions are superimposed as black symbols on the simulations.

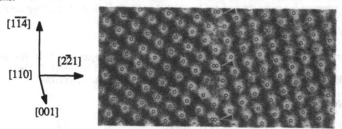

Figure 6: HRTEM micrograph showing the detailed structure of the type B (mirror-plane symmetric) structure. White atomic column image obtained using a defocus of -70nm and for a crystal thickness of about 8nm. Superimposed as open circles are the projected atom positions for both calculated structures of Figure 5.

While this reconstruction effectively destroys the mirror-plane symmetry of the structure, the boundary nevertheless appears to be mirror symmetric in the [110] projection. The image simulation shown in Figure 5d indicates that the "zig-zagged" atomic column is imaged as a single intensity since the projected potential positions are only 0.052 nm apart. All other intensities closely correspond to atom column positions. The excellent agreement between the calculated structure and the intensities in the experimental image is shown in Figure 6. It should be noted that other, more complex reconstructions along the tilt axis are possible and may yield similar images. However, the relatively simple reconstruction shown here leads to a boundary energy which is only slightly higher than for the glide-plane symmetric structures. Thus, the atomistic calculations suggest that this reconstruction is a viable explanation for the appearance of the mirror-symmetric boundary structure. A similar doubling of the periodicity has been observed by Bourret, et al. [11] along the tilt axis of a $\Sigma 3$ twin boundary in germanium.

The reason for the highly facetted boundary morphology seen in Figure 1 is not yet clearly understood. Although the microfacetting maximizes the area of grain boundary that is comprised of the low energy type A and B structures, the presence of the steps increases the total energy of the boundary. Lattice displacements are associated with each of the steps, as can be seen in Figure 1 by sighting along the close packed planes. These long range strains are suggestive of grain boundary dislocations (GBDs). Each GBD will contribute an additional strain and core energy to the total boundary energy. In the case of Fig. 1b, the steps in the boundary accomodate the deviation of the overall boundary plane from the ideal, symmetric orientation. However, the overall boundary plane in Figure 1a is very close to the symmetric (2$\bar{2}$1), yet the boundary is also highly facetted. Thus, the underlying cause of the facetting is not uniquely related to the deviation from the symmetric boundary orientation.

The array of steps or GBDs are probably present to provide an additional tilt disorientation, thereby accounting for the difference between the actual bicrystal tilt angle and the ideal CSL disorientation. Similarly, the small twist component in this boundary may be accomodated by the screw character of the GBDs. The nature of the GBDs are presently being characterized to ascertain whether a relationship exists between their spacing and the measured geometrical parameters of the boundary.

CONCLUSIONS

The structure of a grain boundary in aluminum has been analyzed using HRTEM. Regardless of the overall orientation of the boundary plane, on the atomic scale it is found to be comprised of short microfacets separated by steps. Two distinct microfacet structures have been identified—one with glide-plane symmetry and the other with mirror-plane symmetry. The lowest energy structures calculated using EAM are found to be in excellent agreement with the observed microfacet structures. The comparison between observed image intensities and calculated atom positions is justified based on multislice image simulations. From the initial configurations which have been tested, the only relaxed configuration consistent with the mirror-plane symmetric structure is a reconstructed one in which the periodicity is doubled along the tilt axis.

ACKNOWLEDGEMENTS

The authors would like to thank J. P Scola for the sample preparation, L. A. Brown and G. L. Gentry for their technical assistance. This work is supported by the U. S. Department of Energy, BES-Materials Sciences, under contract number DE-AC04-76DP00789.

REFERENCES

1. T. Watanabe, *Met. Trans.*, **14A**, 531 (1983).
2. A. H. King and D. A. Smith, *Acta Cryst.*, **A36**, 335 (1980).
4. M. S. Daw and M. I. Baskes, *Phys. Rev.*, **B29**, 6443 (1984).
4. S. M. Foiles, *Acta Metall.*, **37**, 2815 (1989).
5. Voter and Chen, *MRS Proceedings*, **82**, 175 (1987).
6. P. Stadelmann, *Ultramicrosopy*, **21**, 131 (1987).
7. V. Vitek, A. P. Sutton, G. J. Wang and D. Schwartz, *Scripta Metall.*, **17**, 183 (1983).
8. C. T. Forwood and L. M. Clarebrough, *Acta Metall.*, **30**, 1443 (1982).
9. R. C. Pond and V. Vitek, *Proc. R. Soc. Lond. B*, **357**, 453 (1977).
10. R. C. Pond, D. A. Smith and V. Vitek, *Acta Metall.*, **27**, 235 (1979).
11. A. Bourret, L. Billard and M. Petit, *Inst. Phys. Conf. Ser.*, **76**, 23 (1985).

ON THE ATOMIC STRUCTURE OF BOUNDARIES IN VERY SMALL METALLIC AND BIMETALLIC PARTICLES.

M. JOSE-YACAMAN,   M. AVALOS,   A. VAZQUEZ,   S. TEHUACANERO,
P. SCHABES AND R. HERRERA.
Instituto de Física, Universidad Nacional Autónoma de México.
Apartado Postal 20-364, 01000 México, D.F., México.

1. INTRODUCTION.

The study of boundaries in small particles is of great
interest in areas such as catalysis, metallurgy and surface
science. A particularly interesting case is that of metallic
particles containing one or several metals. In the present work
we present a number of examples of different mono and
bimetallic systems.

2. EXPERIMENTAL METHODS.

Mono and bimetallic particles were grown by a number of
techniques such as metal evaporation in UHV using several
sources or by precipitation from solutions containing the metal
ions. Some of the Pd samples were prepared in-situ in the
electron microscope under near UHV conditions (1). Samples were
studied in a JEOL 4000EX microscope and with a JEOL 100CX with
STEM attachment. Images were digitized with either a TV-Camera
or a scanning microdensitometer and then processed using
different types of filters (2). The final images were displayed
in an Innovion system attached to a Vax 780 computer.

3. BOUNDARIES IN SMALL PALLADIUM PARTICLES.

A general view of Pd particles obtained by evaporation is
shown in fig. 1a); a processed image of a single particle is
shown in Fig. 1b). The Fourier transform of the particle, shown
in Fig. 1c) clearly indicates the overall five-fold symmetry of
this particle. In a previous paper Gao et. al.(3) have
discussed the structure of what they call Penta-Twinned gold
particles. We have obtained similar results to those of Gao
et. al.(3) in the case of Pd. For instance, it is quite common
to see that the radial twin boundaries do not join along a
common edge, on the decahedral or Penta-Twinned particles, as
shown in Fig. 1b). This contrasts with earlier observations by
Ijima and Ichihasi (4). Most of the particles observed here are
formed by a combination of coherent and shifted boundaries,
assumed "incoherent" by Gao et.al.(3). In fact, the type of
particle shown in Fig. 1b) is the most common type found under
the present experimental conditions. None of the particles
observed in this work had a perfect decahedral shape with all
coherent twin boundaries.
Gao et. al. have explained the non-common edge effect in

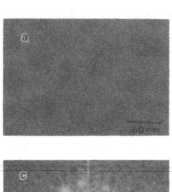

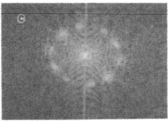

Figure 1. a) General view of Pd particles grown in near UHV in the microscope. b) Computer processed image of a particle with a Penta-Twinned Structure. c) Fourier transform of the previous particle.

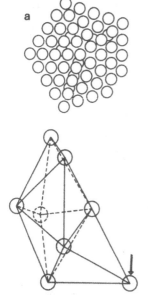

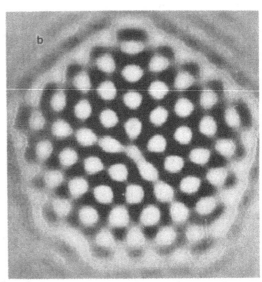

Figure 2. a) Model for the decagonal seed for the Penta-Twinned particle. b) Calculated image at a defocus of Δ=110 nm.

decahedral particles as the result of the minimization of the elastic strain energy of the particles. According to those authors, elastic distortions are necessary to fill out the space on a decagonal particle.

However Herrera et. al. (2) have offered an alternative explanation for the structure of Penta-Twinned particles based on the R-Model introduced by Romeu (5). In this model the initial seed for the particle will be a decagonal arrangement of seven atoms at which an extra atom is added in one of the corners of the decahedron. This results in the growth of a particle with a stacking fault-like defect along the boundary as shown in Fig. 2a). The resulting calculated image is shown in Fig. 2b) which indeed reproduces the experimental images. It appears then that the structure of Penta-Twinned particles is determined by Growth kinetics.

4. BOUNDARIES IN Au-Pd BIMETALLIC PARTICLES.

In the case of particles grown by colloidal methods, it is very easy to produce alloyed particles. In many cases the structure of the bimetallic particles tends to be either single crystalline and then grow very fast or single twinned and then have a rounded shape as shown in Fig. 3a). The Fourier transform of this image shows the single twinned structure of this particle. It is interesting to note that no Penta-Twinned particles were observed at this size level. When particles grow larger they tend to form Penta-Twinned particles. However the shape is more rounded than in the mono-metallic case. Fig. 4 shows the case of a particle of Pt/Rh which has five twins but with an overall rounded shape.

5. MODIFIED BOUNDARIES

A very interesting subject of study is the modification of boundaries by heat treatment in H_2. Fig. 5a) shows a particle of Pt/Rh which was produced by co-evaporation of the metals onto a SiO_2 substrate and then heated in H_2 at $650^{\circ}C$ during several hours (6). The particle clearly shows two portions, one faceted and one rounded. Fig. 5b) and c) shows a processed images of the boundary between the two portions. An optical diffraction study indicates that the upper portion of the particle corresponds to Pt and the lower part to a super lattice produced by the alloying of Pt-Rh; as seen on the figure the boundary between the two portions shows strain and some disorder.

6. ACKNOWLEDGEMENTS

We would like to thank Mr. L. Rendón, Mrs. C. Zorrilla and Mr. A. Sánchez for technical support. We also thank Prof. Lanny Schmidt of University of Minnesota for providing the Pt/Rh

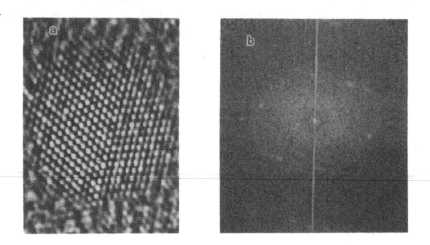

Figure 3. a) Image of Au–Pd small particle showing a single twin. b) Fourier transform of the previous particle.

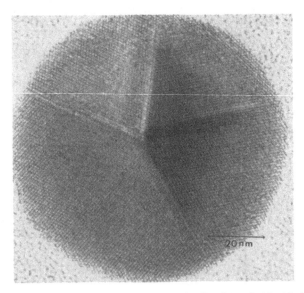

Figure 4. Image of a particle of Pt/Rh Colloidal particle showing five-twins. Note the complex structure of the twin boundaries.

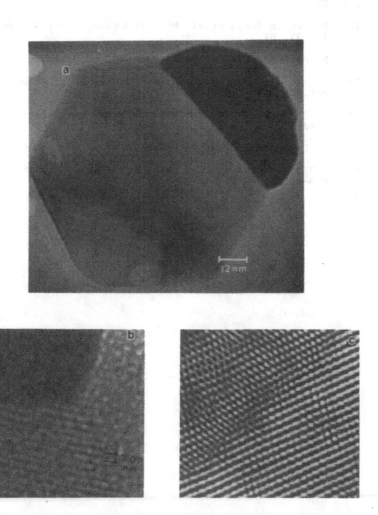

Figure 5. Particle of Pt/Rh produced by co—evaporation and then heated in H₂. a) Bright field image. b) Computer processed image of a portion of the boundary. c) The central part of the boundary between the rounded and faceted parts.

samples. One of us (MJY) was a Guggenheim fellow during the period that this work was made.

REFERENCES

[1]   M. Avalos Borja et.al. MRS Proceedings Series. Vol. $\underline{139}$, (1989), $\underline{81}$.

[2]   R. Herrera et.al. Scripta Met. $\underline{23}$, 1555 (1989).

[3]   Pei-Yu Gao, W. Kunath, H. Gleiter and K. Weiss. Scripta Met. $\underline{22}$, 683 (1988).

[4]   S. Ijima and T. Ichihashi. Phys. Rev. Lett. $\underline{56}$, 616 (1986).

[5]   D. Romeu. Acta Met. In Press.

[6]   T.P. Chojnacki and L.D. Schmidt. Journal of Catalysis $\underline{115}$, 473-485 (1989).

STRUCTURE AND PROPERTIES OF INVERSION DOMAIN BOUNDARIES IN β-SiC

W. R. L. LAMBRECHT, C. H. LEE AND B. SEGALL.

Department of Physics, Case Western Reserve University, Cleveland, OH 44106

ABSTRACT

The structure of inversion domain boundaries in β-SiC (i.e. boundaries between domains with inverted Si and C positions) is investigated by means of a Keating model. For the (110) boundary, the relaxation of the C-C and Si-Si bonds towards the ideal bond lengths which occur in the diamond structure can basically be achieved by a rotation of neighboring Si-C bonds. For the (001) boundary, it is achieved by varying the spacing between the domains. The electronic properties and total energy of formation of the relaxed (110) boundary are studied by means of linear muffin-tin orbital calculations. The interface localized states in the semiconducting gap are mainly due to the Si-Si bonds and lead to a semimetallic situation near the interface.

INTRODUCTION

Compound semiconductor thin films grown epitaxially on a Si(001) substrate often contain inversion domain boundaries (IDB) as a result of the occurence of mono-atomic steps on the substrate. Inversion domain boundaries are boundaries between domains with inverted cation and anion positions. They are also called anti-phase boundaries and sometimes anti-site domain boundaries [1]. For cubic (or β) SiC grown on Si(001) by chemical vapor deposition [2], they were identified by means of convergent beam electron diffraction studies by Pirouz et al. [3]. Their relation to steps on the surface is confirmed by the fact that they can largely be eliminated by growing on a surface which is a few degrees off from from a (001) surface [4]. In that case the surface is dominated by double atomic steps [5]. For SiC, the large lattice-mismatch of 20 % introduces considerable strain due to the interface dislocation along the step line, which exists both for the single and double atomic step. This dislocation has a Burgers vector $\vec{b} = \vec{t}_1 - \vec{t}_2$, where $\vec{t}_{1(2)}$ are the translation vectors of each crystal relating the two terraces separated by the step. These occur in addition to the usual misfit dislocation arrays [6].

Numerous observations by Pirouz and coworkers [3] show that IDBs do not preferentially occur along a specific crystallographic plane, but, rather "meander" through the specimen in an irregular fashion. Sometimes they are facetted, in which case a new type of dislocation occurs at their intersection. Here we present a study of the atomic relaxation near the (110) and two types of (001) boundaries, one with only C-C bonds and one with only Si-Si bonds.

We have studied the relaxation of the structure of these boundaries by means of the Keating model [7], using empirically determined parameters [8]. For the (110) boundary, the structure obtained is close to the bond-rotation model we have proposed earlier [9]. For the (001) boundary the model consists basically of a rigid displacement of the two domains normal to the boundary. In both cases, the major changes in the bond lengths are confined to the layers nearest to the interface. A small but slowly decaying strain field is superposed on the local bond-length relaxation.

In the second part of the work, we present results of electronic structure calculations performed by means of the linear muffin-tin orbital method [10]. Total energies are calculated within the local density functional framework [11]. A full length paper on the electronic structure calculations for the (110) IDB will appear elsewhere [9].

Mat. Res. Soc. Symp. Proc. Vol. 159. ©1990 Materials Research Society

STRUCTURAL RELAXATION

In the Keating model [7], the elastic energy is given in terms of bond-stretching and bond-angle variation terms by the expression

$$\Delta E_{el} = \sum_{i}^{bonds} \frac{3}{8d_i^2}\alpha_i(r_i^2 - d_i^2)^2 + \sum_{i\neq j}^{bond\ pairs} \frac{3}{8d_i d_j}\beta_{ij}(\vec{r}_i \cdot \vec{r}_j - \vec{d}_i \cdot \vec{d}_j)^2. \qquad (1)$$

Here, $\vec{r}_i$ is a vector along a bond pointing from one atom to its neighbor and $\vec{d}_i$ is the corresponding vector in the ideal tetrahedrally coordinated structure, (i.e. $\vec{d}_i \cdot \vec{d}_j = -d_i d_j/3$ since the vectors enclose the tetrahedral bond-angle of 109.4712°). In spite of its simplicity, the Keating model is known to be fairly reliable for systems which deviate only slightly from the ideal tetrahedral coordination. This is the case in the problem studied here. The force-constant parameters α_i and β_{ij} are determined by fitting to the elastic constants and the transverse and longitudinal optical phonon frequency difference as described by Martin [8]. The experimental values used were taken from Ref. [12]. The parameters used are given in Table I. For bond-pair terms involving C-C (or Si-Si) and Si-C bonds, the arithmetic average of the β-s has been used.

TABLE I: Keating model parameters

bond	d(Å)	α (N/m)	β (N/m)
Si-C	1.89	95	22
C-C	1.54	129	85
Si-Si	2.35	48	14

The Keating model using empirical parameters for the pure solids admittedly represents a quite simplified model for the IDB. Our justification for the transferability of the parameters is based on the fact that the coordination is tetrahedral throughout. Furthermore, some tests for the (110) interface indicate that, at least qualitatively, our conclusions are not very sensitive to the precise values of these parameters. As we will see below, the energy of formation of the boundary has an important electrostatic contribution arising from the partially ionic character of SiC and the opposition of like charges near the interface. One might thus need an additional ionic interaction term for the Si-Si and C-C interactions. This point warrants further investigation. Nevertheless, the fact that the Keating model in its present implementation yields a relaxation energy in fair agreement with the value obtained from the electronic structure calculations, gives some a-posteriori evidence that the Keating model is not at all unreasonable.

The minimization of the energy with respect to the atomic positions is carried out within a model with periodic boundary conditions and using a modified Broyden (conjugate gradient) method [13].

The results are shown in Fig. 1. This figure shows the (110) and (001) supercells used in the original, unrelaxed SiC structure with the relaxed bond lengths indicated. Each cell contains two boundaries and in the case of the (110) boundary the bond-length changes are symmetrical because the two interfaces are equivalent.

First, we discuss the (110) interface. We note that the Si-Si and C-C bond lengths at the interface differ significantly from the Si-C bond length (by about 18 %). All the other bond-length relaxations are of the order of 1 % and slowly decay with distance from the interface. This situation is similar to that near a dislocation: there is a core region with strong bond-length changes and a rather slowly decaying strain field, which presumably can be adequately described by means of continuum elasticity theory. The interface bond lengths corresponding to the "wrong bonds" relax very closely to their ideal values given

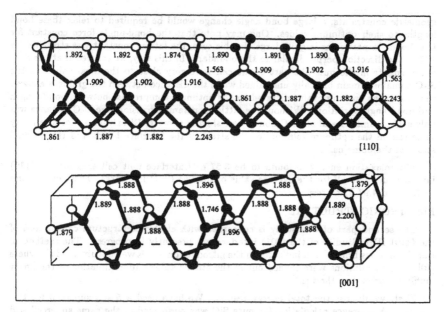

Fig. 1: Structure and bond-length relaxation near (110) and (001) inversion domain boundaries. White circles: silicon, black circles: carbon

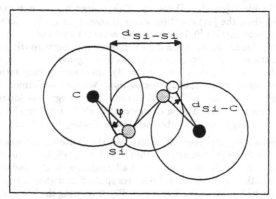

Fig. 2: Bond rotation relaxation model

in Table I. The relaxation mode essentially corresponds to a rotation of the Si-C bonds such that the C atoms approach each other and the Si atoms move away from each other. This model is shown in Fig. 2. The angle of rotation is about 6-7°. One finds that the Si-C bond lengths in the first plane parallel to the interface remain almost unchanged as a result of the opposite movement of the C and Si-atoms. This rigid rotation model was proposed earlier by us [9] and is used in our electronic structure studies presented below. We found that this model is obtained exactly if the bond-angle force constants β are neglected.

In the case of the (001) interface, the deviation of the interface Si-Si and C-C bond lengths from their ideal value is somewhat larger. This is due to the fact that given the geometric constraints, a large bond angle change would be required to relax these bond lengths to their optimum values. One may note that the bond-angle force constant for C-C bonds in particular is quite large. In this case, the lattice constant was stretched by 1 % in the direction perpendicular to the boundary. In this way, we assured that the Si-C bond lenghts in the bulk-like part of the cell were very close to the ideal value, i.e the Si-C domains remain basically unstrained while the distance between the grains is allowed to relax to the optimum value dictated by the optimum compromise between bond-length and bond-angle variations. Without the lattice constant expansion, the C-C distance was only slightly smaller and all Si-C bonds were slightly shorter. Clearly, size effects can play some role in the optimum relation for the strain of the interface bonds and the residual strain in the domains.

The relaxation energy is found to be 3.67 eV/interface unit cell area for the (110) interface and 2.98 eV for the average value of the two (001) interfaces. These correspond to 4.32 and 4.97 J/m^2 respectively.

ELECTRONIC STRUCTURE

The second part of this study is concerned with electronic structure calculations of the (110) IDB. Studies of the (001) interface are presently in progress. The method of calculations has briefly been indicated in the introduction and is well documented elsewhere [10]. The calculations were carried out in the atomic sphere approximation of the linear muffin-tin orbital method [10].

First, we discuss the total energy results. We have used a 5 + 5 supercell for this purpose. A reference calculation for pure SiC was performed in the same supercell and in exactly the same way as for the IDB cell so as to eliminate errors due to inequivalent finite $\vec{k}$-point sampling. The energy of formation per interface unit cell area is found to be 5.91 eV for the ideal (unrelaxed) model. This energy has an important electrostatic component arising from the juxtaposition of like charges near the interface. Using a frozen potential shape approximation [9, 14] where only constant shifts of the bulk potentials per atomic layer are included, we obtain a 3.2 eV interatomic electrostatic component and a 2.1 eV change in the sum of the occupied one-electron eigenvalues. Except for some small second order terms in the change in charge density, these are the only terms which appear in the energy difference within this approximation. We also performed a calculation for the simple bond-rotation model discussed above. The energy was lowered to 3.02 eV/ interface unit cell area. The relaxation energy of 2.89 eV is in reasonably good agreement with the relaxation energy (3.67 eV) obtained in the Keating model, as pointed out above.

Next, we consider the electronic structure. As the C atomic valence levels are about 4 eV lower than the Si levels, one expects that Si-C bonds will have energies intermediate between C-C and Si-Si bonds. We thus expect a broadening of all band structure features near the interface with Si-Si bond related features split-off at higher energies and C-C bond related features pushed down to lower energies. Fig. 3 showing the layer projected densities of states for the bond-rotation relaxed model confirms these expectations. The above effects lead to interface states in the lower bandgap and a peak which splits off above the valence band maximum and is pushed into the bottom of the conduction band. Inspection of the $\vec{k}$ dispersion of these states shows that each interface state peak corresponds to two bands. This is a supercell effect due to the interaction between the two interfaces in each cell.

The fact that the states near the valence band maximum are pushed into the conduction band near the interface means that the system becomes metallic. As the metallicity is due to a band near the center of the Brillouin zone which slightly overlaps with the conduction band, the situation is very similar to a typical semimetal such as Sn. Although the thin superlattices studied here are found to be metallic, a sample with a lower density of IDBs will not become metallic as a whole. Rather, one should think of the IDB as a two-

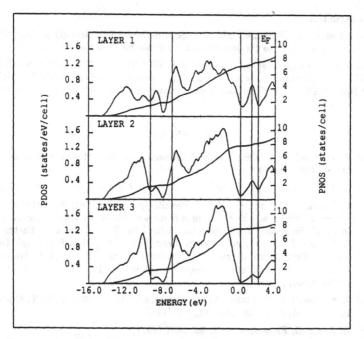

Fig. 3: Layer projected densities of states (PDOS) and its cumulative integral (PNOS) near the (110) inversion domain boundary using the bond-rotation relaxation model. The vertical lines indicate the band edges and the Fermi-level. Layer 1 is closest to the interface.

dimensional semimetal with two Schottky barriers back to back. Near the interface, the densitity of interface states in the semiconductor band gap is very high (one per interface unit cell area). Thus the Fermi level will be pinned at these states near the bottom of the conduction band. Depending on the doping type in the semiconductor and the position of the Fermi level in the bulk of the material, a band bending will occur so as to align the Fermi level throughout the sample.

CONCLUSIONS

Inversion domain boundaries in SiC are found to require a high energy of formation, of the order of 3 eV/interface unit cell area. It is thus unlikely that they would be formed unless by kinetic effects during growth such as the step mechanism mentioned in the introduction.

The Si-Si and C-C bonds near the interface basically try to relax to their normal bond lengths. This is achieved by varying the distance between the domains if only one type of "wrong bonds" occurs or by a rotation of the neighboring Si-C bonds for the (110) interface where both "wrong bonds" occur simultaneously. This local bond-length relaxation is almost perfect for the (110) case, but not for the (001) case since the geometric constraints would require too large a bond-angle variation. A long range slowly decaying strain field accompanies the IDBs analogously to the case of dislocations.

The presence of Si-Si and C-C bonds drastically changes the local electronic structure. The Si-Si bond related features, in particular, occur at higher energies than the corresponding Si-C bond related features and lead to an interface localized band which is pushed up into the conduction band. This turns the IDB in a two-dimensional metal, or, rather a semimetal.

In conclusion, the IDBs are highly energetic defects which strongly perturb the electronic structure and lead to considerable lattice strain fields in the system.

This work was supported by the DARPA-ONR URI at CWRU, Grant No. N000-13-86-K-0773. The electronic structure calculations were performed at the PSC (Pittsburgh) Project No. PHY870036P. Stimulating discussions with P. Pirouz are gratefully acknowledged.

REFERENCES

[1] R. C. Pond, in *Dislocations and Properties of Real Materials*, (London: Inst. Metals 1987), p. 71

[2] S. Nishino, J. A. Powell, and H. H. Will, *Appl. Phys. Lett.* **42**, 460 (1983)

[3] P. Pirouz, C. M. Chorey, and J. A. Powell, *Appl. Phys. Lett.* **50**, 221 (1987); T. T. Cheng, P. Pirouz and F. Ernst, in *Advances in Materials, Processing and Devices in III-V Compound Semiconductors*, edited by D. K. Sadana, L. Eastman, and R. Dupuis, *Mat. Res. Soc. Symp. Proc.* Vol. 144, (MRS, Pittsburgh 1989), in press; P. Pirouz, in *Polycrystalline Semiconductors*, edited by J. H. Werner, H.J. Möller, and H. P. Strunk, Springer Proceedings in Physics, Vol. 35, (Springer, Berlin 1989), p. 200

[4] J. A. Powell, L. G. Matus, M. A. Kuczmarski, C. M. Chorey, T. T. Cheng, and P. Pirouz, *Appl. Phys. Lett.* **51**, 823 (1987)

[5] D. J. Chadi, *Phys. Rev. Lett.* **59**, 1691 (1987)

[6] W. R. L. Lambrecht, B. Segall and P. Pirouz, in *Thin Films: Stresses and Mechanical Properties*, edited by J. C. Bravman, W. D. Nix, D. M. Barnett and D. A. Smith, *Mat. Res. Soc. Symp. Proc.* Vol. 130, (MRS, Pittsburgh, 1989), p. 199

[7] P. N. Keating, *Phys. Rev.* **145**, 637 (1966)

[8] R. M. Martin, *Phys. Rev.* **B 1**, 4005 (1970)

[9] W. R. L. Lambrecht and B. Segall, *Phys. Rev.* **B 40** (1989), to be published

[10] O. K. Andersen, *Phys. Rev.* **B 12**, 3060 (1975); O. K. Andersen, O. Jepsen, and M. Šob, in *Electronic Band Structure and its Applications*, edited by M. Yussouff, (Springer, Heidelberg, 1987)

[11] P. Hohenberg and W. Kohn, *Phys. Rev.* **136**, B864 (1964); W. Kohn and L. J. Sham, *Phys. Rev.* **140**, A1133 (1965); for a recent review see e.g. W. Kohn in *Highlights of Condensed-Matter Theory*, Enrico-Fermi School of Physics, Course LXXXIX, edited by F. Bassani, F. Fumi and M. P. Tosi (North-Holland, Amsterdam 1985), p. 1

[12] *Landolt-Börnstein Numerical Data an Functional Relationships in Science and Technology*, New Series, Edited by O. Madelung (Springer, Berlin 1982) Vol. 17 a

[13] C. G. Broyden, *Math. Comput.* **19**, 577 (1965); D. D. Johnson, *Phys. Rev.* **B 38**, 12807 (1988); D. Vanderbilt and S. G. Louie, *Phys. Rev.* **B 30**, 6118 (1984)

[14] W. R. L. Lambrecht, B. Segall and O. K. Andersen, *Phys. Rev.* **B 40** (1989), to be published

ATOMISTIC SIMULATION STUDY OF INTERFACES IN NANOPHASE SILICON

JAMES A. LUPO AND MICHAEL J. SABOCHICK
Air Force Institute of Technology, Department of Engineering Physics
Wright-Patterson Air Force Base, OH 45433-6583

ABSTRACT

Nanophase silicon was investigated using atomistic simulation. The simulations employed a modified Stillinger-Weber potential appropriate for crystalline and amorphous silicon. Computer "samples" of nanophase material were formed by compressing together three grains of several hundred atoms each, using Fletcher-Powell minimization and external pressures of 0.5 to 13.5 GPa. Relative densities obtained in the samples ranged from 65% to 98% as compared to the perfect crystal. The nanophase materials maintained crystalline order up to the interfaces and no highly disordered interfaces were observed. Calculated bulk moduli exhibited a linear dependence with respect to density, with no significant dependence on structure. The calculated thermal expansion coefficients were up to twice as large as that in the perfect crystal and were structure dependent.

Introduction

In the past few years, a new class of material called *nanophase* or *nanocrystalline* solids has been experimentally investigated. In contrast to polycrystalline substances that consist of relatively coarse grains and in which an overwhelming majority of the atoms see a normal crystalline environment, nanophase materials have a grain size that is small enough that a significant fraction of the atoms are near an interface [1]. It seems reasonable to expect that the properties of the polycrystalline and nanophase materials could be different, and in fact much of the interest in nanophase material has been generated by reports of measurements that support this [2-4]. For example, the specific heat capacity of nanophase Cu is 10% higher than that of polycrystalline Cu, with a corresponding value of 40% for Pd [5]. Unfortunately, many of the properties have only been reported for single materials [2,3], and no metal or compound except perhaps TiO_2 [6-11] seems to have been thoroughly investigated. Thus, the reliability of many of these these measurements has not been verified, and theories explaining the differences in the properties of nanophase and polycrystalline materials are difficult to support. In addition, the nature of the nanophase interfaces themselves--whether they are disordered [12] or are similar to conventional grain boundaries [13], is under investigation.

The purpose of the present work is to use atomistic simulation to investigate these questions. In a previous simulation study of nanophase Cu, the simulated Cu had a relative density of 99+% with respect to the perfect crystal, regardless of compaction pressure [14]. Apparently, the metallic bonding in Cu allowed atoms to easily move and fill in vacancies and voids, and it was impossible to investigate the dependence of various properties on density and compaction pressure. In order to explore this dependence, nanophase Si is simulated in the present work since the covalent bonding in Si should restrict the movement of atoms, and also increase the effects of interface disorder. Small "samples" of nanophase Si are simulated, and the properties of these samples are compared with those of crystalline and amorphous Si.

Method

The computational methods used in the present work were molecular statics and molecular dynamics, using a modified version of the code DYNAMO. The simulations used rectangular volumes whose edge lengths were free to vary with external pressures [15]. Energy minimization in the molecular statics simulations was done with the Fletcher-Powell method, adapted for a large number of independent variables [16]. The interatomic potential function was the Ding and Andersen modification [17] of the Stillinger and Weber Si potential [18]. This potential is appropriate for simulation of solid Si in both the crystalline and amorphous states.

The experiments were conducted on computer-generated nanophase, amorphous and crystalline Si samples. The crystalline or perfect sample consisted of 64 atoms arranged in a diamond lattice structure. The amorphous Si sample was created using the approach of Ding and Andersen [17]. A 216-atom system was melted by heating it to a temperature of 12000K for 1000 steps. The time-step size of 0.002 picosecond for this run did not result in energy conservation, but allowed the system to become disordered quickly. The melted system was then quenched to 0K and zero pressure using Fletcher-Powell minimization. The resulting sample had a relative density of 83% with respect to the perfect crystal density. To assess the effects of density on properties, a second amorphous sample was created by compressing the 83% sample to a relative density of approximately 100% and then relaxing the system to zero pressure. The final relative density of this second amorphous sample was 99%.

The nanophase Si samples were created using techniques analogous to experimental fabrication techniques [10]. Two spherical grains of different sizes were made by removing the corners from a perfect crystal. One was a 105-atom, 7.8A-diameter grain; the other was a 305-atom, 11.3A-diameter grain. Two initial nanophase samples containing 315 and 915 atoms were created by compressing systems consisting of three randomly oriented images of each of these grains to a pressure of 0.5 GPa. The initial systems were then relaxed to zero pressure, resulting in samples with relative densities of 66% for the 315-atom system and 70% for the 915-atom system. The low densities were the result of large voids between the grains. Higher densities were with higher compaction pressures followed by relaxation to zero pressure. The different types of samples are summarized in Table I.

Results

The nature of the interfaces of nanophase materials has been the subject of considerable speculation. On one hand, it has been proposed that the interfaces are structurally similar to a gas-like solid, i.e., highly disordered [12], while others have suggested that the interfaces are similar to common grain boundaries [13]. In the present work, three-dimensional perspective displays of the samples showed that the grains maintained their internal order, even though some distortion must have occurred at the grain boundaries during compression. The order was maintained up to the interface, and the displays were qualitatively identical to electron micrographs of nanophase

Table I. Summary of Calculated Sample Properties

Structure	Number of Atoms	Compaction Pressure [GPa]	Relative Density	B [GPa]	c_p [J/g-K]	α [$\times 10^5$]
Crystalline	64		1.000	103.3	1.20	1.10
Amorphous	256		0.831	63.0	0.85	1.47
			0.989	96.6	0.89	3.02
Nanophase	315	0.5	0.655	28.8	0.88	1.40
		1.5	0.864	70.8	0.89	1.68
		4.5	0.968	107.7	0.83	1.81
	915	0.5	0.696	38.0	0.87	2.42
		1.5	0.723	37.1	0.87	2.88
		4.5	0.754	47.5	0.86	2.83
		13.5	0.888	75.7		2.42

Pd [13]. A cross-sectional view of the 915-atom system is shown in Figure 1. A void is clearly evident in the center of the Figure. Two grains are also in view (the grains are continuous across the edges of the picture due to periodic boundaries), and it is evident that the internal order of the grains is preserved. The pair-correlation functions of the samples at 298K (not shown) indicate that order was preserved to a radius corresponding to the diameter of the grains. This suggests that short range order was maintained for a majority of the atoms. These results are in agreement with the earlier simulation study in Cu [14], but conflict with the interpretation of recent x-ray diffraction measurements, which suggested that systems with grains similar in size to the present work had as many as 50% of the atoms exhibiting neither long nor short-range order [12]. Our results support the theory that interfaces in nanophase materials are similar to grain boundaries in polycrystalline solids [13].

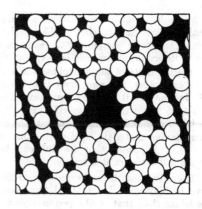

Figure 1. Cross-section of 915-atom nanophase sample. The internal order of the grains is preserved, almost up to the interfaces.

The elastic properties of the samples were investigated by calculating the bulk modulus B. This was done by expanding the zero pressure state by 1%, followed by a minimization fixed at that volume. This was repeated for a compression by 1% of the zero pressure state. The results of these calculations are summarized in Table I and are plotted in Figure 2 as a function of relative density. It is apparent that the bulk modulus is relatively insensitive to structure and depends primarily on density. This observation is emphasized by comparing the amorphous sample at 98% relative density with the perfect crystal. The reduced modulus of the nanophase samples as compared with that of the perfect crystal is consistent with experimental results in nanophase palladium [19]. A recent simulation study of the effect of dimensional changes on grain boundaries in Si showed that expansion of the boundaries resulted in softening of the elastic moduli [20]. Although the results of the present work are similar, the conclusion here is that this softening is due to the effects of voids and not the expansion of interfaces.

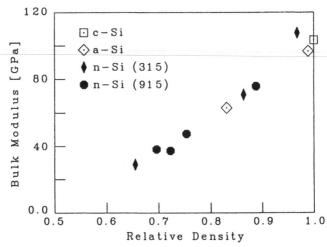

Figure 2. Bulk modulus B as a function of system density. This plot indicates that B primarily depends on density and not structure.

Molecular dynamics simulations at temperatures of 300K and 600K were used to investigate thermal properties of the samples and the perfect crystal. The equilibrium densities and total energies of the samples were calculated at these temperatures in order to determine the specific heat c_p and thermal expansion coefficient α. The calculated specific heats of the nanophase and amorphous samples agreed within a few percent of each other, and were lower than the specific heat of the perfect crystal by about 25%. In contrast, it has been found experimentally that the specific heat of nanophase Cu is 10% larger than the specific heat of polycrystalline Cu; in Pd, the increase is 40% [5]. Rupp and Birringer attribute the difference in specific heats of nanophase and polycrystalline materials to contributions of interfaces that contain no short-range order [5]. It is possible that the difference between experiment and the present work is that the materials are different.

The calculated thermal expansion coefficient varied greatly among the samples, as shown in Figure 3. The thermal expansion coefficients of all the disordered samples were larger than that of the perfect crystal by as much as 200%, in agreement with experiments in nanophase Cu [21]. Unfortunately, no universal trends could be found in the data plotted in Figure 3. For example, the 315-atom nanophase sample had lower thermal expansion coefficients than the 915-atom nanophase sample, even though the interface density should be higher in the smaller system. In general, the thermal expansion coefficient increased with density in the 315-atom and amorphous systems, but this is not supported by the behavior of the 915-atom system. The reason for the scatter in the data is probably that fact that only three grains were used in the nanophase samples, limiting the number of possible boundary orientations. Evidently, more work needs to be done on systems containing more grains.

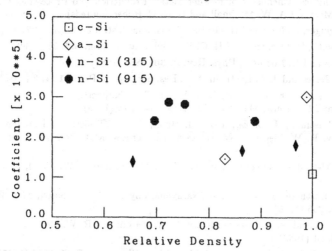

Figure 3. Thermal expansion coefficient α as a function of system density. Although thermal expansion coefficient of the amorphous and nanophase samples is much larger than that of the perfect crystal, no other trends are apparent.

Conclusions

The interface structure and bulk properties of nanophase Si were investigated using computer simulation. No large regions of disorder were observed, and the internal order of the grains was preserved up to the interfaces. The linear dependence of the bulk modulus on relative density for all samples indicates that the bulk modulus is more dependent on density than on structure. The specific heats of the nanophase and amorphous silicon samples were lower than that of the perfect crystal, although no direct correlation with density or structure could be found. The nanophase and amorphous samples had consistently larger thermal expansion coefficients than the perfect crystal, in agreement with experiment.

Acknowledgements

The authors would like to thank R. W. Siegel for continuing interest and encouragement. This work was supported by the Air Force Institute of Technology.

References

1. R. Birringer, H. Gleiter, H.-P. Klein and P. Marquardt, Phys. Lett. **102A,** 365 (1984).

2. C. Suryanarayana and F. H. Froes to be published in the Proceedings of the Symposium on Physical Chemistry of Powder Metals Production and Processing, Oct. 16-18, 1989, St. Mary's, PA, W. M. Small and D. C. C. Robertson (eds).

3. R. Birringer, U. Herr and H. Gleiter, Trans. Jpn. Inst. Met. Suppl. **27,** 43 (1986).

4. J. Horvath, R. Birringer and H. Gleiter, Sol. State Comm. **62,** 319 (1987).

5. J. Rupp and R. Birringer, Phys. Rev. B. **36,** 7888 (1987).

6. R. W. Seigel and J. A. Eastman, Mat. Res. Soc. Symp. Proc. **132,** 3 (1989).

7. J. E. Epperson, R. W. Siegel, J. W. White, T. E. Klippert, A. Narayanasamy, J. A. Eastman and F. Trouw, Mat. Res. Soc. Proc. **132,** 15 (1989).

8. J. A. Eastman, J. E. Epperson, H, Hahn, T. E. Klippert, A. Narayanasamy, S. Ramasamy, R. W. Siegel, J. W. White and F. Truow, Mat. Res. Soc. Proc. **132,** 21 (1989).

9. C. A. Melendres, A. Narayanasamy, V. A. Maroni and R. W. Siegel, J. Mat. Res. **4,** 1246 (1989).

10. J. A. Eastman, Y. X. Liao, A. Narayanasamy and R. W. Siegel, Mat. Res. Soc. Proc. **155,** 255 (1989).

11. C. A. Melendres, A. Narayanasamy, V. A. Maroni and R. W. Siegel, Mat. Res. Soc. Proc. **153,** 21 (1989).

12. X. Zhu, R. Birringer, U. Herr and H. Gleiter, Phys. Rev. B **35,** 9085 (1987).

13. G. J. Thomas, R. W. Siegel and J. A. Eastman, Mat. Res. Soc. Proc. **153,** 13 (1989).

14. M. J. Sabochick and J. A. Lupo, Proc. of Conf. on Diffusion in Metals and Alloys (DIMETA-88), Balantonfured, Hungary, Sept. 5-9, 1988.

15. M. Parrinello and A. Rahman, Phys. Rev. Lett. **45,** 1196 (1980).

16. M. J. Sabochick and S. Yip, J. Phys. F: Met. Phys. **18,** 1689 (1988).

17. K. Ding and H. C. Andersen, Phys. Rev. B **34,** 6987 (1986).

18. F. H. Stillinger and T. A. Weber, Phys. Rev. B **31,** 5262 (1985).

19. H. E. Schaefer, R. Wurschum, R. Birringer and H. Gleiter, J. Less-Common Met. **140,** 161 (1988).

20. S. R. Phillpot, D. Wolf and J. F. Lutsko, Mat. Res. Soc. Proc. **153,** 33 (1989).

21. H. J. Klam, H. Hahn and H. Gleiter, Acta Metall. **35,** 2101 (1987).

THE APPLICATION OF GLANCING ANGLE EXAFS TO STUDY THE STRUCTURE OF PLATINUM-NICKEL MULTILAYERS.

G.M. LAMBLE,[a)]   S.M. HEALD,[b)]   and B.M. CLEMENS [c)]

[a)] North Carolina State University, Box 8202, NC 27695-8202,
[b)] Brookhaven National Laboratory, Upton, NY 11973
[c)] Dept. of Material Science and Engineering, Stanford University, Stanford, CA 94305-2205

ABSTRACT

Recent reports have correlated the mechanical properties of multilayers with a structural expansion of the constituents. Whether this expansion is an interface or bulk effect remains in dispute. We apply the technique of glancing angle EXAFS to obtain information about the local structure and multilayer composition and use these results to shed light on certain aspects of this controversy. The experiments were performed on platinum-nickel multilayers of varying bilayer thickness. Measurements were made by fluorescence detection of EXAFS beyond both the Pt L_{III} and the Ni K edges. Considerable interlayer mixing is evident, as is the presence of order within the interface. We additionally find that the bulk metallic character is retained within the layers.

INTRODUCTION

Observations of the dramatic modifications of the elastic properties of 2-component multilayer metal films, as a function of bilayer thickness, has promoted great interest in this phenomenon and since its first observation [1] there has been much experimental work, using a variety of techniques[2] and many proposed theories as to the origin of this effect. A significant experimental observation is that, along with the changes in the elastic properties of these composition modulated materials, there is a concurrent change in the average lattice parameter in the direction perpendicular to the layer planes, over a critical range of interplanar spacing. It is generally accepted that this change in average lattice parameter is fundamentally related to the observed anomalies in the elastic behaviour of the materials. However, there is still a question as to whether the structural changes extend through the bulk of the layer [3,4,5,6,7,8] or are localized at the interfaces in the multilayer structure [9,10,11]. In particular, Clemens and Eesley [11] have shown that an interface expansion is consistent with the observed structural and elastic behaviour. Recently, Huberman and Grimsditch [8] have proposed that the expansion is electronic in nature and a bulk effect, while Cammarata and Sieradzki [4] have suggested that the observed expansion in the out of plane spacing is due to a bulk biaxial compression resulting from surface stresses acting on each interface.

In this work, we present the results of a Glancing Angle EXAFS study of 3 Pt/Ni multilayers of bilayer period 23, 35 and 100 Å . These particular samples, along with those of two other bimetallic systems, have been previously studied using the techniques of XRD and sound velocity measurements which gave rise to an interface-localized picture of the lattice parameter change and softening of elastic properties[11]. The EXAFS technique probes the local atomic environment around a given type of absorber atom

Mat. Res. Soc. Symp. Proc. Vol. 159. ©1990 Materials Research Society

which provides complementary structural information to that from XRD. In this multilayer study, we have measured the EXAFS above both the Pt L_{III} and Ni K edges. The signal obtained comes from the whole sample and yields the average environment of the Platinum and Nickel atoms. To analyze the data we have used a theoretical curve fitting method [12], based on the EXAFS theory of Lee and Pendry [13], which employs semi-empirical phase shifts obtained from the chosen standard compounds.

EXPERIMENTAL

The multilayer samples were prepared by passing oxidised silicon wafers under shielded sputter sources in a cryopumped dual-source magnetron deposition system. The base pressure was 10^{-7} Torr and the deposition pressure was 2.3×10^{-3} Torr of getter cleaned Ar. The sputter rates were controlled by rate monitors set to produce an equal thickness of each constituent per bilayer. Samples were produced with bilayer periods of 23, 35 and 100 Å and total thickness of around 2000 Å . Electron microprobe analysis showed the composition fluctuation to be less than 2 %.

The EXAFS experiments were performed at the National Synchrotron Light Source (NSLS) on beamline X-11A, by monitoring the fluorescence yield above the Ni K-edge (8333 eV) and the Pt L_{III} edge (11563 eV). The samples were mounted in a perpendicular plane to the direction of the photon beam, with a photon incidence angle of ~ 45 °. A large area ion chamber was used to monitor the fluorescence signal from the samples, situated to collect a large solid angle of emitted radiation. A major problem encountered in the data acquisition was the interception of Bragg peaks in the signal from the diffraction of the photon beam by the single crystal silicon substrates. This problem was surmounted by spinning the samples at a rate of 4 rev/sec to spatially average the diffraction intensity. Bulk materials chosen as reference samples included Ni and Pt metals and Ni_3Pt alloy.

RESULTS AND DISCUSSION

Fig 1 a) b) and c) show the background subtracted raw EXAFS data above the Pt L_{III} edge from the 23, 35 and 100 Å period multilayers respectively. For comparison, the data from bulk Pt metal is shown in d). Data from the Ni K-edge is displayed in analogous form in fig 2, with the layer thickness increasing from a) to c). Bulk Ni data is shown in 2 d). As expected, the EXAFS visually takes on a more metallic character with increase in bilayer thickness.

The Pt and Ni phaseshift and amplitude parameters were obtained from the spectra of the bulk metal standards. Ni k-edge and Pt L_{III} data from the $PtNi_3$ were also used as a reference. Whilst the extent of order in this alloy was not known precisely, the data provided guidelines for the amplitude parameters in the analysis of the mixed phases in the multilayers.

The first Fourier transform peak in each multilayer EXAFS spectrum was filtered and back-transformed into K-space. In every case, three components were required to fit the resulting frequency spectrum, i.e, the Ni edge data shows two backscattering Ni components (one at the bulk f.c.c. distance of 2.48 Å and another at a larger distance of ~ 2.59 Å) with an additional Pt component (at ~ 2.54 Å). The Pt LIII edge EXAFS yields two Pt backscattering components (one at the f.c.c Pt distance of 2.78 Å and one at a shorter distance of around 2.69 Å) with an additional Ni component (at ~ 2.56 Å). Whilst the identification of the three contributing backscatterers was

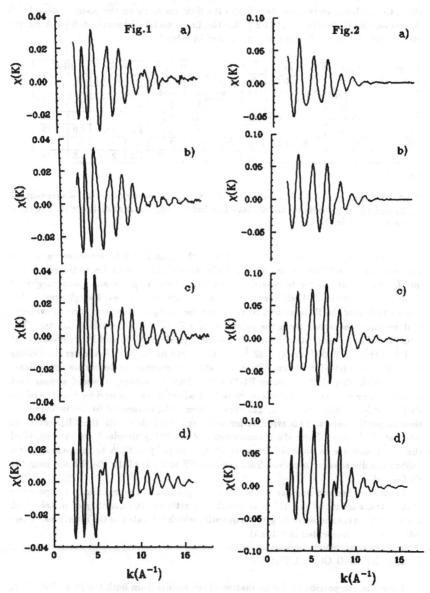

Fig.1 shows the background subtracted raw EXAFS data from the Pt L_{III} edge in Pt-Ni multilayers with bilayer periods of a) 23 Å b) 35 Å and c) 100 Å . The EXAFS data from bulk Pt metal is shown in d).

Fig.2 shows the background subtracted raw EXAFS data from the Ni K edge in Pt-Ni multilayers with bilayer periods of a) 23 Å b) 35 Å and c) 100 Å . The EXAFS data from bulk Ni metal is shown in d).

clear, the inability to resolve them limits the final accuracy on the co-ordination to 35 % and on the distances to ± 0.02 Å for the f.c.c. metal distances, ± 0.04 Å for the others. Results of the fitting are summarized in table I.

Bilayer period (Å)	Edge	Pt-Pt		Pt-Pt		Pt-Ni		Ni-Ni		Ni-Ni	
		N	R(Å)	N	R (Å)	N	R (Å)	N	R (Å)	N	R (Å)
23	Pt L_{III}	5.1	2.76	3.1	2.67	2.2	2.58	-	-	-	-
	Ni K	-	-	-	-	1.8	2.54	4.8	2.59	2.4	2.48
35	Pt L_{III}	8.4	2.77	2.2	2.69	0.9	2.55	-	-	-	-
	Ni K	-	-	-	-	1.7	2.54	5.3	2.59	4.9	2.48
100	Pt L_{III}	10.8	2.78	1.8	2.71	0.6	2.52	-	-	-	-
	Ni K			-	-	0.4	2.53	3.2	2.59	9.5	2.49

Table I.

Table I: Summary of EXAFS results from Pt-Ni multilayer samples with bilayer periods 23, 35 and 100 Å . N denotes the average number of neighbours at distance R. The error bars are quoted in the text.

As expected, the pure metallic content in the multilayers increases along with a decrease in the Pt-Ni co-ordination as the bilayer period increases, i.e, as the number of interfaces per unit volume decreases. Since we are unable to probe any specific region of the multilayer, we cannot tell how the interface composition varies, though presumably it is Pt rich near the Pt layer and Ni rich near the Ni layer. However, there is evidence that mixing does occur, from the existence of a Pt-Ni component in each case. The structure of an ordered stoichiometric PtNi alloy predicts Ni-Ni and Pt-Pt distances at 2.71 Å with a Pt-Ni distance at 2.62 Å i.e, the 'mixed-metal' bond is shorter than either of the Pt-Pt or Ni-Ni distances. From a qualitative comparison to our own multilayer results which also yield a shorter Pt-Ni bond than the others, it would appear that some ordering exists in the interface. Note that whilst we do not find equivalence of the Pt-Pt and Ni-Ni distances, we might expect them to be influenced by the proximity to their respective layers. This evidence for ordering is consistent with the XRD results in reference [11]. Note: From the measurements of the PtNi$_3$ standard, we have adjusted the Ni-Pt distance as observed from the Pt L_{III} edge by -0.03 Å to compensate for a calibration discrepancy observed from the EXAFS analysis of the Pt-Ni bond from the Pt L_{III} edge.

From the observed number of bulk f.c.c. neighbours in the multilayer compared to that in the pure metal (i.e 12) it is possible to estimate the quantity of metal which stays in the layers and accordingly, the quantity which is located in the interfaces. These estimates are represented in table II.

SUMMARY AND CONCLUSIONS

From the comparison and combination of the results from both the Pt and Ni edges evolves a general picture of the structural composition of the multilayer. The EXAFS experiments presented here are able to provide comment and insight on issues in the foreground of the existing controversy about the phenomenological mechanism for the lattice expansion. A bulk biaxial strain responsible for the observed expansion out of the plane would result in a net volumetric dilatory strain of about the same magnitude. Thus, the observed increase in average d-spacing in the growth direction as the bilayer

Bilayer period (Å)	Pt layer	Ni layer	Pt-Ni and Ni-Pt interfaces
23	4.9 Å	2.3 Å	6.6 Å Pt + 9.2 Å Ni
35	12.2 Å	7.1 Å	5.3 Å Pt + 10.4 Å Ni
100	45.0 Å	39.6 Å	5.0 Å Pt + 10.4 Å Ni

Table II

Table II: Pt-Ni multilayer compositions represented in terms of the layer thickness of the constituents.

period decreases would be accompanied by a decrease in the nearest neighbour distances. It is interesting to note that our Pt-Pt and Ni-Ni distances do show a decrease as the bilayer period is decreased from 100 Å to 23 Å. However, the size of this change, while being about the same size as that observed in the out of plane x-ray measurements (1% [11]), is within the error bars of our EXAFS measurement, so we are unable to determine definitively if bulk biaxial compression is the responsible mechanism.

The Pt-Ni multilayers do not exhibit abrupt interfaces. From the extent of Pt-Ni bonding we observe considerable interlayer mixing. It is also clear that the Pt-Ni bond distance is shorter than the corresponding Ni-Ni and Pt-Pt bonds within the interface which would indicate that ordering exists in this region. We can estimate that the two interface regions (per unit cell) are composed of approximately 5 Å Pt and 10 Å Ni. So on average, and without regard to the specific composition, each interface is estimated to be ~ 7.5 Å wide and composed of ~ 2.5 Å Pt and ~ 5.0 Å Ni. An upper limit for the interface width in reference [11] was 5.0 Å thus our own measurements are in agreement with this, within the margin of error for the amplitude determination.

On the basis of what we have learned through this EXAFS study, modelling of the high angle x-ray data, incorporating the influence of interdiffused interface regions, is currently in progress. Additional experiments are planned for multilayers deposited on glass substrates which increases the experimental flexibility. New approaches include, 1) Low temperature experiments would allow extension of the data range beyond 16 A^{-1} beyond which, at room temperature, the EXAFS is indiscernible. Extension of the data range would improve our resolution of the nearest neighbour components. 2) Variation of the photon incidence angle to make use of the polarization vector of the synchrotron beam, may enable the detection of any anisotropic effects within the multilayers. 3) The application of standing wave enhancement of the EXAFS will enable specific regions of the multilayer to be probed[14].

ACKNOWLEDGEMENTS

This work was performed under the auspices of the U.S. Department of Energy under contract Nos. DE-AS05-80-ER10742 and DE-AC02-76CH00016.

REFERENCES

1 W. M. C. Yang, T. Tsakalakos and J. E. Hilliard, J. Appl. Phys. **48**, 876 (1977).

2 I. K. Schuller, Ultrasonics symposium, ed: B. R. McAvoy, IEEE Symposium on Ultrasonics, 1985 (IEEE, New York, 1985) 1093.

3 A. F. Jankowski and T. Tsakalakos, J. Phys. F **15**, 1279 (1985).

4 R. C. Cammarata and K. Sieradzki, Phys. Rev. Lett. **62**, 2005, (1989).

5 T. B. Wu, J. Appl. Phys. **53**, 5265 (1982).

6 I. K. Schuller and A. Rahman, Phys. Rev. Lett. **50**, 1377 (1983).

7 I. K. Schuller and M. Grimsditch, J. Vac. Sci. Technol. B **4**, 1444 (1986).

8 M. L. Huberman and M. Grimsditch, Phys. Rev. Lett. **62**, 1403 (1989).

9 W. R. Bennet, J. A. Leavitt and C. M. Falco, Phys. Rev. B **35**, 4199 (1987).

10 D. Wolf and J. F. Lutsko, Phys. Rev. Lett. **60**, 1170 (1988).

11 Bruce M. Clemens and Gary L. Eesley, Phys. Rev. Lett. **61**, 2356 (1988).

12 S.J. Gurman , N. Binsted and I. Ross, J. Phys. C **17**, 143 (1984).

13 P. A. Lee and J. B. Pendry, Phys. Rev. B **11**, 2795 (1975).

14 S. M. Heald and J. M. Tranquada, J. Appl. Physics **65**, 290 (1989)

THE ELECTRONIC STRUCTURE OF Σ5 GRAIN BOUNDARIES IN CU

ERIK C. SOWA,* A. GONIS* AND X. –G. ZHANG**
* Lawrence Livermore National Laboratory, L356, Livermore, CA 94550
** Physics Department, Northwestern University, Evanston, IL, 60201

ABSTRACT

We present first-principles calculations of the densities of states (DOS's) of unrelaxed and relaxed twist and tilt grain boundaries (GB's) in Cu. The relaxed configurations were obtained through the use of the Embedded Atom Method (EAM), while the DOS's were calculated using the real-space multiple-scattering theory (RSMST) approach recently introduced in the literature. The DOS's of GB's are compared against those of bulk materials as well as against one another. Although the RSMST calculations are still not self-consistent, these comparisons allow us to verify certain expected trends in the DOS's, and to verify the usefulness and reliability of our method.

INTRODUCTION

Electronic-structure calculations require two ingredients: Knowledge of the atomistic structure of a material, and the existence of methods, both formal and computational, for treating the structure at hand. In the case of materials with full translational symmetry, e.g. bulk crystalline solids the atomistic structure may be determined by, e.g., X-ray crystallography. Once this is known, the electronic structure may be obtained with the aid of Bloch's theorem and the associated lattice Fourier transforms, which diagonalize the Hamiltonian in reciprocal space (k-space). The determination of both the atomistic and electronic structure of materials in which translational symmetry is broken by surfaces or internal interfaces is much more problematic. Experimental determination of the atomistic structure of some surfaces is possible through, e.g., low-energy electron diffraction in reciprocal space or scanning tunneling microscopy in real space. There are no techniques available for imaging atomic positions at internal interfaces, although in some cases, partial information may be obtained with electron microscopy. Often one must rely on computer simulations, such as those afforded by the embedded atom method [1, 2, 3, 4].

Knowledge of the atomic coordinates of a surface or interface does not necessarily enable electronic-structure calculations to be performed with conventional methods, because nearly all existing, fully first-principles formalisms for performing such calculations are based on the properties of translational invariance and the use of Bloch's theorem. In order to use these formalisms, it is often necessary to invoke rather severe approximations with respect to the structure of a system, such as the use of slabs of finite thickness to treat surfaces, or of repeating slabs or supercells to study internal interfaces. Such approximations are highly undesirable for a number of reasons: First, they are "uncontrolled", yielding results which should be checked for dependence on slab thickness on a case-by-case basis. Second, they are conceptually unattractive, involving rather severe approximations to the underlying geometry of the system. Third, the extent to which they may yield accurate results may lull one into a false sense of accomplishment, obscuring the existence of a still unsolved problem.

Stated briefly, this problem consists in finding the solution of the one-particle Schrödinger equation, within the Born-Oppenheimer approximation (fixed nuclei) and the local-density approximation to density-functional theory, that satisfies the proper boundary conditions imposed by the structure of a particular semi-infinite material.

In a recent publication [5], a first-principles, multiple-scattering formalism, which allows the exact treatment of the problem just stated, was introduced. As the formal aspects of this real-space multiple-scattering theory (RSMST) method have been reviewed in previous work [5, 6], we shall forgo all but a brief description of this method here. The RSMST is based on the concept of semi-infinite periodicity (SIP), defined as the regular repetition along a given direction of a scattering unit (atom, planes of atoms, etc.), or a set of such units. Systems with SIP possess the property of removal invariance, which states that the scattering properties (scattering matrices) of any such system remain invariant when an integral number of scattering units is removed from, or added to, the free end of the system. Using this property in conjunction with multiple-scattering theory, one can determine the electronic Green function, and hence all one-particle quantities such as the density of states (DOS), directly in real space, bypassing often cumbersome reciprocal-space (k-space) integrations. This formalism provides a unified treatment of the electronic properties of a broad spectrum of systems that includes, but is not limited to, pure elemental solids, compounds, ordered alloys, surfaces and interfaces, and other low-symmetry systems. Only systems with no recognizable periodic structure, e.g. amorphous materials and liquids, fall outside the scope of the RSMST method.

The essence of the method consists in a prescription for the proper renormalization of the scattering properties of the boundary sites of a cluster of atoms. Unrenormalized or "bare" sites in the interior of the cluster describe the region of interest, such as a grain boundary, while the renormalized boundary sites represent the infinite medium surrounding the cluster. This "dressing up" of the cluster is done independently for each part of a system that is characterized by its own SIP, so that grain boundaries between essentially arbitrary crystal structures can be treated. At its present stage of development, our codes can be applied to known atomistic configurations with known electronic one-particle potentials. Work currently in progress is aimed at the incorporation of the often important effects of charge self-consistency, and of total-energy capability, into the program.

In the calculations reported here, we used the self-consistent potentials for bulk Cu given by Moruzzi, Janak and Williams (MJW) [7]. The atomic coordinates of the unrelaxed grain boundaries are easily found through an appropriate twisting or tilting of one half of the underlying lattice. We used unpublished EAM calculations, performed by S. M. Foiles of Sandia National Laboratories, Livermore, to obtain the atomic coordinates of relaxed twist and tilt grain boundaries in Cu.

Figure 1 displays the local electronic DOS's corresponding to a coincidence site (a site which is common to the lattices on both sides of an interface) at both an unrelaxed and a relaxed $\Sigma 5$ (100) 36.9° twist grain boundary in Cu. The bulk Cu DOS, calculated with the RSMST, is also displayed for comparison. (This bulk DOS reproduces the main features of the MJW calculation; small differences may be attributed to our use of a coarser energy grid and a small imaginary component of the energy.) The unrelaxed grain-boundary DOS exhibits considerable smearing of structure compared to the bulk DOS, because of the loss of periodicity and the associated destruction of the Van Hove singularities. The grain boundary DOS is slightly broader than that of the bulk material due to the decreased distance between some of the Cu atoms across the interface. (In fact, simply twisting one half of the crystal with respect to the other results in some of the atoms overlapping across the boundary).

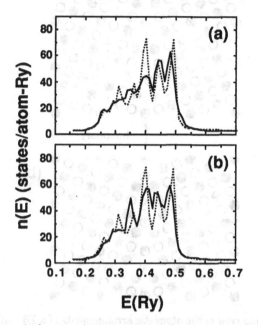

Fig. 1. — Electronic density of states associated with a coincidence site at a $\Sigma 5$ (100) 36.9° twist grain boundary in Cu. DOS's for the unrelaxed and relaxed configurations are shown, respectively, as solid curves in panels (a) and (b), and are compared with those of bulk Cu (dotted curves).

Both the atomistic and electronic characteristics change noticeably in the case of the relaxed grain boundary. The results of the EAM calculations indicate that the interplanar spacing increases by about 20% across the interface from its bulk value, decreases by 2% in the next set of layers, and remains essentially unchanged in layers deeper inside the material. The increase is the result of relieving the overlap conditions across the interface mentioned above, and its effects on the electronic structure are similar to those associated with decreased coordination. In these first calculations, we included only the 20% expansion at the boundary layer, which is the dominant effect. It is seen that the DOS at the grain boundary is indeed narrower than that of bulk Cu. It is also seen that the relaxed grain boundary DOS is shifted slightly toward lower energies compared both to that of bulk Cu and of the unrelaxed configuration, and that it possesses somewhat sharper structure than the DOS at an unrelaxed grain boundary. Although the present, non-charge-self-consistent calculation cannot provide reliable information about the relative energies of the various configurations, both of these effects are consistent with the lower energy of the EAM-relaxed configuration with respect to the unrelaxed one.

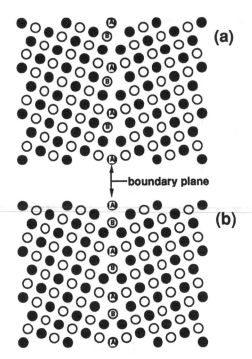

Fig. 2. — Side view of the atomistic arrangements of a $\Sigma 5$ (310) tilt grain boundary in Cu, with filled and unfilled circles representing atoms on two different (100) planes. Unrelaxed configuration, panel (a); relaxed configuration, panel (b).

In order to make clearer the discussion of tilt grain boundaries, it is helpful to refer to the schematic diagram of Fig. 2, which shows the atomic arrangements at a $\Sigma 5$ (310) tilt grain boundary in Cu. The letter A designates a coincidence site, which remains relatively immobile through the relaxation process, while B denotes a site that according to the EAM calculations is displaced the most during relaxation. The relaxations of layer spacings predicted by the EAM for six layers on either side of the boundary were included in this calculation; the primary relaxation is the movement of atom B to a position coplanar with atom A, forming a mirror plane at the boundary. The local DOS's at site A in both the unrelaxed and the relaxed configurations are shown in Fig. 3, and are compared with those of bulk Cu. It is interesting to note that the bulk DOS peak at 0.45 Ry persists in the (100) twist boundary but is absent from the (310) tilt boundary. From this we may conjecture that this structure results primarily from the two-dimensional periodicity of (100) layers. Analogous results for site B are shown in Fig. 4. The changes in the tilt-boundary DOS's upon relaxation are qualitatively similar to those of the twist grain boundary discussed above. (The rather high DOS's at the low-energy region corresponding to site B is most probably an artifact of the scheme used to integrate the cell wave functions, required [5, 6] for calculating the DOS, over the cell polyhedron whose shape in this case deviates substantially from that of the undistorted lattice). We note that these DOS's resemble in structure those obtained in other work [8] where a $\Sigma 5$ (210) tilt grain boundary was studied.

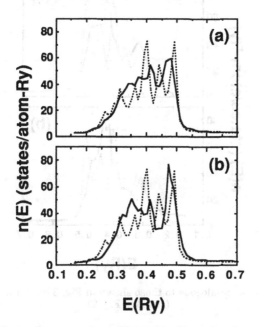

Fig. 3. — Results analogous to those depicted in Fig. 1, but for a coincidence site in a Cu $\Sigma 5$ (310) tilt grain boundary (site A in Fig. 2).

As the calculations just presented illustrate, the RSMST method allows one to obtain the electronic structure of materials with extended defects, such as surfaces and interfaces, on an atom-by-atom basis. Although in its present stage our code is slower than those based on conventional methods when applied to systems of high symmetry, it holds the distinct advantage of being applicable to low-symmetry structures that are not amenable to treatment by these methods. We are currently attempting to increase the efficiency of our code and add charge self-consistency and total-energy capabilities. The results of these efforts will be communicated as they become available.

ACKNOWLEDGMENTS

This work was performed under the auspices of the Division of Materials Science of the Office of Basic Energy Sciences, U. S. Department of Energy, and the Lawrence Livermore National Laboratory under contract No. W-7405-ENG-48.

One of us (SMF) is supported by Director, Office of Energy Research, Office of Basic Energy Sciences, Materials Science Division of the U. S. Department of Energy.

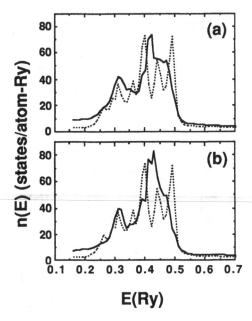

Fig. 4. — Results analogous to those shown in Fig. 3 but for a displaced site (site B in Fig. 2).

REFERENCES

1. M. S. Daw and M. I. Baskes, Phys. Rev. B **29**, 6443 (1984).
2. S. M. Foiles, M. I. Baskes and M. S. Daw, Phys. Rev. B **33**, 7983 (1986).
3. S. M. Foiles, Acta Metall. , to be published (1989).
4. S. M. Foiles, in *Characterization of the Structure and Chemistry of Defects in Materials*, edited by B. C. Larson, M. Rühle and D. N. Seidman (Materials Research Society, Pittsburgh, 1989)
5. X.–G. Zhang and A. Gonis, Phys. Rev. Lett. **62**, 1161 (1989).
6. X.–G. Zhang, A. Gonis and J. M. MacLaren, Phys. Rev. B. **40**, 3694 (1989).
7. V. L. Moruzzi, J. F. Janak and A. R. Williams, *Calculated Electronic Properties of Metals* (Pergamon Press, New York, NY, 1978)
8. S. Crampin, D. D. Vvedensky, J. M. MacLaren and M. E. Eberhart, Phys. Rev. B. **40**, 3413 (1989).

A MECHANISM FOR SOLUTE SEGREGATION TO GRAIN BOUNDARIES

CHU YOUYI AND ZHANG SANHONG
Nonferrous Metals Society of China, B12 Fuxing Road, Beijing 100814, China

ABSTRACT

Based on the local equilibrium among vacancies, solute atoms and vacancy-solute atom complexes, a mechanism for solute segregation to grain boundaries is suggested for a alloy system with binding energy of complex $E_b \gg kT$. A set of dynamic equations for grain boundary segregation is derived, which can describe both equilibrium and nonequilibrium segregations, as the effect of the equilibrium segregation is taken into account in the boundary condition. Theoretical calculation is made by computer for boron segregation at austenite grain boundaries as functions of isothermal holding time, cooling rate and quenching temperature, which agree well with experimental results.

INTRODUCTION

Solute grain boundary segregation can be divided into two categories, equilibrium segregation and nonequilibrium segregations. The former is derived by the binding engergy between solute and grain boundary, which has been clarifyed by McLean[1]. And the latter in quenched alloys has been, so far, attributed to the diffusion of vacancy-solute complexes along vacancy gradients to grain boundaries during cooling and some models have been suggested trying to make quantitative or semi-quantitative descriptions for the phenomenon[2-8]. Recently it has been found that there simultaneously exist equilibrium and nonequilibrium segregations of boron at austenite grain boundaries after quenching from a high temperature[9] and isothermal dynamic curves of boron segregation have been obtained in some boron containing alloys[10]. Based on these latest experimental results, a model for solute grain boundary segregation is developed and a set of dynamic formulas describing both equilibrium and nonequilibrium segregations is derived in the present paper.

MECHANISM OF GRAIN BOUNDARY SEGREGATION

Model

At a temperature T there exists a certain equilibrium concentration of free vacancy C_V^{eq} in a crystal given by:

$$C_V^{eq} = K_V \exp(E_V /kT)$$ (1)

where E_V is the formation energy of a vacancy and K_V the entropy term. For a binary alloy system, vacancy(V)-solute(B) complexes(VB) will be formed by the equilibrium reaction $V + B \rightleftharpoons VB$, when the vacanvy-solute binding energy $E_b \gg kT$. The complex concentration C_{VB} is determined by the equation:

$$C_{VB} = K_o C_V C_B \exp(E_b /kT)$$ (2)

where C_V and C_B are concentrations of vancancy and solute respectively, and K_o is constant. The total solute concentration $C_{B,Tol} = C_B + C_{VB}$. As shown in Fig.1a, at a high temperature T_1 the concentrations of species V, B and VB in a alloy are homogeneously distributed at equilibrium levels. When the alloy is quenched to and isothermally held at a low temperature T (Fig.1b),

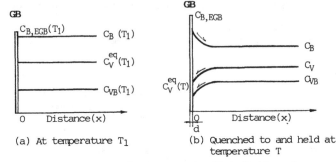

(a) At temperature T_1 (b) Quenched to and held at temperature T

Fig.1 Schematic drawing of concentration distribution of vacany, solute and vacancy-solute complex near grain boundary(GB)

the supersaturated vacancies will be annihilated at grain boundaries, causing the decomposition of complexes there. Thus along with the vacancy gradient a complex gradient is developed driving the complexes moving to grain boundaries and making the enrichment of the solute there. In the meantime a solute gradient is set up at the opposite direction of the complex gradient, the free solute will diffuse away from grain boundaries, which is called back diffusion process. The nonequilibrium grain boundary segregation will arise if the solute enrichment process dominates the back diffusion process. With the prolongation of the isothermal holding time at temperature T(corresponding to the decrease of the cooling rate during a continuous cooling test), the segregation increases at first by consumption of supersaturated vacancies and then turn to decrease when the back diffusion process becomes dominant.

Dynamic Equations

It is suggested that in the process of nonequilibrium segregation, changes in concentrations of V, B and VB in a local region result not only from the diffusion of these species driven by concentration gradients but also from the complex formation and decomposition reactions to keep the equilibrium among these species in the region. Supposing F as the rate of complex decomposition, according to Fick's Second Law and conservation of matter, we find:

$$\frac{\partial C_{VB}}{\partial t} - D_{VB}\nabla^2 C_{VB} = -F$$

$$\frac{\partial C_V}{\partial t} - D_V\nabla^2 C_V = F$$

$$\frac{\partial C_B}{\partial t} - D_B\nabla^2 C_B = F$$

where D_V, D_B and D_{VB} are diffusion coefficients of vacancy, solute and complex respectively. Eliminating F, the above equations become:

$$\frac{\partial C_B}{\partial t} + \frac{\partial C_{VB}}{\partial t} = D_B\nabla^2 C_B + D_{VB}\nabla^2 C_{VB}$$

$$\frac{\partial C_V}{\partial t} + \frac{\partial C_{VB}}{\partial t} = D_V\nabla^2 C_V + D_{VB}\nabla^2 C_{VB}$$

(3)

Assuming the local equilibrium of the complex formation reaction to be maintained in a relatively small region(the reasonableness of this assumption will be discussed latter), the concentrations of vacancy, solute and complex are determined by eq.(2).

Boundary and Initial Conditions

For a specimen heated at a temperature T_1 and then quenched to and isothermal held at a temperature T, the boundary and initial conditions are given as follows(see Fig.1):

(1) The vacancy concentration in the region immediately close to grain boundaries is alway kept at its thermal equilibrium value $C_V^{eq}(T)$ determined by eq.(1) at the temperature T. That is $C_V|_{x=0} = C_V^{eq}(T)$.

(2) Taking account of the equilibrium grain boundary segregation, the solute concentration at grain boundries($C_{B,EGB}$ in Fig.1) can be expressed as $\alpha \cdot (C_B + C_{VB})|_{x=0}$, where α is the equilibrium segregation factor of the solute, which is a function of temperature. Assuming the width of the grain boundary(d) is very small, taken as 10Å in the following calculation, we find:

$$\frac{d}{2}\frac{\partial}{\partial t} [\alpha \cdot (C_B + C_{VB})|_{x=0}] = (D_B \overline{\nabla} C_B + D_{VB} \overline{\nabla} C_{VB})|_{x=0} \qquad (4)$$

(3) Homogeneous distributions of the three species are considered as initial condition. The total vacancy concentration($C_V + C_{VB}$) as well as the total solute concentration($C_B + C_{VB}$) are equal to those at T_1. Using the equilibrium relation of eq.(2) at T, the initial value of C_V, C_B and C_{VB} can be obtained.

Local Equilibrium Assumption

It is assumed, as mentioned before, that the reaction V + B == VB is of equilibrium in a small region of the grain interior. In fact, a certain time is necessary to establish the equilibrium of the reaction, which can be estimated by[8]:

$$\Delta t_o = (\frac{V_a}{C_V} \frac{V_a}{C_B})^{1/3} / D_V D_B$$

where V_a is atomic volume in the crystal. If Δt_o is a negligible quantity as compared with the time for the whole process, the assumption of local equilibrium can be accepted. For the finite difference method used to make numerical calculation, the time increment Δt must be taken large than Δt_o. In the present paper this can be met for the calculation of boron segregation at austenite grain boundaries above temperature $600^{\circ}C$.

CALCULATION RESULTS COMPARED WITH EXPERIMENTAL RESULTS

Calculation Details

According to the mechanism suggested, the segregation of boron at austenite grain boundaries is calculated with computer. The parameters used are:

$D_V(m^2/s)[11]$: $1.4 \times 10^{-5} exp(-1.4/kT)$; $D_B(m^2/s)[12]$: $2 \times 10^{-7} exp(-1.15/kT)$;

$D_{VB}(m^2/s)[4]$: $2 \times 10^{-6} exp(-1.15/kT)$; $C_V^{eq}[13]$: $4.5 exp(-1.4/kT)$;

$E_b(ev)[4]$: 0.5; $K_o[4]$: 12; $\alpha[15]$: $exp(0.42/kT)$.

The initial total B concentration is 20ppm and the grain size 40μm.

The finite difference method is used for the numerical calculation, in which the increment of the distance(x) from a grain boundary is:

$$\Delta x = 0.25 \mu m (\geqslant 720^{\circ}C); 0.07 \mu m (> 720^{\circ}C).$$

To ensure a convergent solution, the time increment is given by:

$$\Delta t = (\Delta x)^2/[2 \cdot max(D_V, D_B, D_{VB})].$$

In order to make comparision with the experimental results by particle tracking autoradiography(PTA), the enrichment factor of grain boundary segregation is defined as $I = (C_{gb} - C_g)/C_g$, where C_{gb} is the total solute concentration at a grain boundary region which includes the grain boundary per se and the area $2.5\mu m$ in the width adjacent to it, and C_g is that in the grain interior away from the grain boundary region.

For simplicity, below $630^{\circ}C$ the diffusion process during cooling is neglected in the calculation since the species considered can hardly move at low temperature. Thus the segregation in the specimen cooled to room temperatures after heating at a temperature lower than $630^{\circ}C$ is same as that at the heating temperature.

Variation of Grain Boundary Segregation with Isothermal Holding Time

The grain boundary segregation of B as a function of isothermal holding time at $1000^{\circ}C$ after quenching from $1200^{\circ}C$ has been calculated(shown in Fig.2a), which is comparable to the experimental result measured by PTA(shown in Fig.2b[10]). It is found that with the prolongation of isothermal holding time the segregation intensifies to a maximum and then declines. The peaks appear at about 3s both on the calculation and experimental curves.

Dependence of Grain Boundary Segregation on Cooling Rate

The calculation result for the grain boundary segregation of B varied with cooling rate after heating at $1000^{\circ}C$ is given in Fig.3, which shows that as cooling rate reduces the segregation increases obviously, but it turns to decrease at relatively low cooling rates, and a maximum appears at about $10^{\circ}C/s$. In fact, by means of SIMS and AP Karlsson[8] found that the stronggest B segregation occures at intermediate cooling rates(arround $13^{\circ}/s$) for 316L austenitic stainless steels.

Influence of Heating Temperature on Grain Boundary Segregation

The influence of heating temperatures on the grain boundary segregation of B in a continuously cooling test has been experimentally measured in Fe-30%Ni alloys(shown in Fig.4b), and the corresponding calculation result is given in Fig.4a. Both results show that at usual cooling rates there exists a minimum segregation at a certain intermediate temperature range referred to as transition temperature. The equilibrium and the nonequilibrium segregations are dominant below and beyond the transition temperature, respectively. In contrary to the equilibrium segregation, the nonequilibrium segregation is enhanced by increasing temperature. Because the latter formed during cooling is sensitive to the cooling rate, the transition temperature goes up as the cooling rate increases. At infinitely high cooling rate the nonequilibrium segregation will be completely inhibited and the segregation exclusively depend on the equilibrium segregation formed during heating.

DISCUSSION

The fit of the theoretical calculations with the experimental results proves the correction of the model suggested for solute grain boundary segregation. Since the equilibrium segregation has been put into consideration in the boundary condition, eq.(4), the segregation behaviour including equilibrium and nonequilibrium segregation is comprehensively described by the dynamic equations derived.

In term of this model, $D_{VB} > D_B$ is a necessary condition for the occurence of the nonequilibrium segregation. In case of $D_{VB} = D_B$, substituting $C = C_{B,Tol} = C_B + C_{VB}$ into eqs.(3) and (4), we find:

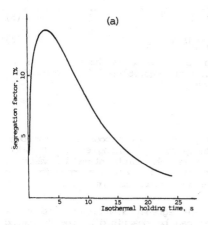

(a)

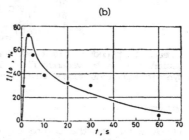

(b)

Fig.2 B segregation at austenite grain boundary as a function of isothermal holding time at 1000°C after quenching from 1200°C.
(a) Calculation result;
(b) Experimental result[10]

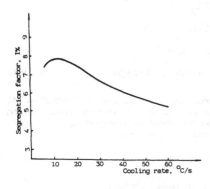

Fig.3 Calculation result of B segregation at austenite grain boundaries as a function of cooling rate after heating at 1000°C, compared with Karlsson's experimental result[8] with maximum at about 13°C/s

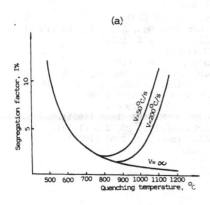

(a)

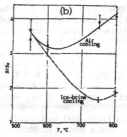

(b)

Fig.4 B segregation at austenite grain boundaries as a function of quenching temperature at different cooling rates.
(a) Calculation result;
(b) Experimental result[9]

$$\frac{\partial C}{\partial t} = D_B \nabla^2 C \tag{5}$$

$$\frac{d}{2} \frac{\partial}{\partial t} (\alpha \cdot C)_{x=0} = D_B \nabla C |_{x=0} \tag{6}$$

It is means that under this condition no influence can be exerted on the segregation process by vacancy flow. In fact, eqs.(5) and (6) are equilibrium dynamic equations suggested by McLean[1].

CONCLUSIONS

(1) A model for the solute segregation to grain boundaries has been suggested. In a alloy system with vacancy-solute binding energy $E_b \gg kT$, there exists local equilibrium among vacancies, solute atoms and vacancy-solute atom complexes in the grain interior. The solute segregation in a quenched specimen will occur by the diffusion of complexes to grain boundaries.

(2) The dynamic process of the solute segregation including both equilibrium and nonequilibrium segregations can be described by the following equations:

$$\frac{\partial C_B}{\partial t} + \frac{\partial C_{VB}}{\partial t} = D_B \nabla^2 C_B + D_{VB} \nabla^2 C_{VB}$$

$$\frac{\partial C_V}{\partial t} + \frac{\partial C_{VB}}{\partial t} = D_V \nabla^2 C_V + D_{VB} \nabla^2 C_{VB}$$

$$C_{VB} = K_o C_V C_B \exp(E_b / kT)$$

The effect of the equilibrium grain boundary segregation is taken into account in the boundary condition.

(3) Theoretical calculations have been made for the boron segregation at austenite grain boundaries as functions of isothermal holding time, cooling rate and quenching temperature with good agreement with experimental results, which proves the correction of the mechanism suggested.

REFERENCES

1. D. McLean, Grain Boundaries in Metals (Oxford, 1957).
2. T.R. Anthony, Acta Metall., 17, 603 (1969).
3. S.J. Bercovici, et al., J. Mater. Sci., 5, 326 (1970).
4. T.M. Williams, A.M.Stoncham and C.R. Harries, Met. Sci., 10, 14 (1976).
5. G.M. Kudinov et al., Phys. Met. Metall., 48, 1244 (1979).
6. R.G. Faulker, J. Mater. Sci., 16, 373 (1981).
7. P. Doig and R.E.J. Flewitt, Acta Metall., 29 1831 (1981).
8. L. Karlsson, PhD thesis, Chalmers Univ. of Techn., Sweden, 1986.
9. Y. Chu, X. He, L. Tang and T. Ko, Acta Metall. Sinica, 23, A169 (1986).
10. X. He, Y. Chu, et al., Acta Metall. Sinica, 23, A291 (1986).
11. A.F. Smith and G.B.Gibbs, Metal Sci. J., 47, 2 (1968).
12. P.E. Busby, M.E. Warya and C. Well, TMS AIME, 197, 1463 (1953).
13. A.F. Rowcliffe and R.B. Nicholson, Acta Metall., 20, 143 (1972).
14. W.F. Jandeska and J.E. Morral, Metall. Trans., 3, 2933 (1972).
15. J.W. Miller, Phys Rev., 188, 1074 (1969).

HIGH-RESOLUTION ELECTRON MICROSCOPY OF OLIVINE-MAGNETITE INTERFACES

STUART MCKERNAN, C. BARRY CARTER, DANIEL RICOULT*, AND A. G. CULLIS**
Department of Materials Science and Engineering, Bard Hall, Cornell University, Ithaca, NY 14853, *Corning-Europe, 7bis Avenue de Valvins, Avon F77210, France, **RSRE St. Andrews Rd. Worcs., WR14 3PS, England.

ABSTRACT

The oxidation of iron-rich olivine to produce magnetite is a model system for the study of phase transitions involving mass transport. High-resolution lattice images of have been obtained from magnetite precipitates in naturally modified iron-rich olivines. The magnetite/olivine interface is shown to be extremely sharp. Steps and misfit dislocations are present at the interface.

INTRODUCTION

The oxidation of iron-rich olivine at elevated temperatures to produce magnetite is a model system for the study of phase transitions involving mass transport. The two structures (olivine and spinel) exist in orientations such that the planar spacing for several low index planes in each phase is very closely matched. The oxygen sublattice of the two systems may be continuous across the phase boundary, although some slight displacement of the atoms occurs in the transformation from one phase to the other. The cation sublattice of both structures contains structural vacancies which facilitates the movement of cations to the transformation front. The transformation from olivine to magnetite requires the transport of Fe ions to the spinel phase, and the transport of Mg and Si ions away from the growing magnetite precipitates. A knowledge of the interface structure and the structure of defects present at that interface, such as dislocations, steps and ledges, is essential in determining the mechanisms for the phase transformation. The morphology of second phase particles in the olivine matrix depends on several factors; the initial iron concentration in the olivine, the time and temperature of the heat treatment and the oxygen partial pressure. An understanding of the precipitate structure may help to deconvolute these different effects, and thus enable the thermal history of naturally occurring olivines to be determined.

High-resolution lattice images have been obtained using a JEOL 4000EX electron microscope from samples of natural iron-bearing olivine from San Carlos. This olivine has been heat-treated at $1000°$ C for 1 hour to internally oxidize the material. Transmission electron microscope images from this material show several different precipitates which heavily decorate the dislocations present in the material. Hematite, magnetite, tridymite, and enstatite are among the minerals which have been identified in he decorated regions [1,3]. In addition to these precipitates nucleated at dislocations, small homogeneously nucleated needle-shaped magnetite precipitates are also observed when the electron beam is oriented parallel to the olivine [010] direction. Similar phase distributions and morphologies have been observed in olivine specimens heat-treated in the laboratory [e.g. 1]. In these studies hematite was found to form at low temperatures although the magnetite spinel may be present in the initial stages of the reaction. The homogeneously nucleated needle-shaped magnetite precipitates were also found in these specimens when annealed at temperatures ~1000°C. Simulated images of magnetite, olivine and some model structures for the olivine/magnetite interface have been calculated using the TEMPAS [4] multislice program.

RESULTS

From electron diffraction patterns of the magnetite precipitates, the orientation relation between the olivine and the magnetite is a_{ol} // $(11\bar{1})_{mag}$, b_{ol} // $(11\bar{2})_{mag}$ and c_{ol} // $(1\bar{1}0)_{mag}$. The olivine orthorhombic unit cell is given by a=4.75 Å, b=10.21 Å, c=5.98 Å, and the magnetite cubic unit cell is given by a=8.394 Å. This orientation relationship means that the olivine (200) planes are parallel to the magnetite $\{11\bar{1}\}$ planes; that is the oxygen 'close-packed' planes in each

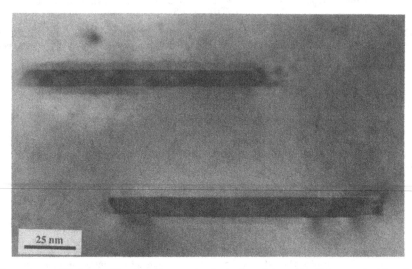

Figure 1   High-magnification micrograph of spinel precipitates in olivine matrix. Note the regions of darker contrast around each precipitate.

system are parallel. When viewed along the olivine **b** axis, the diffraction spots from the olivine structure and those from the spinel structure are almost co-incident. The presence of the second phase is then only detectable by a change in the relative intensity of the diffraction spots. The morphology of these magnetite particles is very striking. Figure 1 is a relatively high magnification, bright-field image showing two of the magnetite precipitates. The edges of the particle are extremely sharp and highly planar, suggesting that they are parallel to the [010] electron beam. The ends of the precipitate are more rounded. The precipitates are ~ 6-10 nm wide and 100-150 nm long.

The precipitates appear to extend completely through the TEM foil, giving them a minimum thickness of a few tens of nm. The long sides of the precipitates are parallel to the 'close-packed' oxygen planes, which appear end-on at this orientation. It can also be seen that around each precipitate there is an irregular region approximately 5 nm wide which has a darker contrast than the matrix. This darker region has been observed in association with all of the precipitates. High-resolution images of this darker contrast region do not show any large differences from olivine regions away from the precipitates, however the pattern does appear to be less distorted than images taken away from the magnetite precipitates.

Figure 2 shows a boundary between the olivine (above) and a magnetite precipitate (below). The 'close-packed' oxygen planes in both systems lie horizontally and are 'edge on' in this image. The olivine has been almost totally thinned away, leaving a small layer attached to the spinel. This lip may result from the geometrical shadowing effect of the adjacent, more resistant spinel, or it may result from the different physical properties of the layer of material around the precipitate which appears darker in figure 1. The interface is obviously atomically abrupt, and very uniform (although there is some variation in specimen thickness across the image). There appears to be a relatively large amount of amorphous material visible at the edge of the olivine; presumably a result of ion-milling damage.

The schematic view of a projected atomic arrangement is shown in figure 3. The two structures are obviously very similar in this configuration, the oxygen sublattice, in particular, is continuous across the phase boundary. The octahedrally co-ordinated ion positions in the spinel project onto the cation positions in the olivine structure, and (in this particular configuration) the silicon positions in the olivine structure project onto the tetrahedrally co-ordinated cations in the spinel structure. There is, however, a difference in the distribution of the structural vacancies in

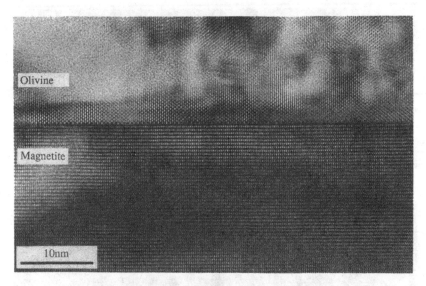

Figure 2   High-resolution image of an olivine/magnetite boundary in thin crystal. The boundary is atomically smooth and abrupt.

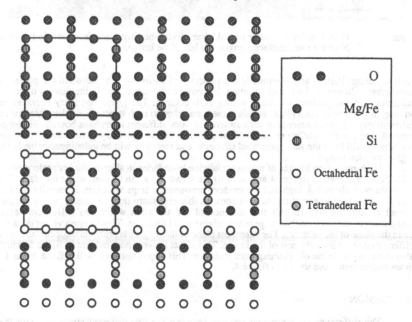

Figure 3   Schematic diagram of the projection of one possible Olivine-Magnetite interfaces. The different shadings of the atom columns represents structural vacancies at different heights in the columns. Projected unit cells are outlined.

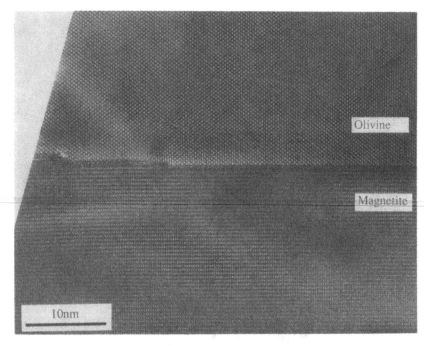

Figure 4   High-resolution image of a thicker olivine-hematite interface than in figure 2.
Note the two interface steps on the left of the image.

each structure. Particularly obvious is the unoccupied cation column between the projected
tetrahedral columns in the spinel structure, but other differences exist in the density of cation
columns, and the position of atoms within each column. The phase boundary is shown as
passing through an oxygen plane, as this plane is common to each structure. As there are several
different oxygen environments in each structure several different interphase boundaries can be
created, each with a different cation configuration at the boundary plane. Some of these
boundaries will have the same projected structure, and therefore will be indistinguishable in the
high-resolution images.

High-resolution images of magnetite/olivine boundaries in slightly thicker regions of the
TEM foil are shown in figures 4 and 5. In figure 4 two steps can be seen in the interface. The
steps are both about 5 Å high and the terraces between the steps are again atomically flat and
abrupt. The lattice image at the steps is considerably less sharp than elsewhere, and a Burgers
circuit around the step reveals the presence of a dislocation which has a Burgers vector
component in the plane of the image in the olivine [101] direction. This Burgers vector does not
lie in the plane of the interface. The outermost plane of the magnetite precipitate appears to have a
different contrast from the rest of the image. This effect is even more evident in figure 5, which
also shows a step in the olivine/magnetite interface. This step is not at all well defined, but is 1.5
times higher than those shown in figure 4.

DISCUSSION

The difference in interplanar spacing between the two different structures with this
orientation relationship is set out in table 1. From this table it would appear that the lowest energy
shape for the magnetite precipitates would be platelets parallel to the olivine (100) plane. This is

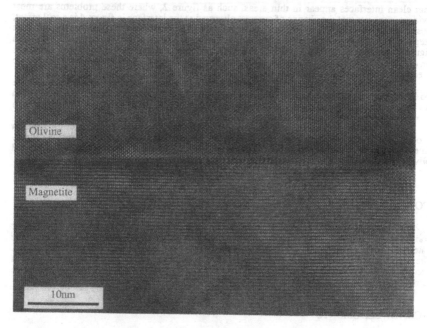

Figure 5   High-resolution image of a thicker olivine-hematite interface than in figure 2.
            Note the gradual change in appearance of the interface step.

TABLE 1

Trace of plane	Olivine (Å)		Magnetite (Å)		Misfit
Horizontal	(100)	4.75	(111)	4.85	2%
Vertical	(001)	5.98	(110)	5.94	0.7%
Plane of figure	(010)	10.21	(112)	3.43	
			x3 = 10.28		0.7%

also the 'close-packed' oxygen plane, which would be a low energy plane. Assuming that the misfit shown above is accommodated by perfect dislocations, then the width of the platelets is about half of the separation between misfit dislocations. This may be because the long edge corresponds to a low-energy surface, and growth normal to this surface is very slow. The platelets are also about twice as long as the misfit dislocation separation. Thus we do not expect to see dislocations at the platelet edge, but they may occur along the platelet. The occurrence of dislocations at steps on the interface, as shown in figure 4, enables large terraces of low energy interface to exist. The small size of the Burgers vector of these dislocations means that if the misfit is accommodated by these dislocations, several of them will be present along the interface. If the misfit dislocations form a regular array in the interface plane then there will also be dislocations in the interface parallel to the plane of the specimen. This will affect the appearance of the interface in the high-resolution images [4,5]. The interface contrast may also be affected by various artifacts introduced during specimen preparation; different thicknesses across the interface due to different ion-milling characteristics, grain boundary grooving due to preferential milling and the possible filling of such a groove with amorphous, sputtered material. The fact that clean interfaces appear in thin areas, such as figure 2, where these problems are most evident, suggests that the change of contrast observed at the interface in figure 4 is a real effect, possibly related to the stoichiometry at the interface. The variation of contrast at the interface steps in figure 5 appears to arise from an inclined step. The step itself is larger than those in figure 4, and appears to be 1.5 unit cells high. This should lead to different interfacial structures on either side of the step. The difference in specimen thickness at either end of the inclined step precludes direct interpretation of the image.

The relative thinness of the precipitates, the existence of steps at the interface, and the possible non-stoichiometry of the outermost magnetite layer suggest that the growth mechanism for these platelets is the migration of steps along the interface; this migration may be assisted with movement of dislocations associated with the steps.

The band of dark contrast around the precipitates may represent a buildup of the reaction products following the phase transformation, or a concentration gradient of the diffusing cations. In either case there appears to be no structural difference between this zone and the unaffected olivine. Further work is in progress to determine the stoichiometry of this region.

CONCLUSIONS

Magnetite spinel precipitates in olivine can form as platelets (which appear as needles or rods in projection) with a high degree of lattice matching between the corresponding planes. The interfaces between the magnetite and the olivine in this case are atomically smooth and abrupt. Steps and interfacial dislocations are present on the interface which may indicate the growth mechanism for the platelets. There appears to be differently structured steps which may give rise to multiple atomic structures of the interface plane.

ACKNOWLEDGMENTS

The authors would like to thank Mr R. Coles and Ms M. Fabrizio for technical support. The electron microscopes are part of a central facility provided by the Materials Science Center at Cornell and are supported, in part, by National Science Foundation. This research has been supported by the Materials Science Center at Cornell University.

REFERENCES

1. P. E. Champness, Mineral.Mag. **37**, 790 (1970).
2. D. L. Kohlstedt and L. B. Vander Sande, Contrib. Mineral. Petrol. **53**, 13 (1975).
3. C. B. Carter, Ber. Bunsenges. Phys. Chem. **90**, 643 (1986).
4. R. Kilaas, Proc. **45**th EMSA meeting, p66 (1987).
5. D. R. Rasmussen, Y. K. Simpson, R. Kilaas and C. B. Carter, Proc. 46th EMSA, p.610 (1988).
6. D. R. Rasmussen, S. McKernan, and C. B. Carter, Proc. 47th EMSA, p130, (1989).

ATOMIC SCALE STRUCTURE OF TWIN BOUNDARY IN Y-Ba-Cu-O SUPERCONDUCTORS

Yimei Zhu, M.Suenaga, Youwen Xu, and M.Kawasaki*
Department of Applied Science, Brookhaven National Laboratory, Upton, NY 11973
*JEOL USA, Inc. Peabody, MA 01960

ABSTRACT

High resolution electron microscopy of the twin boundary layers in $YBa_2(Cu_{1-x}M_x)_3O_{7-\delta}$ for $x=0$ and 0.02 and $M=$Zn, Fe and Al showed that the boundary widths are ~ 1 nm for the pure and the Zn substituted $YBa_2Cu_3O_7$, 2.5~3 nm for the Fe and Al substituted oxide. It was found that the lattice plane is shifted across the twin boundary by $(1/3 \sim 1/2) \cdot 2d_{(110)}$ along the boundary. The broadening of the layer for the cases of Fe and Al is also thought to be associated with a reduction of the twin boundary energy, which also leads to an increased twin density.

INTRODUCTION

Since the discovery of high T_c superconductivity in $YBa_2Cu_3O_7$, twinning has been presumed to be important in determining superconducting properties, particularly critical current density. For example, it has been argued that the twin boundary is an area of weak superconductivity[1,2] or flux pinning sites.[3,4] On the other hand, some reports suggest that the twin boundary may act as a diffusion path for oxygen.[5,6] It is therefore of interest to study the twin boundary in detail to understand the correlation between the structure and superconducting properties. Our previous electron diffraction studies[7] on twin boundaries in $YBa_2(Cu_{1-x}M_x)_3O_{7-\delta}$, where $M=$Zn, Al, Fe and Ni, indicated the existence of non-orthorhombic twin boundary layers. The layer thickness, which depends on the doping element and oxygen content, varied from $\sim$1 nm for a pure $YBa_2Cu_3O_7$ and $M=$Al, and Zn and $x=0.02$, to $\sim$2.5 nm for $M=$Ni and Fe and $x=0.02$. Furthermore, direct observation of a twin boundary by high resolution electron microscopy and optical diffraction from a single twin boundary supported the existence of such a twin boundary layer.[8,9] In the present paper, we report a further investigation of the structure and of the nature of strain associated with the twin boundary in $YBa_2Cu_3O_7$. In particular, the structure of the boundary for a pure oxide will be compared with those for which 2% of Cu is replaced with Zn, Fe and Al.

EXPERIMENTAL

High-quality bulk samples of Y-Ba-Cu-O were produced by sintering the powders as described elsewhere.[10] The HREM specimens were prepared by crushing bulk samples into fine fragments in acetone, and then depositng onto a holey carbon film. Electron microscopes employed were *JEOL JEM-2000FX, JEM- 2000EX* and *JEM-2010*, all operated at 200kv.

RESULTS

1. Pure $YBa_2Cu_3O_7$

Fig 1(a) is a typical multi-beam image of a twin boundary seen edge on in $YBa_2Cu_3O_7$ at a direct magnification of 100,000 x. Thin lines of contrast at twin boundaries are clearly visible. Unless the twin boundary had been damaged (including the loss and disordering of oxygen at twin boundary) by the electron beam irradiation, such sharp and straight twin boundary contrast was always observed under multi-beam imaging conditions in $YBa_2Cu_3O_{7-\delta}$ ($\delta \sim$0). This suggests that the crystal structure at the boundary may be different from the orthorhombic twin matrix, and the thickness of this region is 1 to 1.5 nm as measured from the multi-beam images.

The selected area diffraction (SAD) pattern was formed from an area of about 1 micron diameter which included several twin boundaries. The pattern shows streaks in [110] orientation which are parallel to the direction of twinning of the reciprocal spots and perpendicular to the (110) twin boundary [see the inset in Fig.1(a)]. Also, the streaks are sharp and straight, and equally extended in all the diffraction spots including the origin. Although the intensity of the streaks is not strong, strongly overfocusing the second condenser lens of the electron microscope renders the streaks easily observable. As discussed earlier,[7] careful measurements showed that the length of the streak is inversely proportional to the thickness of the twin boundary and indicated that the streak in reciprocal space is a geometric effect of the thin layer at the twin boundary in real space. The thickness of the boundary determined from the multi-beam bright field image was 1 to 1.5 nm, in agreement with the value previous obtained from the length of the streaks. In addition,

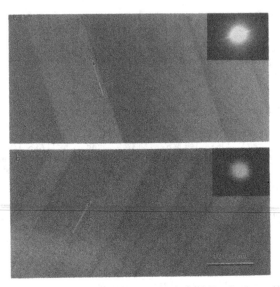

Fig.1 Multi-beam images of twin boundaries viewed along the [001] direction from a $YBa_2Cu_3O_7$ sample (a), and a $YBa_2(Cu_{0.98}Al_{0.02})_3O_7$ sample (b). The insets show two 110 SAD spots from $YBa_2Cu_3O_7$ and $YBa_2(Cu_{0.98}Al_{0.02})_3O_7$, respectively. Note that the twin boundary image from $YBa_2(Cu_{0.98}Al_{0.02})_3O_7$ is wider and fuzzier than that of $YBa_2Cu_3O_7$, the streak is longer in (a) than that in (b).

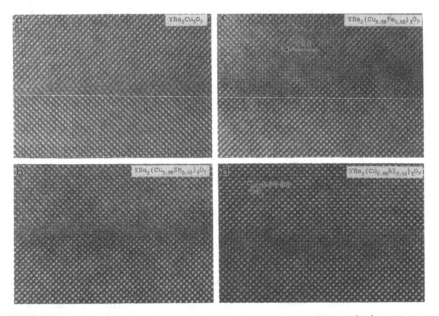

Fig.2 High resolution structure images of twin boundaries observed along the [001] direction for $YBa_2Cu_3O_7$ (a), $YBa_2(Cu_{0.98}Zn_{0.02})_3O_7$ (b), $YBa_2(Cu_{0.98}Fe_{0.02})_3O_7$ (c) and $YBa_2(Cu_{0.98}Al_{0.02})_3O_7$ (d). The horizontal distorted region in the middle of each micrograph is a twin boundary.

in order to confirm that the streaks are due to the boundary layer, an optical diffractogram using a laser beam and a digitized diffractogram using a Eikomix camera from a single twin boundary of a HREM micrograph were performed. The results unambiguously confirmed that these streaks arise from the twin boundary.[8,9] Similar streaks were also observed with the same techniques in a diffraction pattern from a small twin boundary region of a two-dimensional schematic model drawing with only oxygen-copper coordination in CuO basal plane.

A high resolution image of a twin boundary taken from an area such as that shown in Fig.1 at a direct magnification of 400,000 x is shown in Fig.2. The horizontal distorted region in the middle of the each micrograph is a twin boundary, which is a transition area of adjacent twin domains with a width of several unit cells. In the twin domain on both sides of the boundary, the bright dots arranged in a rectangular unit cell correspond to the fundamental perovskite unit cells. The approximate 3.82Å and 3.88Å period dot patterns correspond respectively to the a and b lattice parameters of the Y-Ba-Cu-O oxides. Both the geometry and the scale of the patterns allow us to associate the white dots with the columns of either Ba(Y) or Cu(O) atoms. This can not be the column of O(vacancy), since one of the sides of the O(vacancy) unit cell should be parallel to the twin boundary. Also, our primary computer simulation of the image contrast of the twin domain using the multislice method showed that white dots are likely to be heavy atoms, i.e. Ba(Y) atoms. (Unfortunately, this is not conclusive, because of the lack of knowledge about the specimen thickness and defocus.) Interpretation is tentative, since for very thin crystals, a many beam image can be interpreted as a two-dimensional projection of the charge density. Thus, in that case the white dots could be potential valleys and might be assigned to arrays of lighter atoms such as Cu(O) [since these could not be the O(vacancy) columns as argued above]. In either case the arguments below still hold true.

Several observations can be made about the details of the interchange of the a and b planes across the twin boundary. The interchange is accomplished not only by the rotation of the planes in the left twin with respect to the right by an angle of $90° - \phi$, where $\phi = 2(b - a)/(a + b)$, but also by a translation of the a (or b) plane by approximately $(1/3 \sim 1/2) \cdot 2d_{(110)}$ along the twin boundary [i.e. the ($\bar{1}$10) direction]. This displacement, or shift, is most clearly seen by tilting the micrograph and observing the series of the dots along the [100] or [010] direction across the boundary. This can be clearly seen in Fig.3(a) which was taken by placing an optical camera at a near-glancing angle from a print having a large area of the same boundary shown in Fig. 2(a). Also, the camera was pointed along the [100] or [010] direction on the image. Such displacement of the lattice at the twin boundary is readily observable in all specimens, although the detailed image of the twin boundary layer is not always resolved. Our experiments show that a satisfactory image depends both on the Scherzer focus of the particular twin boundary and at least a 2.0 Å point-to-point resolution imaging condition of the microscope, and as well as on the strain distribution at the twin boundary, i.e., boundary structure itself. For all the specimens we observed, the transition region (shown in Table 1) is in the range of $4 \sim 6$ times the (110) interplanar spacing ($\sim$1 to 1.5 nm) which is consistent with our low-magnification observation discussed above.

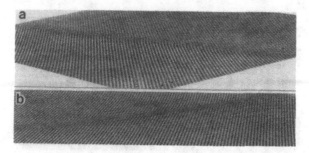

Fig.3 Photographs [a: $YBa_2Cu_3O_7$, b: $YBa_2(Cu_{0.98}Al_{0.02})_3O_7$] taken at a near-glancing angle from prints with a large area of the same boundary shown in Fig.2(a) and Fig.2(d). It can be clearly seen that the a and b planes interchanging at the twin boundary involves a shifting along the boundary as well as a rotation across the boundary.

Table 1. Orthorhombicity ϕ_{100} and the twin boundary thickness

Samples	X-ray	SAD Pattern	Thickness of T.B.*
$YBa_2Cu_3O_7$	0.0171	0.0170	$1.0 \sim 1.5$ nm
$YBa_2(Cu_{0.98}Zn_{0.2})_3O_7$	0.0173	0.0166	$0.9 \sim 1.3$ nm
$YBa_2(Cu_{0.98}Fe_{0.2})_3O_7$	0.0117	0.0115	$2.0 \sim 3.0$ nm
$YBa_2(Cu_{0.98}Al_{0.2})_3O_7$	0.0119	0.0114	$2.6 \sim 3.9$ nm

* Measured from mult-beam images of individual twin boundary (T.B.) and the length
of the streak from selected area diffraction pattern involving several twin boundaries.

2. Effects of Alloying

In order to study the effect of replacing Cu by other elements in $YBa_2Cu_3O_7$ on the structure
of the twin boundaries, we have studied $YBa_2(Cu_{0.98}Zn_{0.02})_3O_7$, $YBa_2(Cu_{0.98}Fe_{0.02})_3O_7$ and
$YBa_2(Cu_{0.98}Al_{0.02})_3O_7$. It is generally agreed that Fe, Al and Co substitute predominately for
Cu on the chain site, while Zn and Ni tend to the plane site (at least at low concentrations, ~2%).
Furthermore, our previous SAD study noted that the substitutions of Cu by Zn and Ni do not
cause the thickness to vary significantly from that the pure oxide while Fe and Al cause a widened
layer. As shown in Fig.2(b), 2(c) and 2(d) these high resolution structural images confirm the
earlier observation by electron diffraction that the thickness of the distorted region is essentially
unchanged from the pure specimen with Zn substitution [Fig.2(b)], but is widened considerably
with Fe [Fig.2(c)] and Al [Fig.2(d)] substitution. We also noted that the image of the boundary in
$YBa_2(Cu_{0.98}Zn_{0.02})_3O_7$ is very similar to the pure $YBa_2Cu_3O_7$, i.e. straight and narrow. On the
other hand, the addition of Fe and Al cause the thickness not only to widen but also to be diffuse
and wobbly. The specimens with a wider twin boundary layer have lower orthorhombicity than
those with a narrow one. Measurements of the orthorhombicity from SAD patterns are consistent
with x-ray diffraction results (see Table 1).

DISCUSSIONS

The twin boundary structure in $YBa_2Cu_3O_7$ has been studied extensively. However, very
few clear atomic-scale images of twin boundary were reported. In order to obtain a high quality
structure image, one must identify a very thin crystal (edge) with an undamaged twin boundary
and protect the structure from damage (from excessive beam heating and irradiation) until the
HREM image is recorded. Great care is necessary during the observation, since the main effect
of electron irradiation in $YBa_2Cu_3O_7$ is to induce the fading of the twin due to the loss and/or
disordering of the oxygen.

Fig.4(a) is a schematic [001] projection of twin boundary structure. The atomic position is
consistent with a highly resolved structure image of the twin boundary shown in Fig.4(b). Since
the Cu-O coordination is most important for constructing the twin boundary, here we only draw
the CuO$_5$ pyramids (▨) and Cu-O chain (double line). The black dots are related to the projected
columns of Cu(O); the columns of oxygen atom are at the corner of the pyramid. The projection
of the unit cell is shown in inset.

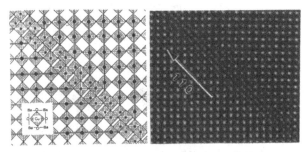

Fig.4 A schematic [001] projection of twin boundary structure (a). The black dots denote Cu atoms,
the oxygen atoms are at the corners of the pyramid. The unit cell is shown in the inset. The atomic
position is consistent with a highly resolved twin boundary image from $YBa_2(Cu_{0.98}Zn_{0.02})_3O_7$ (b).

The sketch clearly shows a lattice displacement of about half a unit cell along the twin boundary. In addition, if the atomic position is measured perpendicular to the boundary, from one side of the twin domain to the other, the same amount of the displacement is also found. Such a configuration can be either due to a shear along the twin boundary or an expansion/contraction perpendicular to the boundary.

In order to determine the nature of lattice displacements at the twin boundary described above, two-beam studies of $YBa_2(Cu_{0.98}Zn_{0.02})_3O_7$ were also carried out using both analysis of boundary fringe contrast and $\vec{g} \cdot \vec{R} = 0$ invisibility criterion.

First, it was found that the outer fringes of the tilted twin boundary are symmetric in bright-field image and alternatively symmetric and asymmetric in dark-field image. From the criteria given by Gevers et al.[11] and Amelinckx et al.[12], such fringes are neither from α-interface (where $\alpha = 2\pi\vec{g} \cdot \vec{R}$) nor from δ-interface (where $\delta = \xi_{g1}s_1 - \xi_{g2}s_2$, see the reference for definition), it is from a mixture of α-δ interface. The δ characteristic could arise from a slight misorientation of the twin domains whereas the α character could arise from the displacement. Thus, this also clearly indicates that the twin boundary consists of not only a rotation but also a translation of the respective lattice plane at the boundary.

Secondly, we observed the same twin boundaries by tilting $\pm 35°$ from [001] through $[\bar{1}\bar{1}1]$ to $[\bar{3}\bar{3}2]$ zone axes. When $\vec{g}$=110, 112 and 113 is operated in these zone axes, respectively (i.e. $\vec{g}$ perpendicular to the boundary), the boundary fringes are out of contrast. This implies that the phase angle, $\alpha = 2\pi\vec{g} \cdot \vec{R}$, if such $\vec{R}$ exists, is either 0 or $2n\pi$. Then unless the component of $\vec{R}$ which is perpendicular to the twin boundary is a multiple of $d_{(110)}$, this two beam contrast study suggests that the displacement at the twin boundary is along the $(\bar{1}10)$ boundary and not perpendicular to the boundary.

Although a detailed investigation of atomic arrangement at the twin boundary distorted area would be of significant interest due to its possible role in determining irreversible superconducting properties of $YBa_2Cu_3O_7$, such a study is not now possible without extensive image simulations and model calculations of various possible boundary structures. However, in the above discussions, it has been clearly shown that there exists a distorted layer at the twin boundary. Also, the lattice planes on one side of the boundary are shifted along the (110) twin plane by $(1/3 \sim 1/2) \cdot 2d_{(110)}$ with respect to the planes on the other side. Although the exact cause for this distortion is not clear, it is certain that the oxygen in the chain plays a dominant role in causing this shear strain and in determining the width at the boundary. An ideal twin boundary is produced by cutting an orthorhombic structure along the (110) plane at the Cu site and attaching it to a region of otherwise identical material (except for a $\sim 90°$ rotation of the a and b axes). It is likely that the chain-oxygen at the boundary region would repel each other to minimize the energy due to oxygen-oxygen repulsion. Such repulsion may lead to a local volume expansion of the boundary region, and such an expansion could schematically reproduce similar bent (100) or (010) planes across the boundary. However, it was shown here that the displacement is along the (110) twin plane, not perpendicular to it.

Another possibility for the distorted boundary is that some of the chain oxygen at the boundary region could be missing. In fact, this possibility was proposed by Jou and Washburn.[8] However, a specific model does not lead to the type of twin boundary distortion which was observed here. (A careful examination of their twin boundary image by high resolution electron microscopy also shows a similar distortion and shear displacement at the boundary although they did not note the latter.)

A clear indication of the importance of the chain-oxygen in determining the nature of the distortion is seen by the effects of the substituting elements Zn, Fe and Al on the width of the boundary layer. As pointed out above, Zn (at a level of 2%) is thought to substitute for Cu in the CuO_2 plane and thus is likely not to influence the location of the chain oxygen. For Fe (which substitutes into the chain layer) additional layer is incorporated in the chain region ($\sim \frac{1}{2}$ oxygen per Fe). We anticipate a similar situation for Al. This extra oxygen is likely to be accompanied by additional disordering of the chains at the boundary region and to create a diffuse boundary.

It is also interesting to note that the addition of Al and Fe reduce the twin spacing.[10,13] This reduction was argued to be due to the decreased interface energy, γ, of the boundary.[14] This was deduced from the relationship for the spacing d,[15]

$$d \cong \sqrt{\frac{g\gamma}{cM\phi^2}} \tag{1}$$

and from the measurements of d, g, and ϕ for pure $YBa_2Cu_3O_7$, Fe and Al substituted specimens. Here, $\phi = 2(b-a)/(a+b)$, c is a constant of the order of unity, M is the elastic shear modulus and g is the grain size measured along the twin boundary. Then, based on the observed broadening and diffuseness of the boundary due to the Fe and Al substitutions, we suggest that a large fraction of the interface energy γ is strain energy which is reduced by the Fe and Al additions. Thus, the sharp boundaries, which were seen in the pure and the Zn substituted $YBa_2Cu_3O_7$, are highly strained and this strain leads to a high surface energy.

In contrast, the wider twin boundaries observed when substituting Fe and Al for Cu are often associated with a proposed ordering of the substitution atoms along the boundary leading to a reduced boundary energy.[16,17] However, to the present, no conclusive evidence for such a picture is available. On the contrary, we argue that, since the twins are formed at relatively low temperature compared with the temperature required for substantial diffusion of Fe and Al, such a segregation of the atoms to the twin boundary is unlikely. Thus, the broadening of the boundary in the Fe and Al substituted $YBa_2Cu_3O_7$ is probably associated with the distortion of the oxygen-copper chain due to the Fe or Al incorporation in the chain. In support of this argument, a similar broadening in the boundary was also observed in an oxygen reduced specimen, $YBa_2Cu_3O_{6.67}$.

Finally, it is puzzling that the displacement along the (110) twin plane which was observed in Fig.2 and Fig.4(b) is not limited to or near the distorted boundary region. It appears that the entire material is shifted across the twin boundary, by a significant amount along the (110) plane. Such a shift might be expected to be a very high energy process and unlikely to take place. However, it may be that the displacement occurs slowly during twin formation (a process which also involves movement of the atoms). In this way, an apparently high energy process can be accommodated.

ACKNOWLEDGMENTS

This research was performed under the auspices of the U.S. Department of Energy, Division of Materials Sciences, Office of Basic Energy Sciences under Contract $No.DE - AC02 - 76CH00016$.

REFERENCES

1. G. Deutscher and A. K. Muller, *Phys. Rev. Lett.* **59**, 1745 (1987).

2. M. Daeumling, J. Seuntjens and D. C. Larbalestier, *Appl. Phys. Lett.* **52**, 590 (1988).

3. T. Matsushita, K. Funaki, M. Takeo and K. Yamafuji, *Jpn. J. Appl. Phys.* **26**, L1524 (1987).

4. G. J. Dolan, G. V. Chandrashekhar, T. R. Dinger, C. Field and F. Holtzberg, *Phys. Rev. Lett.* **62**, 827 (1989).

5. K. N. Tu, N. C. Yeh, S. I. Park, and C. C. Tsuei, *Phys. Rev. B.* **39**, 304 (1989).

6. C. J. Jou and J. Washburn, *J. Mater Res.* **4**, 795 (1989).

7. Y. Zhu, M. Suenaga, Y. Xu, R. L.Sabatini, and A. R. Moodenbaugh, *Appl. Phys. Lett.* **54**, 374 (1989).

8. Y. Zhu, M. Suenaga, R. L. Sabatini, and Y. Xu, *Proc. 47th Annual Meeting of Electron Microscopy Society of America, San Antonio, Aug.1989* G.W.Bailey, ed., p.168, (1989).

9. Y. Zhu, M. Suenaga, and Y. Xu, *Philos. Mag. Lett.* **60**, 51 (1989).

10. Y. Xu, M. Suenaga, J. Tafto, R. L. Sabatini, A. R. Moodenbaugh, and P. Zolliker, *Phys. Rev. B* **39** 6667 (1989).

11. R. Gevers, J. Van Landuyt, and S. Amelinckx, *Phys. Status Solidi* **11**, 689 (1965).

12. S. Amelinckx and J. Van Landuyt, in *Electron Microscopy in Mineralogy*, J. M. Christie, J. M. Cowley, A. H. Hener, G. Thomas and N. J. Tighe, editors, p.68 (1976).

13. R. Wordenweber, G. V. S. Sastry, K. Heinenmen, and H. C. Heryhardt, *J. Appl. Phys.* **65** 1649 (1989).

14. N. Chandrasekar, D. O. Welch, and M. Suenaga, presented at *1988 Fall Meeting of the Materials Research Society, Boston*, (1988).

15. D. O. Welch, unpublished and see Ref.10.

16. P. Bordet, J. L. Hodeau, P. Strobel, M. Marezio, and A. Santoro, *Solid State Commun.* **66**, 435 (1988).

17. J. L. Hodeau, P. Bordet, J. J. Capponi, C. Chaillout, and M. Marezio, *Physica C* **153-155**, 582 (1988).

MODELING OF NANOPHASE CONNECTIVITY IN SUBSTANCE-VOID COMPOSITE BY OBLIQUE DEPOSITION

T.MOTOHIRO, S.NODA, A.ISOGAI AND O.KAMIGAITO
Toyota Central Research & Development Laboratories, Inc., 41-1, Nagakute-cho, Aichi-gun, Aichi-ken, 480-11, Japan

ABSTRACT

Obliquely vapor deposited thin film is characterized by its unique inclined columnar structure. Recently one of the authors developed thin film optical quarter-wave plate by oblique deposition. SEM observation revealed the inclined columns of ~10 nm in diameter and ~3 microns in length. Its birefringence indicates those columns are less closely spaced in the plane of vapor incidence (PVI) than normal to PVI, composing alternatively stacked substance layer (columns laterally connected with each other)-void layer(residual space) nanophase composite with 2-2 connectivity. The growth mechanism of the inclined columnar structure has been successfully explained by the self-shadowing effect in 2D-space computer simulation in PVI. However, the connectivity development perpendicular to PVI is not self-evident. In the present work, we performed simple 3D-space simulation of oblique depositipon and observed substantial feature of the connectivity development and related features on this nanophase structure.

INTRODUCTION

New types of solids called cluster-assembled materials(CAM) or nanophase composites(NPC) has been receiving considerable attention recently[1]. The CdS superclusters in zeolites[2] is a typical example of this class of materials. Discrete $(CdS)_4$ cubes, spontaneously being isolated within the regularly arranged sodalite cages and stabilized by the interaction between Cd atoms and framework O atoms, form an interlocking three-dimensional supercluster structure with peculiar optical properties. From this example emerge three intriguing ideas: (1)isolation of clusters, (2)cluster arrays and (3)self-organization of clusters. The isolation of clusters(1) is absolutely essential for CAM or NPM to prevent cluster fusion. The cluster arrays (2) may add new aspects of properties to CAM or NPC through interlocking of the arrayed clusters or their anisotropy. The self-organization(3) is necessary because it is very difficult to define nanometer length scale three-dimensional structure (not one dimensional in-depth structure) artificially, or even if it is realized, there is no doubt that the process is of high cost which prevents the CAM or NPM from prevailing in applications. In addition, in such nanometer scale inhomogeneous materials, there is a reasonable prediction that stable property will be expected only in the system spontaneously formed by self-organization. Therefore, the ideas (1)-(3) are very suggestive as guidelines for searching new CAM or NPM.

Recently one of the authors developed thin film optical quarter-wave plate by oblique deposition, and realized that the film has nanometer scale structure which is essential to its function: birefringence. In the CdS superclusters in zeolites, the three ideas (1)-(3) are realized by the interaction between clusters(CdS) and the support(zeolite). The aim of this paper is to characterize the atomic scale structures formed on substrates by oblique deposition with a view to consider them as another candidate of the support for synthesis of new types of CAM or NPC.

NANOPHASE STRUCTURE IN OBLIQUELY DEPOSITED THIN FILMS

Figure 1 shows a fractured section of a double-layered Ta_2O_5 thin film

Substrate   Film   Film surface

Fig.1 Fractured section of a thin film
optical quarter-wave plate.

Fig.2 Blue patches growing in
an obliquely deposited WO_3 film.
The sample size is 50x50 mm.
The plane of vapor incidence is
in parallel with the horizontal
edge of the sample.

quarter-wave plate by oblique deposition observed with a field emission type
scanning electron microscope. What surprized us in this photo is its densely
packed fine fibrous structures of ~10 nm in diameter in spite of its large
thickness ~3 microns because we are accustomed to thicker and coarser columnar structures ~100nm or more in diameter in intensively studied obliquely
deposited metallic alloy thin films for magnetic use[4]. In our obliquely
deposited oxide thin film for optical use, coarsening of the fine fibrous
structure has been confirmed to lead to hazy or opaque appearance which
reduce applicability of the film very much. The fine fibrous structures
formed at high deposition rate in a high vacuum have been very stable at
least for two years in room temperature until today.

    As for structure information perpendicular to this cross section, many
previous works[4],[5] support a view that the fibrous structures or columns
are less closely spaced in the plane of vapor incidence (PVI) than normal to
PVI, composing alternatively stacked substance layer (columns laterally connected with each other)-void layer(residual space) nanophase composite with
2-2 connectivity[6]. Our observation on the growth of elliptical blue patches
in [Al/SiO_2/obliquely deposited WO_3/glass substrate] system also supports
this view[7]. According to our speculation, exposure of the sample system to
the air triggers proton diffusion from the Al layer through pin holes in the
SiO_2 isolator layer into the obliquely deposited
WO_3 layer, forming blue tungsten bronze regions.
The elliptical shapes of the blue patches with
their major axes perpendicular to the PVI as
shown in Fig.2 suggest difference of proton diffusion rate caused by anisotropy of the packing
condition of the fibrous structures.

    Here, we have a sample of nanometer length
scale structure which fits the following condition: "Birefringence may arise when there is
an ordered arrangement of similar particles of
optically isotropic material whose size is large
compared with the dimensions of molecules, but
small compared with the wavelength of light"
(Born and Wolf [8]). Actually, the film shows
birefringence of half as large as that of
calcite at the deposition angle $\delta=70°$ as shown
in Fig.3.

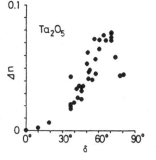

Fig.3 Birefringence vs the
deposition angle δ

MODELING OF THIN FILM GROWTH BY OBLIQUE DEPOSITION

Modeling of thin film growth by oblique deposition has been successfully done by computer simulation[4],[5], and we followed the similar method.

Figure 4 shows the method of simulation. Particles are supposed to be homogeneously generated on a square S whose surface normal n is inclined from the surface normal of the substrate Z by a deposition angle δ. Then particles are supposed to drift in parallel with n until they make their first contacts with the substrate plane or the previously deposited particles. No relaxation process is considered because we deal with low mobility oxide adparticles, whose validity is discussed later.

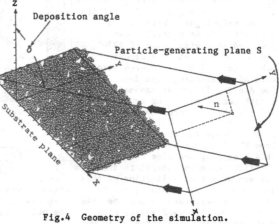

Fig.4 Geometry of the simulation.

Figure 5 shows calculated deposits-patterns after depositions of 10000 particles at δ=70° and δ=30°. It is shown that the particles are arranged inhomogeneously and form many clusters leaving voids in between at large δ. As can be estimated from the cluster size, the deposited structure is of nanometer length scale.

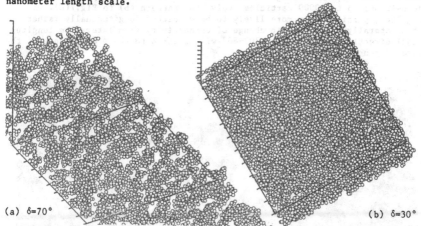

(a) δ=70°   (b) δ=30°

Fig.5 Caluculated deposits-patterns after depositions of 10000 particles.

Figure 6 shows the same deposits-pattern that in Fig.5(a), but 'fluoroscoped' from the direction at a polar angle of 54° with respect to Z-axis in parallel with PVI. There exists an anisotropic distribution of particles, being more likely to be connected toward the lateral direction rather than the longitudinal direction. The two hatched figures attached longitudinally (a) and laterally (b) are the distributions of the length of the vectors connecting arbitrary pairs of two projected particles in the main figure; (a) for those vectors whose angular deviations from the longitudinal line are within ±9°, (b) for those vectors whose angular deviations from the lateral line are within ±9°. The existence of the depleted zone only in the distribution (a) certifies the anisotropic arrangement of particles in the main figure. This lateral connnectivity of particles can be observed at a wide range of δ>30°, indicating that oblique deposition makes inclined quasi-layered structure of alternating high density and low density layers at nanometer length scale. According to Born and Wolf[8], this lateral connectivity of particles is the origin of the birefringence for the light entering from the 'fluoroscoped' direction. This birefringence also occurs for the light entering normal to the film surface when the inclination of this quasi-layered structure is not so large. However, this is not the case at large δ. Figures 7(a),(b) show patterns after depositions of 10000

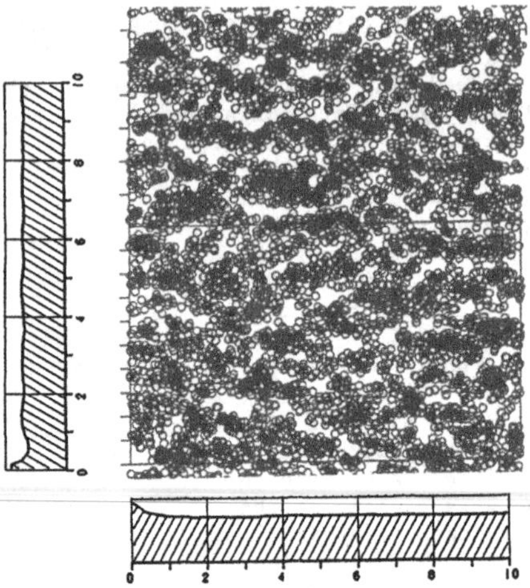

Fig.6 Deposits-pattern at δ=70° 'fluoroscoped' from the direction at the polar angle of 54°,10000 particles.

particles at δ=85° 'fluoroscoped' from the direction in parallel with PVI at a polar angle of (a) 80° and (b) 0°, respectively. Here again we got the quasi-layered pattern in (a). However, in spite of the same particles' arrangement viewed from the different angle, the pattern (b) is entirely different from (a), indicating the arrangement is very anisotropic. Succeeding depositions up to 40000 particles evolve the pattern (b) to Fig.7(c). In Fig.7(c), particles are more likely to be connected longitudinally rather than laterally. This drastic change of connectivity from lateral to longitudinal direction with δ across 70°-80° will relates to the δ dependence of the birefringence shown in Fig.3.

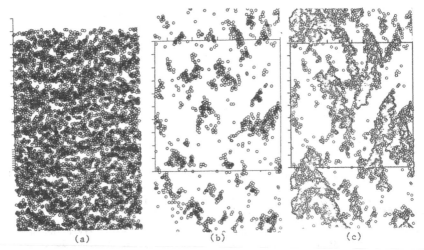

(a)   (b)   (c)

Fig.7 10000 particles' deposits-patterns at δ=85° 'fluoroscoped' from the direction at the polar angle of (a)80° and (b)0°. (c) for 40000 and 0°.

DISCUSSION

Figure 8 shows a typical elementary process of the present modeling. A particle B is supposed to rest at its first contact with a previously deposited particle A. A succeeding particle C cannot reach the substrate but a partcle D can. Here appears a hatched shadowed region which is to remain as voids. This is what is well-known as self-shadowing effect. However, in case the sticking force of A to B is not strong, and B hits A at the peripheral region of collision cross section with a large momentum, it

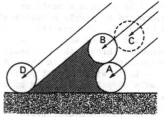

Fig.8 Typical elementary process of oblique deposition.

is likely that B does not rest at its first contact with A but is scattered or forwarded into the shadowed region. This is a kind of relaxation process which has not been taken into consideration in the present model. Suppose an extreme case in which B never sticks to A unless it is a head-on collision, the growth direction of columns (represented by a polar angle α from the substrate normal hereafter) evidently does not deviate remarkably from the direction of vapor incidence: δ. However, our SEM obsevation of the obliquely deposited Ta_2O_5 films deposited at various δ revealed a large deviation of α from δ, and even a deviation from the well-known 'tangent rule' : $2\tan\alpha = \tan\delta$ at large δ values as shown in Fig.9 [3].

There is another type of relaxation scheme used by Henderson et al[9], which is very popular in these types of modeling: each captured particle is relaxed into the nearest triangular pocket formed by three hard spheres. This relaxation is not directional and more related to the substrate temperature rather than the momentum of particles before deposition. This relaxation is known to fatten columns and increase density of simulated structure[5]. Figure 10 shows our experimental results on the effect of the substrate temperature on birefringence of obliquely deposited WO_3 films[3]. It is shown that birefringence decrease with the substrate temperature, indicating the Henderson's relaxation scheme is not essential for forming the nanophase structure which cause birefringence. The very fine fibrous structures shown in Fig.1 also don't support any remarkable fattening of fibrous structures caused by the Henderson's relaxation process.

Thus, so far as our deposition process of oxides to form artificial birefringent substance is concerned, the present simple model with very sticky particles with no relaxation depicts the essence of the phenomena.

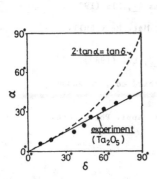

Fig.9 Structure inclination α of Ta_2O_5 film vs. deposition angle δ.

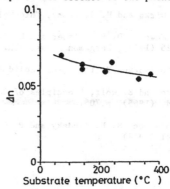

Fig.10 Birefringence of obliquely deposited WO_3 filmes at $\delta=70°$ vs. the substrate temperature.

Now we have experimental evidences that nanophase structure can be formed by oblique deposition, and the mechanism of formation is reasonably understood by simple geometrical phenomena: self-shadowing. As clearly shown in Fig.5(a), the formation of clusters and their isolation is spontaneously performed by the self-shadowing effect. The cluster arrays also appears spontaneously through the stochastic process of deposition. The origin of the periodicity appeared in the quasi-layered structure somewhat resembles that of periodic wind-wrought patterns on a sand dune which is also a typical example of patterns formed by self-organization. Thus, involving the three idea (1)isolation of clusters, (2)cluster arrays and (3)self-organization of clusters, the atomic scale structure formed on substrates by oblique deposition can be another possible candidate of the support for systhesis of new types of CAM or NPC. In addition to this, it is very advantageous that the process can be comprehended by a simple model because varieties of modifications of this nanophase structure can be designed on the model.

CONCLUSION

The nanophase structure found in the thin film retardation plate by oblique deposition was analysed with the aid of computer modeling. Anisotropic connectivity development was observed, which can cause the birefringence which actually takes place in the thin film retardation plate. Through the modeling, it is shown that the self-shadowing effect by oblique deposition causes (1) isolation of clusters, (2) cluster arrays and (3)self-organization, indicating the possible development of new types of CAM or NPC on the base of this oblique deposition technique.

REFERENCES

1. Panel Report on Clusters and Cluster-Assembled Materials, Monterey, California, January 1988, COUNCIL ON MATERIALS SCIENCE, DIVISION OF MATERIAL SCIENCE, U.S. DEPARTMENT OF ENERGY.

2. Y. Wang and N. Herron, J. Phys. Chem. 91, 257 (1987).

3. T. Motohiro and Y. Taga, Applied Optics 28, 2466 (1989).

4. S. Keitoku and K. Nishioka, Jap. J. Appl. Phys. 20, 1249 (1981).

5. A. G. Dirks and H. J. Leamy, Thin Solid Films 47, 219 (1977).

6. R. E. Newham, D. P. Skinner and L. E. Cross, Mat. Res. Bull. 13, 525 (1978), Pergamon Press, USA.

7. T. Motohiro and Y. Taga, Thin Solid Films 72, L71 (1989).

8. M. Born and E. Wolf, Pronciples of Optics, third ed., Pergamon, Oxford, (1965), p.705.

9. D. Henderson, M. H. Brodsky and P. Chaudhari, Appl. Phys. Lett., 25, 641 (1974).

HIGH-RESOLUTION ELECTRON MICROSCOPY OF INTERFACES IN AlN - BRAZE METAL ALLOY SYSTEMS

A. H. CARIM
Center for Micro-Engineered Ceramics & Department of Chemical and Nuclear Engineering University of New Mexico, Albuquerque, NM 87131

ABSTRACT

The microstructural aspects of active brazing of AlN with a Ag-Cu-Ti alloy have been investigated. A series of reaction product layers are formed. TiN is produced in contact with the polycrystalline, bulk AlN. High-resolution transmission electron microscopy and microdiffraction demonstrate that some of the TiN grains at the interface display specific orientation relationships with respect to the adjoining AlN crystallites. As has sometimes been observed in studies of epitaxy in other systems, these relationships are not necessarily those that provide the minimum geometrical mismatch between one or more sets of lattice planes. Farther from the substrate, an η-type nitride phase with composition $(Ti,Cu,Al)_6N$ occurs as a reaction product. High-resolution images confirm the absence of amorphous or crystalline intervening phases at the TiN-η interface and at η grain boundaries.

INTRODUCTION

Interfaces between ceramics and metals are present in many circumstances, including electronic packaging, structural joining, contacting of electronic ceramics, and composites [1-4]. An understanding of ceramic - metal bonding requires knowledge of the structures at such boundaries. This work describes the interfacial microstructures associated with active brazing of AlN with a Ag-Cu-Ti alloy, as observed by transmission electron microscopy (TEM). Further information on phase formation, identification, and composition is presented elsewhere [5,6]. A layered series of reaction products are present: TiN is found adjacent to the polycrystalline bulk AlN, and a complex η-type phase $(Ti,Cu,Al)_6N$ is found farther from the substrate. High-resolution TEM and microdiffraction are used to provide information on AlN-TiN orientation relationships. HREM images also confirm the absence of amorphous or crystalline intervening phases at the TiN-η interface and at η grain boundaries, and illustrate the crystallography of twins in the recently-identified η phase [5].

EXPERIMENTAL PROCEDURE

AlN substrates were stacked together with intervening layers of Ag-Cu-Ti active braze alloys containing 1.7 wt.% Ti (Cusil ABA) and 4.5 wt.% Ti (Ticusil). These "sandwiches" were bonded at 900°C for 5 to 30 minutes in an Ar ambient under a pressure of 1.38 MPa (200 psi). For microscopy, slices were cut perpendicular to the interfaces and then mechanically ground, dimpled, and argon ion milled to produce TEM samples [7]. Structural and chemical analyses were carried out on a JEOL JEM-2000FX microscope operated at 200 kV equipped with a Tracor-Northern 5500 energy-dispersive spectroscopy system. Higher-resolution images were obtained on a Philips CM30ST at 300 kV; this system has a point-to-point resolution of better than 0.20 nm.

LAYER STRUCTURE AND DEVELOPMENT

The microstructure of the reaction zone nearest to the substrate is shown in Fig. 1. A layer of fine-grained TiN, as identified by selected-area diffraction ring patterns, adjoins the AlN. Small (d≤10 nm) Cu-containing particles are also present within this reaction product layer. Beyond the TiN is a region of large (d>1 μm), nearly defect-free grains of $(Ti,Cu,Al)_6N$. The TiN layer grows in thickness with increasing time at 900°C, and the AlN - TiN interface becomes more irregular due to penetration and reaction at grain boundaries. The implication is that this interface is where further reaction between the substrate and Ti supplied

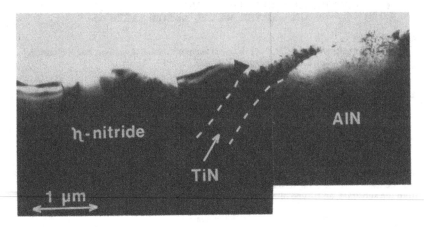

Fig. 1. Layered microstructure in the AlN substrates brazed with Ag-Cu-Ti. The TiN layer is fine-grained and includes small copper-containing precipitates; the η-phase layer beyond it has an approximate composition of $(Ti,Cu,Al)_6N$.

from the braze alloy takes place. The morphologies seen were similar for samples brazed with Cusil ABA and for those brazed with Ticusil. Microstructural evolution took place more rapidly in the latter since more of the active element Ti was available.

AlN-TiN INTERFACES AND ORIENTATION RELATIONSHIPS

High-resolution images demonstrate that there is no atomic-scale amorphous layer or intervening crystalline phase at the AlN-TiN interface. Although at many places there is no clear orientation relationship across this boundary, occasional cubic TiN grains [8] do exhibit registry of lattice planes with the larger hexagonal AlN crystals [9] across the interface. Figs. 2 and 3 show such cases; the associated microdiffraction patterns on each side of the interface are inset for Fig. 2 and the selected-area diffraction pattern over the boundary is shown in Fig. 3. The interfaces are incompletely determined in each case since the actual plane of the boundary is not parallel to the viewing direction (note Moiré fringes indicating overlap) and could not be readily identified. Since the interfaces are ill-defined and appear to be nonplanar, a comparison was made with geometrically "optimal" relationships defined on the basis of a volumetric mismatch, represented here by planar mismatches along three orthogonal axes. These "ideal" alignments, which also preserve three-fold symmetry elements, are not consistent with those observed in the samples (see Table I), suggesting that the arrangement of near-boundary chemical bonds may also play a role in orientation relationship development.

THE TiN-η INTERFACE

At the interface of the TiN layer with the $(Ti,Cu,Al)_6N$ η-type phase, there were again no intervening phases. Lattice fringes in the two materials extended all the way to the boundary between them. At this boundary, however, no particular orientation relationships between low-index planes in the TiN and those in the η phase were observed. Slightly preferential ion thinning occurred in some places along this interface, suggesting a possible stoichiometry variation in one or both phases, but this has not yet been confirmed.

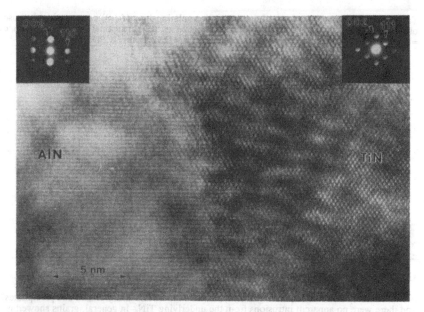

Fig. 2. HREM image of the AlN-TiN interface; at this location the AlN grain is aligned along the [1$\bar{1}$0] zone axis and TiN is aligned along [110]. Microdiffraction patterns are inset.

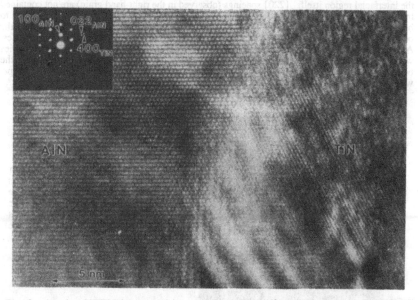

Fig. 3. Another HREM image and inset selected-area diffraction pattern at the AlN-TiN interface; AlN is oriented along [01$\bar{1}$] and the TiN (200) planes are aligned with AlN (011).

Table I. Lattice matching for optimal orientation relationships (A, B) compared with those observed in Figs. 2 and 3. Lattice parameters are taken as: AlN(hexagonal), a = 0.31114 nm, c = 0.49792 nm (JCPDS card #25-1133); TiN(cubic), a = 0.4240 nm (JCPDS #6-0642).

case	parallel planes	ratio of d-spacings	% mismatch (with respect to TiN planes)
A (ideal)	$(002)_{AlN}$ ‖ $(111)_{TiN}$	2.490 / 2.448	+1.7
	$(1\bar{2}0)_{AlN}$ ‖ $(2\bar{2}0)_{TiN}$	1.556 / 1.499	+3.8
	$(010)_{AlN}$ ‖ $(10\bar{1})_{TiN}$	2.695 / 2.998	-10.1
B (ideal)	$(002)_{AlN}$ ‖ $(111)_{TiN}$	2.490 / 2.448	+1.7
	$(110)_{AlN}$ ‖ $(2\bar{2}0)_{TiN}$	1.556 / 1.499	+3.8
	$(1\bar{1}0)_{AlN}$ ‖ $(10\bar{1})_{TiN}$	2.695 / 2.998	-10.1
Fig. 2	$(\bar{1}10)_{AlN}$ ‖ $(110)_{TiN}$	2.695 / 2.998	-10.1
	$(110)_{AlN}$ ‖ $(2\bar{2}0)_{TiN}$	1.556 / 1.499	+3.8
	$(002)_{AlN}$ ‖ $(002)_{TiN}$	2.490 / 2.120	+17.5
Fig. 3	$(011)_{AlN}$ ‖ $(200)_{TiN}$	2.370 / 2.120	+11.8

GRAIN BOUNDARIES AND TWINS IN THE η PHASE

Within the η-nitride phase, grain boundaries were free of glassy phases and precipitates, and there were no apparent intrusions from the underlying TiN. In general, grains showed no specific orientation relationships across boundaries. Fig. 4 is an HREM image of a typical grain boundary in the η, with inset optical diffractograms from the regions on each side and one from the interfacial region itself. {220} spacings (observed in the right-hand grain) are not contained in the [123] zone of the left-hand grain, ruling out a simple tilt boundary, and there appears to be no particular alignment of low-index directions. Slight distortion of lattice planes near the boundary may be indicative of interfacial dislocations which accommodate lattice mismatches. Although no linear defects or stacking faults were observed in $(Ti,Cu,Al)_6N$, twin boundaries were occasionally seen as shown in Fig. 5. A selected area diffraction pattern over the boundary is inset and indexed in the accompanying sketch. The boundary plane is {111}-type, with the axes $[011]_A$ ‖ $[011]_B$ aligned along the viewing direction.

Fig. 4. Grain boundary in the η phase. The left-hand grain is viewed along a <123> axis and the adjoining grain shows {220}-type spacings. Optical diffractograms from each grain and over the interface are included.

429

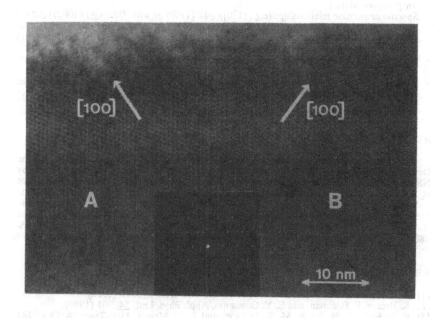

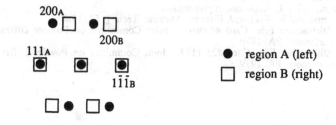

Fig. 5. High-resolution image of a twin boundary in the η phase along a common {111}-type plane. The selected-area diffraction pattern across the boundary is inset and indexed in the accompanying schematic diagram.

SUMMARY

- In the layered sequence of reaction products that form at the interface between AlN and Ag-Cu-Ti braze alloys, there are no amorphous or ultrathin crystalline phases at heterointerfaces or grain boundaries.
- Specific orientation relationships (e.g., $[1\bar{1}0]_{AlN} \parallel [110]_{TiN}$ with $(002)_{AlN} \parallel (002)_{TiN}$) are seen at some AlN-TiN interfaces; these are not those which produce minimum (three-dimensional) mismatch, implying that chemical effects may need to be considered in interfacial development in addition to geometrical considerations.
- No particular orientation relationships were observed at the TiN-η interface or at η grain boundaries.
- Twin boundaries are occasionally present in the cubic $(Ti,Cu,Al)_6N$ η-phase, characterized by a $\{111\}$ interface plane and alignment of <011>-type directions.

ACKNOWLEDGEMENTS

TEM work was carried out at the Electron Microbeam Analysis Facility in the Department of Geology and Institute of Meteoritics at UNM, which is supported in part by NSF, NASA, DOE-BES, and the State of New Mexico, and at the Center for Materials Science at Los Alamos National Laboratory. The assistance of M. Harrington (Sandia National Laboratories), C. Hills (SNL), and T. Wawrzyniec (UNM) with sample fabrication and preparation is gratefully acknowledged. I also thank R. E. Loehman (SNL) and J. F. Smith (LANL) for helpful conversations. This work was funded by SNL under contract numbers 05-0802 and 75-9825. Portions of this work have appeared previously elsewhere [Ref. 5].

REFERENCES

[1] R. E. Loehman and A. P. Tomsia, Am. Ceram. Soc. Bull. 67, 375 (1988).
[2] R. Brener, F. Edelman, and E. Y. Gutmanas, Appl. Phys. Lett. 54, 901 (1989).
[3] E. L. Hall, Y. M. Kouh, M. R. Jackson, and R. L. Mehan, Met. Trans. A 14A, 781 (1983).
[4] P. Martineau, R. Pailler, M. Lahaye, and R. Naslain, J. Mater. Sci. 19, 2749 (1984).
[5] A. H. Carim, J. Mater. Res. 4, 1456 (1989).
[6] A. H. Carim and R. E. Loehman, in preparation.
[7] J. C. Bravman and R. Sinclair, J. Electron. Microsc. Tech. 1, 53 (1984).
[8] Powder Diffraction File, Card #6-0642. Joint Committee on Powder Diffraction Standards, Swarthmore, PA, 1979.
[9] Powder Diffraction File, Card #25-1133. Joint Committee on Powder Diffraction Standards, Swarthmore, PA, 1979.

PART X

Novel Experimental Techniques

NOVEL COMPACT AND HIGH-RESOLUTION ION BACKSCATTERING ANALYSIS SYSTEM,CHIRIBAS

YOSHIAKI KIDO, AKIRA KAWANO, AND ICHIRO KONOMI
Toyota Central Research and Development Laboratories, Inc.,
Nagakute-cho, Aichi-gun, Aichi-ken, 480-11, Japan

ABSTRACT

The design and performance of CHIRIBAS(Compact and High-Resolution Ion Backscattering Analysis System) are presented. The system consists of a 60 kV duoplasmatron and unique energy analysis systems connected to a microcomputer. The 5-60 keV H, He, and He^+ beams are backscattered to 180° and their energies are analyzed by a time-of-flight technique or with an electrostatic deflector combined with a position sensitive microchannel plate. The probing depth and depth resolution are estimated to be up to 50 nm and better than 0.5 nm, respectively. The computer-simulated spectrum analysis allows rapid and accurate determination of surface and interface structures such as depth profiles of elemental compositions and lattice defects. CHIRIBAS also makes it possible to measure excited state populations using Stark ionization for 5-10 keV H neutralized at the exit surface and thus provides the possibility to determine the energy and spatial distributions of the valence electrons extending outward from the top surface. The present article includes the beam parameters achieved and the preliminary data on reactive-ion etched Si surfaces.

INTRODUCTION

Ion scattering has attracted much attention as a unique and versatile tool for determining the depth profiles of composite elements and of lattice disorder near the surface regions. Up to now, ion scattering analysis has been classified into following three categories upon the probing beam energy; (1) LEIS(Low Energy Ion Scattering: ≲ a few keV), (2)MEIS(Medium Energy Ion Scattering: 100-300 keV), and (3)RBS(Rutherford Backscattering Spectrometry: 1-3 MeV). RBS has been widely used for analyzing elemental compositions and crystalline structures as a simple and reliable technique[1,2]. The probing depth reaches up to 1 μm and the detection efficiency obtained with a solid state detector(SSD) is 100%. In addition, one can employ the simple Rutherford scattering cross sections and the energy-independent Bohr straggling formula. The weak point of RBS resides in its poor depth resolution of about 10 nm and low scattering yields because the noise level of SSD is more than 10 keV and the scattering cross section is proportional approximately to $1/E^2$(E: ion energy). In order to realize good depth resolution, it is essential to use low energy ion beams. Recently, MEIS has demonstrated its excellent depth resolution better than 1 nm and has been applied to the analysis of the initial growth process of metal-semiconductor islands and of semiconductor surface regions altered by reactive-ion etching[3,4]. MEIS is coupled with a Cockcroft accelerator and an electrostatic energy analyzer. Therefore, it takes a large area to be set up and needs high cost. In addition, its narrow dynamic range, corresponding to the energy resolution, requires long measurement time and frequently results in sample deterioration such as interface mixing and defect formation. LEIS provides a unique tool for analysis of impurities and atomic structures at surfaces[5,6]. However, its application is limited to special cases because of low penetration depth(≲ a few monolayers) and multiple scattering effect.

In this paper, we propose a novel compact and high-resolution ion backscattering analysis system, CHIRIBAS. The system consists of a duoplasmatron coupled with a pair of extraction electrodes(up to 60 kV) and an einzel lens and of two types of energy analyzers connected to a microcomputer. CHIRIBAS provides typically 50 keV He and He^+ and 5-10 keV H beams and allows energy analysis with accuracies of 5 X 10^{-3}. We can employ, case by case, a time-of-flight(TOF) technique and a novel detection system of neutral beam incidence and backscattered ion deflection(NIBID). The above two detection systems have wide dynamic ranges and therefore the dose of the probing beam is considerably lower than that in MEIS analysis, typically of the order of 1-2. In order to realize rapid and accurate data analysis, we have extended the computer programs to simulate the random and aligned backscattering spectra for a wide range of beam energy(10-2000 keV)[7]. We employ the Molière interatomic potential, Ziegler's stopping power formula[8], and the extended Lindhard-Winther straggling theory[9]. The aligned spectrum analysis is made according to the Feldman and Rodgers analytic treatment[10]. As another capability of CHIRIBAS, the populations of the excited states with the principle quantum number, $n \geq 8$ can be determined using Stark ionization. This technique suggests the possibility to determine the energy and spatial distributions of the valence electrons extending outward from the top surface(e.g. dangling bond). The fixed scattering angle of $180°$ makes it easy to set up double-alignment conditions and to design the system with very simple and compact structure. In fact, the size of CHIRIBAS is at most 2.5 m in length. Thus it realizes low cost and allows *in situ* observation of epitaxial layer-by-layer growth by combining it with an ultra-high vacuum chamber for molecular beam epitaxy(MBE) etc.

DESIGN AND PERFORMANCE OF CHIRIBAS

As mentioned earlier, good depth resolution is obtained in the backscattering analysis if one uses low energy ion beams. The depth resolution at the near surface region is represented by

$$\Delta t = (\Delta E/E)(E/(K S(E)/\cos\theta_1 + S(K E)/\cos\theta_2)) \qquad (1)$$

where $\Delta E/E$ is the energy resolution, K the kinematic scattering factor, $S(E)$ the stopping power, and θ_1 and θ_2 the incidence and emergence angles, respectively. Figure 1 shows the energy dependence of the depth resolution for the He → Ge and He → Si cases, where we assume the scattering angle of $180°$, the target tilt angle of $45°$, and the energy resolution(FWHM) of 5×10^{-3} for MEIS and CHIRIBAS and the system resolution of 15 keV(FWHM) for RBS. It is also interesting to estimate the probing depth dependence of the depth resolution. Figure 2 is the typical calculations for the 50, 200, and 2000 keV He^+ incidence on Ge. The energy resolution and the geometrical condition are the same as that in the previous case. As the energy straggling, the extended Lindhard-Winther theory is used for CHIRIBAS(50keV) and MEIS(200 keV) and the Bohr straggling values are used for RBS(2 MeV). As clearly seen, CHIRIBAS and MEIS provide the excellent depth resolution at the near surface region up to 20-30 nm and thus they are especially suitable for *in situ* analysis of the initial growth process of thin films and interface structures.

The schematic diagram of CHIRIBAS is illustrated in Fig. 3. The system consists of mainly three components; (1)ion source and extraction, (2)beam transport and charge neutralizer, and (3)particle detection.

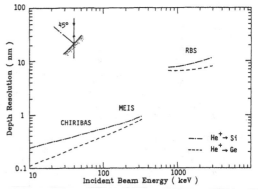

FIG. 1. Energy dependence of depth resolution (standard deviation).

A duoplasma ion source is e-quipped in order to obtain intense H^+ and He^+ beams with good emittance and narrow energy spreading($<$ a few eV). The ions generated in the ion source are extracted with the first extraction electrode and then accelerated typically up to 50 keV after being focussed with the einzel lens. The focal point is adjusted to the center of the aperture with a diameter of 0.5 mm, where an annular microchannel plate(AN-MCP) is placed just 1 m apart from the target position. A retarding aperture and a X-Y deflector are added after the einzel lens for 5-10 keV H^+ extraction and fine beam-adjustment. The accelerated ion

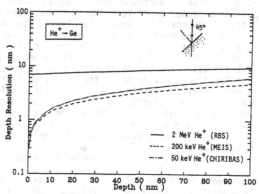

FIG. 2. Probing depth dependence of depth resolution(standard deviation) for the He^+ incidence on Ge.

beam is deflected to $20°$ to eliminate impuritiy, neutral, and molecule components. If one confines the ion species to He^+ only, one may omit the analyzing magnet. Impurity concentrations can be kept less than 10^{-3} for He^+ beams. Thus more compact and low-cost CHIRIBAS can be realized and used for He^+ backscattering analysis.

It is the notable point that CHIRIBAS provides both ion and neutral beams. The accelerated ion beam passes through the neutralizer canal with the size of 3 mm in diameter X 10 mm in length into which Ar gas is introduced. If one sets the pressure of the neutralizer to 1×10^{-4} Torr, the neutral fraction after passing through the canal is about 1% for the 10-20 keV/amu H^+ and He^+ beams. Now, we estimate the energy loss and the beam energy spreading from experimental data. In the case that 100 keV He^+ ions pass through Ar gas in 1×10^{-4} Torr, the energy loss and straggling(FWHM) are at most a few eV and 30 eV, respectively. In the case of Ar pressure less than 5×10^{-4} Torr, the energy loss and straggling is negligibly small compared with the system resolution(FWHM) of CHIRIBAS detection system, typically 200-300 eV.

CHIRIBAS allows two types of analyses; (1)backscattering with 50 keV He^+ and He beams and (2)resonance electron capture and subsequent Stark ionization with 5-10 keV H beams. In the case of the 50 keV He^+ beam, a backscattering energy spectrum is obtained as a time spectrum using the AN-MCP. The acceler-

CHIRIBAS

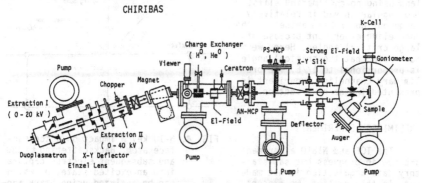

FIG. 3. Schematic diagram of CHIRIBAS.

ated He$^+$ beam is chopped at the chopping plate which is supplied with fast clock pulse(amplitude: 0-100 V, width: 1-10 ns, rise and fall times: 0.4-2 ns,repetition frequency:0-100 kHz). The elapsed time between backscattering and impinging on the AN-MCP ranges from 0.6 to 1 X 10^{-6} s for 20-50 keV He$^+$ ions. Thus we expect the time resolution $\Delta t/t$ of 2-4 X 10^{-3} and the resultant energy resolution of 4-8 X 10^{-3}. The TOF technique needs no correction upon charge neutralization and the irradiation dose is very low level.

For the 50 keV He beam, the energy analysis of the backscattered He$^+$ ions is made using the electrostatic deflector coupled with the position sensitive MCP(PS-MCP), as illustrated in Fig. 3. Its sensitive area is 20 X 80 mm^2 with the spatial resolution of 130 μm. The incident He beam is collimated with the aperture, 0.5 X 5 mm^2 fixed on the top of the PS-MCP holder. The backscattered He$^+$ ions are collimated to 0.5 X 5 mm^2 with the movable X-Y slit and then detected with the PS-MCP after passing through the deflector. The characteristic energy resolution is expected to be 200-300 eV.

The populations of the excited states of the backscattered H atoms are measured using Stark ionization. Very strong electric field up to 200 kV/cm is generated by applying the voltages $\pm$30 kV on the parallel plates whose gap is 3 mm. Thus it makes it possible to ionize the excited states n$\geq$8 of H. Figure 4 illustrates the situation of H impact on a top surface atom and subsequent 180° backscattering. The incident H atom is decelerated by the repulsive Coulomb force of the target nucleus and stopped at the position of the closest approach, typically 0.001-0.01nm from the nucleus. In this process, the bound electron is stripped off by the strong Coulomb force with high probability. Then the bare H$^+$ is accelerated backward and passes through the cloud of the valence electrons extending outward from the top surface. If an electron capture with a sharp resonance occurs at the exit surface(Massey's criterion), one can determine the energy and spatial distributions of the valence electrons at the surface(e.g. dangling bond).

CHIRIBAS has been set up this September and now starts on operation. The beam current and shape are monitored with the movable viewer placed in front of the neutralizer. We have obtained the 50 keV He$^+$ beam current of 150-200 μA with the spot size of 10 mm^2 in diameter. After passing through the neutralizer canal, 3 mm^2 in diameter and 10 mm long and the small aperture, 0.5 mm in diameter, the final beam current on the target is 200-300 nA. The beam spot size can be changed continuously from 0.5 up to 5 mm in diameter with the einzel lens using no collimating slits. The angular spread is relatively large but no problem arises in the channeling experiment because of large critical angles. Here, we must note that the critical angle is proportional to $1/\sqrt{E}$. The preliminary channeling experiment is presented in the next section.

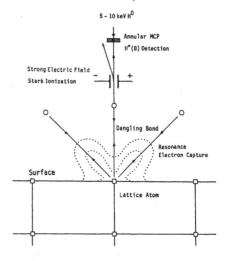

FIG. 4. 5-10 keV H impact on a top surface atom. Backscattering to 180° and subsequent electron capture into an excited state, n, which can be analyzed using Stark ionization.

PRELIMINARY EXPERIMENT

The TOF and NIBID experiments are now in progress but satisfactory data acquisition is not made yet. In this section, we present the preliminary experimental data on reactive-ion etched Si surfaces. We used the 50 keV He$^+$ beam provided by CHIRIBAS and an electrostatic toroidal analyzer, which was fabricated according to the work reported by Smeenk et al.[11]. The incident He$^+$ beam was collimated to 0.5 X 1 mm^2 with the X-Y slit and the beam current on the target was 200-300 nA. The thermal oxide layer with thickness 100 nm was grown on a chemically etched Si wafer. The oxide layer was completely removed off by reactive-ion etching(RIE). The sample was placed on the powerd electrode and supplied with 300 W RF power for 7 min. A self-biased voltage of 400 V was applied in a CF$_4$/H$_2$ gas mixture(CF$_4$/H$_2$ ratio = 45/15). The RIE was stopped just when the whole SiO$_2$ layer was etched away(no overetching). In the case of the CF$_4$/H$_2$ ratio less than 45/15, a thin fluorocarbon layer is formed.

Figure 5(a) shows the random and [100]-aligned CHIRIBAS spectra from the just-etched sample measured with the toroidal analyzer. The best-fitting of the front edge of the random spectrum leads to the energy resolution of 300 eV(FWHM), which corresponds to the depth resolution of 0.54 nm(standard deviation) for Si. Only surface damage is seen but significant oxide and fluorocarbon layers are not observed. The present result is consistent with the no overetching condition. The computer-simulated spectrum analysis reveals that an amorphous layer with thickness about 4 nm is formed on the surface(Fig.5 (b)). This is possibly caused by energetic ion bombardment because the ions are accelerated by the applied DC voltage of 400 V and hit the sample surface. For comparison, a commercially obtained Si wafer was also analyzed and the [100]-aligned spectrum is depicted in Fig. 5(a). The present spectrum analysis shows the presence of a natural oxide layer(SiO$_x$: x=1.3-1.5) with thickness about 2 nm. It is seen that the oxide layer with thickness more than 0.5 nm is detectable and the probing depth reaches up to 30 nm. However, as mentioned previously, the toroidal analyzer needs a long measurement time and high dose and therefore target deterioration becomes pronounced.

The TOF measurement has the advantages that charge neutralization correction is unnecessary and the probing ion dose is very low level. We expect that a good TOF spectrum would be obtained by altering the chopping plate structure for suppressing the jitter of the applied DC pulse and by adopting constant fraction timing for both the MCP signal and the start pulse generated simultaneously with the chopping pulse. For further improving the time resolution, we are now setting an aperture of 7-8 mm in diameter in front of the chopping plate.

CONCLUSION

The design and performance of CHIRIBAS with the unique detection systems are presented.

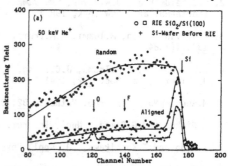

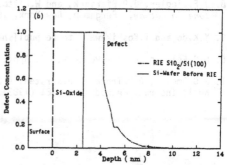

Fig. 5(a) Observed random and [100]-aligned spectra from the just-etched(○,□) and non-dry-etched(+) Si samples. The solid curves correspond to the best-fitted spectra.
(b) The near-surface structures determined from the above best-fitting.

The preliminary measurements using the toroidal analyzer reveal the Si surface structures. The computer simulation of random and aligned spectra allows rapid and accurate determination of the depth profiles of elemental compositions and lattice disorder near surface regions. The present results show that the amorphous layer with thickness about 4 nm is formed on the just-etched Si surface and the natural oxide layer(SiO_x: x=1.3-1.5) with thickness about 2 nm exists on the top surface of a Si wafer. The preliminary experiment using the 50 keV He^+ beam and the toroidal analyzer certifies the excellent depth resolution better than 0.5 nm and the probing depth up to about 30 nm.

ACKNOWLEDGMENTS

We would like to thank Dr. Y.Miyake of Nippon Physitech Corporation for valuable discussion and comment on the design of CHIRIBAS. Mental support of Dr. J.Kawamoto and Dr. T.Hioki is also acknowledged.

REFERENCES

1. W.K.Chu, J.W.Mayer, and M-A.Nicolet, *Backscattering Spectrometry*, (Academic, New York, 1978).

2. L.C.Feldman and J.W.Mayer, *Fundamentals of Surface on Thin Film Analysis* (North-Holland, New York, 1986).

3. E.J. van Loenen, M.Iwami, R.M.Tromp, and J.F. van der Veen, Surf. Sci. 137, 1(1984).

4. G.S.Oehrlein, R.M.Tromp, J.C.Tsang, Y.H.Lee, and E.J.Petrillo, J. Electrochem. Soc. 132, 1441 (1985).

5. M.Aono, Nucl. Instrum. Methods, B2, 374 (1984).

6. H.Niehus and G.Comsa, Nucl. Instrum. Methods, B15, 122 (1986).

7. Y.Kido and T.Koshikawa, J. Appl. Phys. in press.

8. J.F.Ziegler, J.P.Biersack, and W.Littmark, *The Stopping and Range of Ions in Matter*, (Pergamon, New York, 1985).

9. Y.Kido and T.Koshikawa, to be published in Phys. Rev. A.

10. L.C.Feldman and J.W.Rodgers, J. Appl. Phys. 41, 3776 (1970).

11. R.G.Smeenk, R.M.Tromp, H.H.Kersten, A.J.H.Boerboom, and F.W.Saris, Nucl. Instrum. Methods 195, 581 (1982).

SIMULATION AND QUANTIFICATION OF HIGH-RESOLUTION Z-CONTRAST IMAGING OF SEMICONDUCTOR INTERFACES

D. E. JESSON, S. J. PENNYCOOK, AND M. F. CHISHOLM
Solid State Division, Oak Ridge National Laboratory, Oak Ridge, TN 37831-6024

ABSTRACT

Incoherent characteristics of Z-contrast STEM images are explained using a Bloch wave approach. To a good approximation, the image is given by the columnar high-angle cross-section multiplied by the s-state intensity at the projected atom sites, convolved with an appropriate resolution function. Consequently, image interpretation can be performed intuitively and quantitative simulation can be implemented on a small computer. The feasibility of 'column-by-column' compositional mapping is discussed.

INTRODUCTION

Z-contrast STEM is a new technique for generating chemically sensitive, unambiguous atomic images of crystals. Remarkably, the images exhibit no contrast reversals with specimen thickness or objective lens defocus. These incoherent characteristics provide a clear advantage for the materials scientist investigating unknown crystal structures but may run contrary to the intuition of researchers more familiar with conventional phase contrast imaging techniques. The purpose of this paper is therefore to clarify the origin of these effects and show that, to a good approximation, the image is given by a convolution of the incident probe intensity profile with a strongly Z-dependent object function peaked at the atom sites. Consequently, images of perfect crystals may be predicted intuitively and information on a scale below the resolution limit can be interpreted by deconvolution. We also show how quantitative image simulation can be performed very quickly on a small computer, and consider the extension of these methods to the simulation of semiconductor interfaces. Finally, utilizing these results, the feasibility of 'column-by-column' composition mapping using Z-contrast STEM is discussed.

EXPERIMENTAL GEOMETRY

In a typical Z-contrast experiment using a demagnified field emission source, the coherence width of the illumination in the plane of the objective aperture is considerably greater than its diameter so that to a good approximation the focused probe is perfectly coherent. For each position of the probe on the specimen surface, a convergent beam electron diffraction (CBED) pattern is formed in the detector plane. If we consider a low-index zone axis with the probe diameter smaller than the lattice spacing, the CBED discs will overlap and the intensity of the overlap regions (except at special midpoints) will depend on lens aberrations, probe defocus and position [1].

For the purpose of our discussion, it is convenient to partition the CBED pattern into four separate components (Fig. 1a). Firstly, a strong dynamical central region consists of the bright field disc surrounded by low-order reflections which are close to or at the Bragg condition within the angular range defined by the objective aperture. This is enclosed by a weak annulus of medium to higher order zero-layer reflections associated with large excitation errors. Eventually, the Ewald sphere construction intersects higher order Laue zones (HOLZ's) of the reciprocal lattice, which provides a third component visible as concentric rings of intensity. Finally, a complicated distribution of pseudo-elastic thermal diffuse scattering is superimposed on the pattern.

The optimum detector configuration to capitalize on the detailed information available in CBED patterns has been discussed by several authors[1,2]. For Z-contrast STEM we are concerned with the special case of a wide annular detector possessing a large inner angle α as proposed by Howie [3]. Assuming α is sufficiently large to exclude the strong dynamical central region of the pattern, the signal detected for a given probe position will typically consist of a large thermal diffuse component, some weak zero-layer diffraction, and the first HOLZ ring. Clearly, the details of this interpretation will depend on the particular zone-axis geometry under consideration although the above imaging conditions relevant to Si [110] are applicable to numerous other low-index zone axes of technological importance.

440

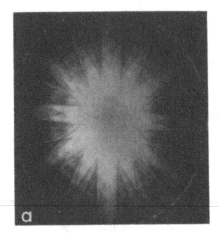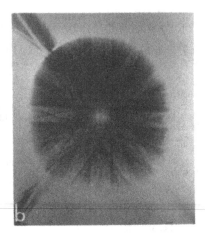

Fig. 1. (a) CBED and (b) large-angle channeling patterns for Si [110] at 100 keV.

THE HIGH-ANGLE SIGNAL

At first sight, the calculation of Z-contrast images is very complicated, requiring a full dynamical simulation involving multiple elastic and inelastic scattering. However, initial attempts at such calculations [4] have produced predictions of strong image contrast over a wide range of objective lens defocus (~160 nm) which we would not expect to observe experimentally (see later). This could be due to the sensitivity of the high-angle signal to the scattering models employed and/or efforts to reduce computing time. On the other hand, similar calculations by Shin et al. [5], which neglect three-dimensional diffraction and thermal vibrations, appear to produce image contrast in excellent agreement with experiment . This is perhaps surprising, but we shall see later from simple analytical arguments why the calculations should give the correct contrast although the wrong thickess dependence.

Our approach to the interpretation of Z-contrast images is motivated by the simplicity of the experimental results and the realization that black-box multislice methods give little physical insight into the origin of high-angle scattering phenomena. At the outset of the analysis, it is important to identify two simplifying constraints imposed by the geometry of our detector. First, the large inner detector angle (eg. 75 mrad) means that our measured signal originates at or close to the atomic sites. Second, its width (~75 mrad) ensures that in most cases of practical importance, dynamical effects on the outgoing electrons coupled with depth-dependent coherence effects are washed out. To identify the mechanism responsible for the incoherent nature of the images we have therefore concentrated on a Bloch wave analysis of the problem, which is both predictive and readily applicable to practical specimens.

We first consider the special case of HOLZ diffraction. Such effects are relatively discrete and may give rise to phase contrast associated with overlapping contributions from different discs. An interesting possibility therefore exists of using a special mask (perhaps combined with specimen cooling) to exclude all but a particular HOLZ ring contribution and facilitate high-angle phase contrast imaging of the relevant conditional projected potential [6]. We would expect such images to display the usual contrast reversals with specimen thickness and defocus and it should be possible to perform computer simulation using a simple pseudo-kinematic program combined with Bloch wave calculations. The contribution of these phase-contrast effects to Z-contrast imaging using a wide annular detector has recently been a matter of some debate [7]. In order to estimate the magnitude of the HOLZ contribution, a visual analysis of a CBED pattern is often misleading since the eye is extremely good at detecting differences in intensity and relatively poor at integrating over a diffuse background. What we really require is the ratio of HOLZ intensity (or more precisely that contained in the overlap regions) to the integrated thermal diffuse scattering component. A much improved experimental basis for making such judgments is provided by the large-angle channeling pattern (LACP) result presented in Fig. 1b for Si [110]. Such patterns effectively map electron intensity at the atom sites as a function of incident beam orientation [8]. The absence of discrete bright lines toward the pattern center associated with diffraction to HOLZ's, is a clear indication that, in this case, 3D diffraction is a negligible component of the integrated signal for the angular range relevant to Z-contrast STEM. Similar conclusions still apply to much stronger HOLZ diffraction and can be conveniently verified by the LACP method.

Having discounted the HOLZ contribution, we now proceed to discuss the thermal diffuse component. Under the reasonable assumption that this component dominates the combined HOLZ and high-angle zero-layer scattering, Pennycook and Jesson [9,10] have shown how the incoherent characteristics of Z-contrast images arise. In particular, a Bloch wave analysis of the coherent probe propagation implicates atomic orbital-like s-states as the dominant component at the atom sites. This is both a consequence of an angular integration associated with the incident electron probe and a depth integration which eliminates dependent cross terms and justifies the use of an independent Bloch wave model [11]. To a good approximation, tight binding theory then gives the image as a convolution of the surface probe intensity with a sharply peaked object function located at the projected atom sites. At currently available resolutions, it is justifiable to regard the object function for each column as a delta function weighted by a product of the relevant s-state intensity and a partial cross section σ_i for scattering into the high-angle detector. σ_i, which is readily derivable from atomic cross-sections [12], is strongly composition dependent since it approaches the Z^2 dependence of unscreened nuclear scattering. Furthermore, with some practice it is possible to guess the arrangement and approximate strength of s-states for a given projected potential. We note that for very heavy strings at 100 keV, 2s-states may introduce a thickness modulation to the images of such columns. Fortunately, such effects may also be anticipated and incorporated directly into our analysis. Although a simple tight-binding picture may require modification under special circumstances [13], the intimate relation between s-states and the parent crystal projection is central to the unambiguous character of Z-contrast images. Consequently, qualitative image simulation can often be performed intuitively with no dynamical computation and, as we shall demonstrate in the following section, quantitative simulation requires only simple calculations of Bloch states and a convolution.

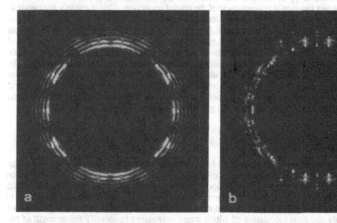

Fig. 2. Kinematic simulation of the outer portion of a CBED pattern for Si [110]. In (a) the probe is located over the dumbbell center and in (b) over a channel. The crystal thickness is 20 nm, the inner detector angle is 50 mrad, and the objective lens defocus is -70 nm.

To conclude this section we consider the coherently scattered zero-layer component at the detector. Figure 2 shows a kinematic simulation of the CBED pattern contribution for Si[110] at 100 keV. As expected, the disc overlap regions exhibit constructive and destructive interference in (a) and (b) with the probe located over a dumbbell and channel respectively. It is customary to describe such effects using the theory of STEM imaging developed by Cowley [2]. However, the convenient assumption of a phase object (which neglects the curvature of the Ewald sphere) becomes invalid for large-angle scattering, even in very thin crystals. With the aim of avoiding unpredictable numerical solutions, we have therefore adopted an alternative analytical strategy which takes into account both the dynamical diffraction of the incident probe and three-dimensional coherence effects. This approach will be fully explained in [14] and offers a new way to quickly simulate and interpret coherent CBED patterns from thin crystals. In real space, the analysis shows how the detected signal may be resolved into a transverse convolution of the surface probe intensity with a thickness (or z) dependent object function. The object function can be resolved into a set of highly localized, compositionally sensitive columnar object functions which are coherently integrated over the column lengths to account for the defocusing effects of atoms located at different heights in the column. The reciprocal space manifestation of the columnar defocusing is clearly evident in Fig. 2(a) as a system of radial fringes. For very thin crystals, these Fresnel interference effects produce a rapid increase in the coherent zero-layer signal, which is greatly suppressed as the column length exceeds a few nm in response to the contraction of the radial fringes toward the pattern center. The enhanced large-angle diffraction with increasing thickness is therefore compensated by the loss of the intense inner fringe from the detector, which produces an oscillatory thickness dependence with no overall linear increase. This is the reason why the thermal diffuse component quickly dominates the zero-layer coherent component at large scattering angles in specimens above a few nm's in thickness.

Our analysis of large-angle electron scattering has revealed several important implications for the atomic scale investigation of materials. Firstly, although incoherent imaging theory has little or no relevance to conventional atomic resolution electron microscopy, our experimental arrangement for Z-contrast STEM closely approximates a two-dimensional incoherent imaging system. Hence, the detected signal $I(\underline{R})$ for a probe at the surface coordinate $\underline{R}$ is given by the standard optical equation for incoherent imaging [15]

$$I(\underline{R}) = P(\underline{R}) * O(\underline{R},z) \tag{1}$$

where $P(\underline{R})$ is the instrumental response equal to the surface probe intensity (apart from a scaling factor) and $O(\underline{R},z)$ is a crystal object function. Compared with the equivalent expression for coherent imaging, Eq. (1) offers a number of intrinsic advantages including an improved resolution limit and the retrieval of object function information by deconvolution. However, the usefulness of Eq. (1) depends on the extent to which the object function is characteristic of the crystal projection. This is entirely analogous to the inverse scattering problem of phase contrast microscopy in which the aim is to determine the crystal potential from the exit surface wave function. It is therefore a significant advantage for Z-contrast microscopy that $O(\underline{R},z)$ remains intuitively linked to the composition and structure of the crystal projection even in thick crystals (~20–50 nm). The dominant effect of crystal thickness is just to scale the columnar contrast which although highly nonlinear in thin crystals is readily calculable [14]. In thicker crystals (≥ 10 nm) the dominance of the diffuse scattering builds incoherence into the third dimension so that $O(\underline{R},z)$ simply varies as the thickness integrated s-state. We emphasize that this almost 'classical' view of Z-contrast imaging is derived from a rather complicated dynamical scattering situation [9,10,14] and is a consequence of the experimental geometry rather than simplification of theory.

IMAGE SIMULATION OF PERFECT CRYSTALS

In crystals of a few tens of nm's thick used for Z-contrast experiments we have seen why large-angle zero-layer reflections are suppressed by z-coherence. Furthermore, although HOLZ diffraction scales with thickness like the thermal diffuse component, such effects are generally very small when compared with the total integrated signal. It is therefore justifiable under typical imaging conditions to include an object function appropriate for thermal diffuse scattering in Eq. (1) so that for quantitative image simulation only one axial calculation of Bloch states is required to identify the strength of the relevant s-states. We have checked this procedure with exact dynamical calculations and found excellent agreement for several projected potentials. An immediate consequence of Eq. (1) is the relative ease with which one can execute a Z-contrast through focal series. Figure 3, for example, shows such a sequence for the [110] axis of Si at 100 and 300 keV. In the 100 keV case, a well-defined defocus close to –70 nm optimizes both the contrast and resolution. In real space this corresponds to the Scherzer defocus condition for the most compact probe which in this case is about 2.2 Å FWHM intensity. We note that this is totally different from the stationary phase condition advocated by Spence et al. [7], which is really only applicable to point detectors. Surface probe calculations show that at lower defocus values the probe broadens, whereas at higher values the central peak sharpens as more intensity is transferred into the wings of the probe. These effects produce a blurring of the image as observed experimentally. At 300 keV, the 1.38 Å separation of the two Si columns comprising a [110] dumbbell exactly matches the Scherzer resolution limit for incoherent imaging. The predicted contrast at the optimum defocus of –50 nm, therefore closely agrees with the Rayleigh definition of the resolution limit. At lower defocus values, the simulations blur like the 100 keV results. However, a more interesting variation is observed at higher values with improved resolution occurring away from the Scherzer condition in response to the sharpening of the central probe peak. Notice significant intensity now present in the channels due to the enhanced wings. At very high values of defocus, reversals occur but with much weaker contrast. In practice, such effects are swamped by the experimental background and the inherent image noise so that the image is only visible in the high-contrast region around Scherzer defocus.

Experimental evidence of the convolution present in Eq. (1) is displayed in Fig. 4 for [110] axes of Si and InP. The accompanying simulations match the experimental results closely and clearly reproduce the elongated Si [110] 'rugby' balls corresponding to individual dumbbells. Scattering from InP is however dominated by the In columns so that the image consists of 'soccer' balls. Thus, information below the resolution limit of the microscope can be extracted by deconvolution. To avoid confusion it should be emphasized that the slight deviation from circular symmetry associated with the simulation of the In columns is not meant to infer the polarity of the lattice in the experimental image, although such information might be retrievable through image processing.

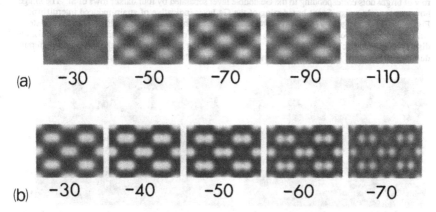

(a)   −30   −50   −70   −90   −110

(b)   −30   −40   −50   −60   −70

Fig. 3. Simulated Z-contrast through focal series for Si [110] at (a) 100 and (b) 300 keV. Defocus values are given in nm and C_s = 1.3 mm. Atoms always appear 'white' in Z-contrast images.

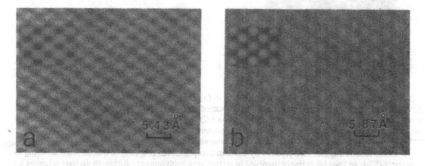

Fig. 4. Z-contrast images and accompanying simulations for (a) Si [110] and (b) InP [110] at 100 keV.

SIMULATION OF SEMICONDUCTOR INTERFACES

In extending the above discussion to the interpretation of Z-contrast images of semiconductor interfaces it is important to verify the integrity of any simplifying approximations. Formally, this can be achieved by supercell calculations, which we are currently implementing. For the moment we therefore present more intuitive arguments which focus on our experimental results and preliminary Bloch wave calculations.

The qualitative interpretation of images from semiconductor interfaces can be predicted from our discussion of s-states in perfect crystals. Such states are least sensitive to the strengths of neighboring strings so that the interface object function should be representative of the local composition. This is not in general true for conventional phase contrast images, however, which usually reflect interference patterns from several less localized states. The localization of the Z-contrast image is then simply given by the probe intensity profile. For Si [110] at 100 keV, our calculations show that for the probe located over a dumbbell, 83% of the image originates from that dumbbell. Consequently, we would expect Z-contrast images to at least qualitatively map interface structure and chemistry.

Figures 5(a) and (b), for example, show Z-contrast images obtained from MBE-grown ultra-thin superlattices $(Si_mGe_n)_p$, where m and n refer to alternating {400} monolayers of Si and Ge repeated p times. The aim in (a) was to produce a $(Si_8Ge_2)_{100}$ superlattice which should appear as one vertical row of bright dots corresponding to the Ge double layer separated by four darker rows of Si. The image however reveals a significant broadening of the Ge layer indicative of strain-induced interdiffusion. Figure 5(b) is an image of a $(Si_2Ge_6)_{100}$ superlattice and in this case the observed contrast is consistent with the expected layer periodicities. The vertical waviness in the image is mimicked by the simulation displayed in Fig. 5(c), which links two regions where the Si atoms are in the same and different dumbbells [Fig. 5(d)] indicating that the images are sensitive to {400} monolayers.

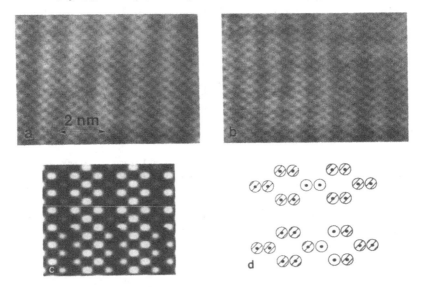

Fig. 5. Z-contrast images of ultra-thin multilayers. In (a) a $(Si_8Ge_2)_{100}$ superlattice exhibits monolayer diffusion and in (b) the image displays the contrast expected for a $(Si_2Ge_6)_{100}$ multilayer. The apparent waviness in (b) is simulated in (c) and attributed to {400} monolayer steps as shown in (d). The open circles represent Si atoms and the shaded circles represent Ge atoms.

These results suggest that column-by-column compositional mapping is indeed feasible using Z-contrast STEM, and we are actively involved in quantifying such images. Although various schemes are possible, it is clear that difficulties associated with an inherently weak signal-to-noise ratio must be overcome. Furthermore, the magnitude of an absorbative plane-wave component as well as the effects of strain on s-states should be considered. An additional possibility is to complement the dark-field signal by enhancing the contrast between weak columns using the bright field detector. Such complementary information coupled with the realization of 1.3 Å resolution at 300 keV will have important implications for many areas of materials science. For semiconductor interfaces, this would allow all atomic columns to be resolved at major zone axes with the chemical sensitivity and image interpretability of the Z-contrast technique (Fig. 6).

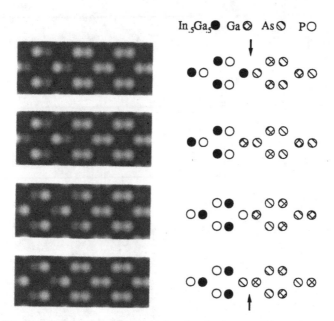

Fig. 6. Simulated Z-contrast images illustrating that the four possible configurations of an $In_{0.5}Ga_{0.5}P$/GaAs interface should be distinguishable at 300 keV.

ACKNOWLEDGMENTS

We gratefully acknowledge C. W. Boggs, T. C. Estes, and J. T. Luck for technical assistance. This research was sponsored by the Division of Materials Sciences, U.S. Department of Energy under contract DE-AC05-84OR21400 with Martin Marietta Energy Systems, Inc. One of us (DEJ) was supported by an appointment to the U.S. Department of Energy Postgraduate Research Program at Oak Ridge National Laboratory administered by Oak Ridge Associated Universities.

REFERENCES

1. J. C. H. Spence and J. M. Cowley, *Optik* **50**, 129 (1978).
2. J. M. Cowley, *Ultramicroscopy* **2**, 3 (1976).
3. A. Howie, *J. Microsc.* **117**, 11 (1979).
4. Z. L. Wang and J. M. Cowley, *Ultramicroscopy* (in press).
5. D. H. Shin, E. J. Kirkland and J. Silcox, *Appl. Phys. Lett.* **55**, 2456 (1989).
6. W. Cochran and H. B. Dyer, *Acta Cryst.* **5**, 634 (1952).
7. J. C. H. Spence, J. M. Zuo, and J. Lynch, *Ultramicroscopy* **31**, 233 (1989).
8. D. E. Jesson and S. J. Pennycook, in preparation.
9. S. J. Pennycook, D. E. Jesson, and M. F. Chisholm, in *Proc. 6th Conf. on the Microscopy of Semiconducting Materials*, Inst. Phys. Conf. Ser. No. 100, Section 1, p. 51 (1989).
10. S. J. Pennycook and D. E. Jesson, *Phys. Rev. Lett.* **64**, 938 (1990).
11. D. Cherns, A. Howie, and M. H. Jacobs, *Z. Naturforsch.* **28a**, 565 (1973).
12. S. J. Pennycook, S. D. Berger, and R. J. Culbertson, *J. Microsc.* **144**, 229 (1986).
13. R. Vincent, D. M. Bird, and J. W. Steeds, *Philos. Mag. A* **50**, 765 (1984).
14. D. E. Jesson and S. J. Pennycook, in preparation.
15. S. G. Lipson and H. Lipson, *Optical Physics*, Second Edition, (Cambridge University Press; Cambridge) (1981).

ATOMIC STRUCTURE AND CHEMISTRY OF Si/Ge INTERFACES DETERMINED BY Z-CONTRAST STEM

M. F. CHISHOLM, S. J. PENNYCOOK, AND D. E. JESSON
Solid State Division, Oak Ridge National Laboratory, Oak Ridge, TN 37831-6024

ABSTRACT

The technique of Z-contrast STEM provides a fundamentally new and powerful approach to determining the atomic scale structure and chemistry of interfaces. The images produced do not show contrast reversals with defocus or sample thickness, there are no Fresnel fringe effects at interfaces, and no contrast from within an amorphous phase. Such images are unambiguous and intuitively interpretable. In this paper, the technique has been used to directly image sub-nanometer interdiffusion in ultrathin $(Si_mGe_n)_p$ superlattices. The Z-contrast image of a $(Si_8Ge_2)_p$ superlattice grown by MBE at 400°C clearly shows significant broadening of the Ge-rich layer. Also, film formation and misfit accommodation in epitaxial Ge films on (001)Si produced by implantation and oxidation of Si wafers was studied. It was found that the Ge films, which are constrained to grow layer-by-layer, remain completely coherent with the Si substrate to a thickness of 5–6 nm. This is 3 to 6 times thicker than the observed critical thickness for Ge films grown on Si by MBE. It is observed that misfit accommodating dislocations nucleate at the film surface as Shockley partials. The Z-contrast images show these partials can combine to form perfect dislocations whose cores are found to lie entirely in the elastically softer Ge film.

INTRODUCTION

Until fairly recently, all high-resolution electron microscopy images have been obtained using phase contrast which is due to the interference of diffracted beams. These are "lattice fringe" images unless special conditions for true structure images apply. In general, though, there is no simply interpretable connection between the observed intensities in the fringe patterns and the crystal structure. This is because the relative phases of the diffracted beams are strongly influenced by dynamical scattering effects. This is demonstrated in Figure 1, which is a montage of simulated phase contrast images of Si viewed in the <110> direction. Complete contrast reversals and all states in between the extremes are produced depending on defocus or sample thickness.

Z-contrast imaging is a completely new approach to high-resolution electron microscopy. This technique is based on scanning an electron probe of atomic dimensions across the sample and collecting the compositionally sensitive high-angle scattered electrons with an annular detector. Diffraction at high angles is effectively kinematic [1]. Therefore, the high-angle scattering can be viewed as being generated by each atom in proportion to the electron intensity close to the atomic sites. Such images are unambiguous and can be interpreted intuitively. With Z-contrast, the obtained image is not a reconstruction of the object from the Bragg diffracted beams as in phase contrast, but instead a map showing at atomic resolution the scattering power of the sample. Also, at high angles the scattering is largely unscreened and the cross section approaches the full Z^2 dependence of Rutherford scattering [2]. There are no contrast reversals with defocus or sample thickness and no Fresnel fringe effects at interfaces, so observed rigid shifts are independent of the sample thickness and objective lens defocus. These characteristics combined with the strong chemical sensitivity make Z-contrast imaging a powerful technique to study the atomic structure and chemistry of interfaces.

Two examples involving interfaces between Si and Ge are used to demonstrate the characteristics of high resolution Z-contrast imaging. The first involves strained layer superlattices of Si/Ge. These heterostructures consist of a few monolayers of Si alternating with a few monolayers of Ge. When the superlattice period is this short it changes the symmetry of the unit cell.

Defocus (Å)

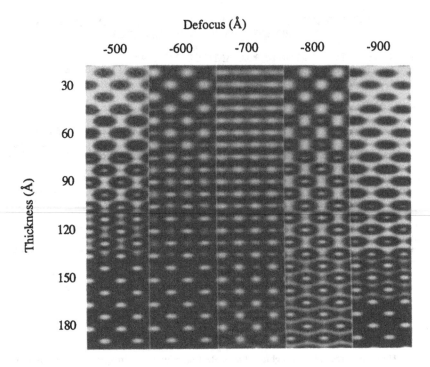

Fig. 1. Simulated phase contrast image of the <110> projection of Si as a function of sample thickness and objective lens defocus. (C_s = 1 mm, spread of defocus = 4.0 n m, divergence halfwidth = 1 mrad, 400 keV, objective aperture radius = 0.59 Å^{-1}).

The principal observable effect of this change in unit cell symmetry on the electronic structure is the creation of new, structurally induced energy levels [3]. The abruptness of the superlattice interfaces is then an important concern because the deviation of only one atomic layer results in large relative thickness changes in the superlattices. In principle, with Z-contrast imaging it should be possible to obtain a column-by-column composition map across such interfaces.

The second example discussed is epitaxial films of Ge grown on (001) Si. These Ge films were produced by an implantation and oxidation technique first described by Fathy et al. [4], whereby (001)Si wafers are implanted with Ge ions. A thin layer of Ge is formed during steam oxidation of the implanted wafer. This is the result of complete rejection of Ge by the growing oxide without any appreciable diffusion of Ge into the Si substrate. A range of Ge film thicknesses, formed by varying the Ge implant dose, can be produced to study the various stages of misfit accommodation in these epitaxial films. Electron microscopy is well suited for the study of the initial stages of dislocation nucleation and dislocation reactions. High resolution Z-contrast imaging has the added advantage of producing intuitively interpretable images with chemical sensitivity. Thus, determinations of the interface and dislocation core structure, as well as the dislocation core location, can be made.

EXPERIMENTAL

The ultrathin $(Si_mGe_n)_p$ multilayers were grown by MBE. Details of the MBE growth of these samples have been published elsewhere [5]. The epitaxial Ge films used in this study were produced by implanting (001)Si wafers with 35 keV Ge^+ ions at fluences from 1×10^{16} cm^{-2} to 3×10^{16} cm^{-2}. The wafers were then steam oxidized at 900°C for 30 min. Cross sections were mechanically polished and then ion milled with 6 keV Ar until perforation. The samples were finished using 2 keV I for 5 min to reduce the surface amorphous material [6]. The samples were examined using a VG Microscopes HB501 STEM operating at 100 keV and equipped with a high-resolution pole piece (resolution limit of 0.22 nm).

RESULTS AND DISCUSSION

Figure 2a shows a high-resolution Z-contrast image from part of a $(Si_8Ge_2)_{100}$ super-lattice viewed along the <110> direction. This sample was grown by MBE at 400°C. The image consists of a roughly hexagonal array of white spots. Each spot corresponds to a column of atom pairs aligned along the beam direction since the 0.14 nm separation between {400} planes cannot be resolved. If there was no interdiffusion, there would be two possible images; one where the two monolayers of Ge comprise a single row of spots or where they are shared between two rows of spots. The image should have one or two row(s) of bright spots (Ge Z=32) alternating with four or three rows of darker spots (Si Z=14). The image shows the Ge-rich layer actually extends over two or three spots. This is the result of interdiffusion and demonstrates that Z-contrast imaging provides sufficient resolution and chemical sensitivity to detect subnanometer interdiffusion in these artificially layered crystals. Figure 2b shows a high-resolution Z-contrast image from part of a $(Si_2Ge_6)_{100}$ superlattice grown at 350°C. At this reduced growth temperature, superlattices with sharp interfaces and the expected periodicities were produced.

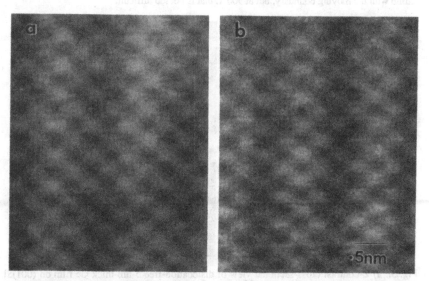

Fig. 2. Z-contrast images of Si_mGe_n superlattices in cross section along a <110> direction. (a) Image of part of a $(Si_8Ge_2)_{100}$ superlattice grown at 400°C showing broadening of the Ge-rich layers; (b) part of a $(Si_2Ge_6)_{100}$ superlattice grown at 350°C with the expected periodicity.

Figure 3a shows a cross-section view of a dislocation-free 5-nm thick Ge film on (001)Si produced by oxidation of a Ge (2×10^{16} ions/cm^2) implanted Si wafer. This is three to six times thicker than observed critical thickness for Ge films on Si produced by MBE [7]. Using conventional deposition techniques, Ge normally grows as islands on Si and after the equivalent of six monolayers of Ge is deposited, dislocations are introduced at the island perimeters to relieve the strain. With the implantation and oxidation technique, the Ge film is constrained to grow layer-by-layer. Dislocations, formed by plastic deformation of the film, are nucleated at a free surface and glide to the interface to accommodate the misfit. This process is delayed compared with island growth mode films due to kinetic considerations. Only three defects were found in ~2 cm of thinned cross sectional interface in this sample. These defects are the earliest stages of misfit relief in these films. In all three cases the Ge-rich film bulged at the dislocation (Fig. 3b). The defect is a 30° Shockley partial dislocation ($\bar{b}$ = a/6[2$\bar{1}$1]), which has glided to the interface on a (111) plane leaving a stacking fault. Figure 4 is a defect in a more advanced stage. A 90° Shockley partial ($\bar{b}$ = a/6[11$\bar{2}$]) has nucleated and followed the 30° partial, annihilating most of the stacking fault. When combined, these two partial dislocations produce a 60° dislocation ($\bar{b}$ = a/2[10$\bar{1}$]). This mechanism has been proposed earlier to explain the presence of stacking faults in Si films grown on GaP [8] and residual diffraction contrast around dislocation lines in Co films grown on Cu [9] and is conclusively demonstrated here.

A dose of 3×10^{16} Ge ions/cm^2 produced a 7 to 8 nm thick film with a fairly regular array of interfacial dislocations spaced every ~17 nm. It is observed that the dislocation cores are entirely in the Ge layer, which is the elastically softer material (Fig. 5). This is an important observation and one that could not be made so easily with phase contrast imaging. Although the expected 60° dislocations are present at the interface (Fig. 5a), most of the observed interfacial dislocations are in the edge orientation (Fig. 5b). It is proposed that these edge dislocations ($\bar{b}$ = a/2[$\bar{1}$10]) are the result of the moving interface, which cause the dislocations, which glide on (111) planes, to sweep the interface as it moves into the wafer. They react with other dislocations when favorable to produce a more efficient misfit accommodating dislocation. The created edge dislocations, with their Burger's vectors in the plane of the interface, must now climb with the moving boundary, but at 900°C that is not too difficult.

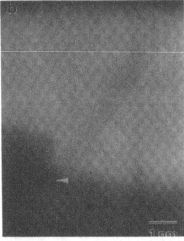

Fig. 3. a) Z-contrast cross-sectional view of dislocation-free 5-nm-thick Ge film on (001)Si produced by oxidation of a Ge(2×10^{16} ions/cm^2) implanted wafer. b) Area of the same film where a 30° Shockley partial dislocation ($\bar{b}$ = a/6[2$\bar{1}$1], arrowed) has glided from the surface to the interface producing a stacking fault.

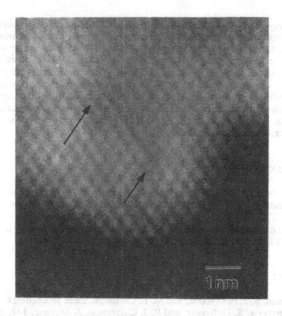

Fig. 4. Z-contrast image of dissociated 60° dislocation with a stacking fault between the two partial dislocations (arrowed).

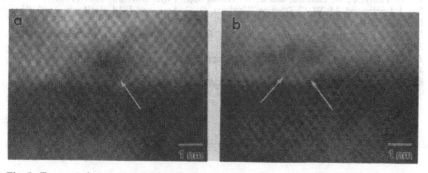

Fig. 5. Z-contrast image of the Si/Ge interface showing the dislocations cores lie entirely on the Ge film. a) 60° dislocation; b) edge dislocation with its Burger's vector parallel to the plane of the interface. Extra half plane indicated by arrows.

CONCLUSIONS

Z-contrast imaging with its atomic resolution, chemical sensitivity and intuitively interpretable images is an extremely powerful tool for the study of interfaces. There is sufficient resolution and chemical sensitivity to detect subnanometer interdiffusion in an artificially layered Si/Ge crystal grown by MBE at 400°C.

Ge films on Si produced by implantation and oxidation are constrained to grow layer-by-layer instead of the normally observed island mode. Films less than 6-nm thick are coherent, chemically abrupt, and planar. This is three to six times thicker than the observed critical thickness for Ge films grown on Si by MBE. It is observed that misfit accommodating dislocations nucleate at the film surface as Shockley partials. The 30° partial dislocation nucleates first and glides to the interface producing a stacking fault. Subsequently, a 90° partial can nucleate and follow the 30° partial annihilating the stacking fault. These two partials can combine to form a 60° dislocation. In supercritical thick films, the perfect dislocation cores lie entirely in the elastically softer Ge film. The interfacial dislocations are now predominantly in the edge orientation. This is believed to be the result of the moving Ge/Si boundary during the oxidation step of the growth process. The dislocations gliding on (111) planes sweep the interface and react to produce the more efficient pure edge dislocations.

ACKNOWLEDGMENTS

The authors gratefully acknowledge O. W. Holland for provision of the Ge-implanted Si, D. C. Houghton and J.-M. Baribeau for provision of the MBE-grown superlattices, and C. W. Boggs, T. C. Estes, and J. T. Luck for technical assistance. This research was sponsored by the Division of Materials Sciences, U.S. Department of Energy under contract DE-AC05-84OR21400 with Martin Marietta Energy Systems, Inc.

REFERENCES

1. D. E. Jesson, S. J. Pennycook, and M. F. Chisholm, these proceedings.
2. S. J. Pennycook, S. D. Berger, and R. J. Culbertson, *J. Microsc.* **144**, 229 (1986).
3. See ref. in T. P. Pearsall, *CRC Critical Reviews in Solid State and Materials Science* **15**, 551 (1989).
4. D. Fathy, O. W. Holland, and C. W. White, *Appl. Phys. Lett.* **51**, 1337 (1987).
5. J. Lockwood, M. W. C. Dharma-Wardona, G. C. Aers, and J.-M. Baribeau, *Appl. Phys. Lett.* **52**, 2040 (1988).
6. N. G. Chew and A. G. Cullis, *Appl. Phys. Lett.* **44**, 142 (1984) and *Ultramicroscopy* **23**, 175 (1987).
7. G. Abstreiter, K. Eberl, E. Friess, W. Wegscheider, and R. Zachai, *J. Cryst. Growth* **95**, 431 (1989).
8. W. A. Jesser and J. W. Matthews, *Philos. Mag.* **17**, 461 (1968).

SIMULATED IMAGE MAPS FOR USE IN
EXPERIMENTAL HIGH-RESOLUTION ELECTRON MICROSCOPY

MICHAEL A. O'KEEFE, ULRICH DAHMEN and CRISPIN J.D. HETHERINGTON
National Center for Electron Microscopy, University of California, Lawrence Berkeley
Laboratory, 1 Cyclotron Road, Berkeley, CA 94720.

ABSTRACT

A "map" of all possible high-resolution images may be simulated for a
crystalline specimen in a chosen orientation for any particular transmission electron
microscope (HRTEM). These maps are useful during experimental high-resolution
electron microscopy and make it possible to locate optimum imaging conditions even
for foil thicknesses beyond the weak-phase object limit. Although defects such as
grain boundaries are not generally periodic, image maps of perfect crystal can be
used to optimize defect contrast during operation of the microscope by reference to
the image of the perfect crystal neighboring the defect.

INTRODUCTION

The current availability, at reasonable cost, of high-speed computing has made
feasible the rapid generation of simulated images. A useful application [1] of these
advances is the production, before the experiment, of all possible images (rather than
a select few after the event). For perfect periodic crystals, image contrast conditions
repeat with both crystal thickness and microscope defocus, allowing a map to be made
that includes all possible HRTEM images. Inclusion of partial coherence limits the
map to the range of defocus that can be used experimentally. For a crystal with repeat
distance d of 2 to 3 Ångstrom units (in projection), the Fourier image defocus period
$2d^2/\lambda$ is small enough to allow a useful map to be produced as an array of about ten to
fifteen defocus values by ten to twenty thickness values; typical steps might be 100Å
in defocus and 20Å in thickness.

THEORY

The ranges of values of microscope defocus and specimen thickness over which
an image map should extend may be determined by consideration of Fourier-images
and dynamic extinction distance. Fourier-image theory [2] has been shown to be
useful in experimental measurement of defocus values in high-resolution electron
microscopy [3], and the concept of Fourier-image defocus period provides a means of
defining the focus range over which changes in defocus can produce new image
contrast. The phenomenon of near-repetition of image contrast with increasing
crystal thickness is due to the effect of dynamic extinction distance [4], and depends
upon both specimen structure and orientation, as well as the electron wavelength
used.

The overall phase of a diffracted beam may be written as

$$\phi(k) = \pi/2 + \phi_{dyn}(k) + \chi(k) \qquad (1)$$

where $\pi/2$ is the phase change on kinematic scattering, $\phi_{dyn}(k)$ is the additional
change due to dynamical effects, and $\chi(k)$ is the phase change imposed by the
objective lens [3]. The dynamical term, $\phi_{dyn}(k)$, varies with crystal thickness, and, for
simple metals and semiconductors is near-periodic with the period of the dynamic
extinction distance if absorption is neglected [4]. Therefore, for a given lens defocus,
and for HRTEM specimens thin enough for absorption to be negligible, $\phi(k)$ can be
periodic in crystal thickness to quite a good approximation. Similarly, for any given
specimen thickness, the lens term, $\chi(k)$, [and hence $\phi(k)$], is periodic with objective
lens defocus with a period of $2d^2/\lambda$, where d is the (projected) unit cell length [2].

Mat. Res. Soc. Symp. Proc. Vol. 159. ©1990 Materials Research Society

EXPERIMENTAL

To determine optimum viewing conditions for an interface in aluminum, (in this case a Σ99 {557} <110> tilt boundary), we considered images of perfect aluminum up to 440Å thick, over one Fourier-image period of $2d^2/\lambda$ for experimental conditions corresponding to the NCEM Atomic-Resolution Microscope (JEOL ARM-1000).

Figure 1. Model of aluminum in [110] projection showing the original cell and the cell of size |a| by |a|/√2.

In [110] projection, the smallest orthogonal cell in perfect aluminum (fig.1) is |a| by |a|/√2, or 4.04Å by 2.86Å. These dimensions yield Fourier-image periods of 3178Å and 1589Å respectively for an electron wavelength corresponding to the experimental energy of 800keV; the overall period is the lowest common multiple, or 3178Å. A plot of contrast-transfer functions (CTFs) at three values of defocus differing by 3178Å confirms that $\chi(k)$ values for all three are identical at spatial frequencies equal to √n/|a| (fig.2).

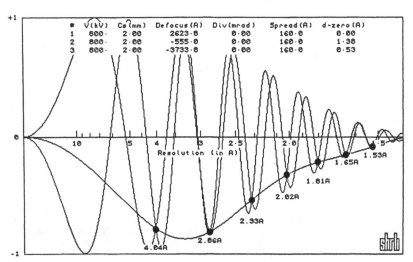

Figure 2. CTFs for defocus values separated by one Fourier image period for ARM conditions (listed at top) and |a| = 4.04Å. Positions where all three curves intersect are marked and labelled with real-space distances equal to |a|/√n for n=1 to 7.

RESULTS

Images simulated using the NCEMSS programs [5] over the defocus range +381Å to -3175Å show a defocus period of 3175Å, corresponding to the Fourier-image value. Four types of high-contrast images occur (fig.3). At Scherzer defocus (-500Å), thin-crystal images have black spots at atom positions (S-type images). Near -900Å defocus, white spots appear at atom positions (W images). Images near defocus values of -1500Å and -2500Å are white-spot images like that near -900Å, but shifted by a half unit cell ($\overline{W}$ images). Similarly, the image at -2000Å defocus is a black-spot image displaced by a half cell ($\overline{S}$ image). Images at 0Å and -3175Å (a Fourier pair) are W images.

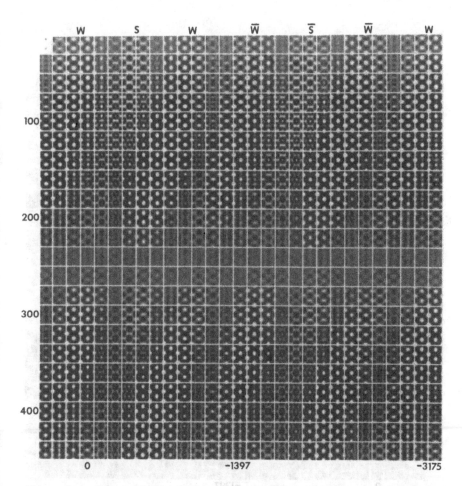

Figure 3.   Images simulated for ARM conditions covering the defocus range horizontally from +381Å to -3175Å in 29 steps of 127Å and the crystal thickness range vertically from 20Å to 440Å in 22 steps of 20Å. Display contrast is held identical for all images, allowing the loss of contrast at the extinction thickness of 240Å to be seen clearly. For comparison with the images, the positions of atoms within the cell are displayed as black dots in the projected potential plot (top left).

Figures 2 and 3 were computed without the effect of incident electron beam convergence (spatial coherency) in order to demonstrate the periodity of images with defocus. When convergence is included at the level corresponding to the condenser aperture commonly used in the ARM, it is found that the range of useful images is less than one complete Fourier-image period. The beam-dampening effect of convergence is weak near Scherzer defocus, but stronger as defocus is increased or decreased [6], resulting in very low contrast far from Scherzer defocus (fig.4). These lower-contrast images are not experimentally useful and may be neglected, condensing the required image map to a defocus range of approximately one half of the Fourier-image period.

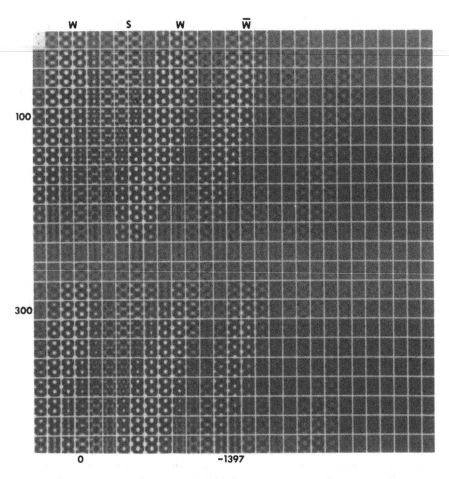

Figure 4. Images computed for the conditions of figure 3, but including an incident beam convergence of 0.6millirad. halfangle, corresponding to the measured experimental value. As defocus becomes larger, images lose resolution, then contrast.

Because underfocus conditions generally result in higher resolutions [7], a useful image map for [110] aluminum imaged in the ARM at 800keV need cover only the range from zero to approximately -1400Å. This map (fig.5) clearly shows the 260Å repeat in thickness, and the high-contrast regions that become skewed in the direction of positive defocus as crystal thickness is increased [8]. From the image map (fig.5), the white-spot (W) images near -800Å defocus and 100Å thickness show good fidelity (each white spot is centered on an atom position) and possess higher contrast than the black-spot (S) images near Scherzer defocus (-500Å). For these reasons, the "best" conditions of -800Å defocus and 100Å thickness were selected when an experimental image of a Σ99 {557} <110> tilt boundary in aluminum was obtained. Figure 6 shows a comparison of the experimental image with one simulated from a grain-boundary model [9].

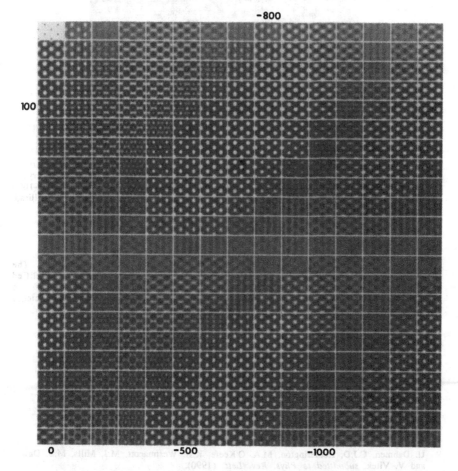

Figure 5. Final image map for [110] aluminum under ARM conditions. Defocus values (horizontal) are from 0 to -1400Å in steps of -100Å. Thickness values (vertical) are from 20Å to 440Å in steps of 20Å. Atom positions (black dots) are shown top left.

Figure 6. Comparison of experimental image (left) of Σ99 {557} <110> tilt boundary in aluminum with image (right) simulated at a defocus value of -800Å and 103Å crystal thickness. The simulation conditions correspond to those of the JEOL ARM-1000 operated at 800keV: C_S = 2.0mm; spread of focus = 160Å; beam convergence = 0.6millirad. A Gaussian vibration corresponding to 0.5Å halfwidth at the specimen is included.

ACKNOWLEDGEMENTS

The authors wish to thank J.H. Turner for assistance with the figures. The NCEMSS simulation programs and other NCEM facilities are available to qualified microscopists at no cost for non-proprietary research; they are supported by the Director, Office of Energy Research, Office of Basic Energy Sciences, Materials Science Division of the U.S. Department of Energy under contract no. DE-AC03-76SF00098.

REFERENCES

1. J.M. Cook, M.A. O'Keefe, D.J. Smith and W.M. Stobbs, *J. Micros.* **129**, 295 (1983).
2. J.M. Cowley and A.F. Moodie, *Proc. Phys Soc.* **70**, 486 (1957).
3. S. Iijima and M.A. O'Keefe, *J. Micros.* **117**, 347 (1979).
4. J.C.H. Spence, M.A. O'Keefe and H. Kolar, *Optik* **49**, 307 (1977).
5. R. Kilaas, *Proc. 45th EMSA*, 66 (1987).
6. G.R. Anstis and M.A. O'Keefe, *Proc. 34th EMSA*, 480 (1976).
7. O. Scherzer, *J. Appl. Phys.* **20**, 20 (1949).
8. D.F Lynch, A.F Moodie and M.A. O'Keefe, *Acta Cryst.* A31, 300 (1975).
9. U. Dahmen, C.J.D. Hetherington, M.A. O'Keefe, K.H. Westmacott, M.J. Mills, M.S. Daw and V. Vitek, *submitted to Phys. Rev. Lett.* (1990).

SPECTROSCOPIC ELLIPSOMETRY AS A NON-DESTRUCTIVE TECHNIQUE FOR CHARACTERIZATION OF ATOMIC-SCALE INTERFACES

J.L. STEHLE *, J.P. PIEL *, J.H. LECAT *, C.PICKERING **, L.C. HAMMOND ***
* SOPRA, 26/68 rue Pierre Joigneaux, 92270 BOIS-COLOMBES, FRANCE
** Royal Signals and Radar Establishment, St Andrews Road, Malvern, Worcs WR14 3PS, U.K
*** ARIES/QEI, Concord MA 01742, USA.

ABSTRACT

Analysis of oxide interfaces with semi-conductor substrates, such as crystalline silicon, gallium arsenide, or indium phosphide is critical in processing and electrical performances. Interfaces can be characterized by spectroscopic ellipsometry (SE), which has a wide spectral range (1.3 to 5.3 eV) allowing an optical penetration depth of 10 nm to a few microns.

A multilayer stack can be characterized in terms of its layer thicknesses and composition. These physical parameters must be calculated through a mathematical model. Linear regression analysis is used to minimize the differences between the measured spectrum and the calculated model. If necessary, an interlayer can be introduced into the model to enhance the fit. This can be complemented by a new method involving calculation of apparent index values which amplifies interface sensivity allowing the thickness to be measured to better than 2 Angstroms. Examples will be given.

I : INTRODUCTION

In semi-conductors, interfaces have an important role. Their presence can seriously modify the electrical properties of many electronic and opto-electronic devices. Therefore, the need for appropriate characterization techniques is important. Spectroscopic ellipsometry has the advantage of being a sensitive and non-destructive procedure, which can be adapted to in-situ measurements (1). Detailed examples of the sensitivity of this technique will be given.

II : BASIC PRINCIPLES - INSTRUMENTATION

The principle of spectroscopic ellipsometry is the analysis of the polarization state of a collimated light beam for each wavelength reflected from the surface to be characterized. The light, emitted by a Xenon arc source, is collimated by a spherical mirror and then modulated by a rotating polarizer. After reflection on the sample, the state of polarization of the light is modified. This effect is related to the sample properties : thickness and composition of each layer.

The reflected beam passes subsequently through a second polarizer. Spectroscopic scanning is achieved by using a double monochromator with a grating and a prism. The detector (photomultiplier) produces a sinusoidal signal and its phase and amplitude are obtained from an Hadamard transform. From these values, the computer calculates and displays for each wavelength the two ellipsometric parameters Tan Ψ (amplitude ratio) and Cos Δ (phase difference) .

III : BASIC PRINCIPLES - THEORY

As spectroscopic ellipsometry is an indirect technique, a model must be chosen to interpret the change of state of polarization of the light created by the surface to be characterized (2).

Two models have been here used . The first one considers the sample as an ideal three-phases system (ambient - overlayer -substrate). When the substrate composition and the thickness of the overlayer are known, it is possible to calculate both real (n) and imaginary part (k) of the index of the upper layer. A transparent layer (like SiO2) with an abrupt interface will have k = 0 for any wavelength. The presence of an interface between the substrate and the layer will create some disturbances on the apparent values of n and k, calculated from the ideal model.

The second model considers a more complete multilayer system, and is based on matrix computation (3). Thickness and composition of each layer are given as parameters for the fit. A least-squares regression algorithm has been developed, in order to select the best set of parameters to describe the depth profile of the sample (4),(5).

In this approach, an interface is a very thin intermediate layer, which is a mixture of the materials of the two closest layers. The index of this layer is calculated using the Bruggeman effective medium theory (6). The introduction of this intermediate layer should improve the agreement between the measurement and the model. This agreement is quantified by the standard deviation σ between measured and calculated values (7).

IV : EXAMPLES

The technique has been applied to three different thicknesses of SiO2 on silicon : 42.2 nm, 113.1 nm, 205.4 nm, with measurements made at an incidence angle of 75.07 degrees.

The apparent n and k values have been calculated from the results using thicknesses obtained from fitting single SiO2 layer models to the data (see table 1). These results have been compared with simulated n,k spectra (fig 1,2,3).

Beside this data reduction, the results of least squares fitting are shown on table 1. For the 42.2 nm sample, the sigma value is improved, first by introducing a void fraction in the oxide layer. A negative value of the void fraction is usually physically interpreted as a greater density of the oxide layer, in comparison with the Malitson SiO2 values (8) used in the index files. A further improvement is obtained by addition of an intermediate layer. The index of this layer is calculated with a physical mixture containing 50% SiO2 and 50% of silicon.

Upper Layer (nm)	Void Fraction	Interface (nm)	Sigma
43.7	0.	-	0.0314
42.0	- 0.0675	-	0.0116
42.2	- 0.0349	1.0	0.0088
113.6	0.	-	0.0853
113.3	- 0.0043	-	0.0392
113.1	- 0.0041	0.2	0.0372
205.8	0.	-	0.0113
205.4	- 0.0029	-	0.0107
205.4	0.	0.2	0.0112
205.0	- 0.0025	0.2	0.0107

Table 1 : *Best fit parameters and mean square deviations obtained for three different thicknesses of SiO2 on Si .*

The thickness value of the interface (1.0 nm), the thickness of the upper oxide layer (42.2 nm) and its void fraction (-0.035), are close to the values used to calculate the curve on fig 1 (1.0 nm, 42.5 nm and -0.02 respectively). The simulated curves reproduce the form of the experimental values, including the marked discontinuity in n . Differences may indicate some information on the nature of the interface and further work is in progress to understand these.

For the 113.1 nm oxide layer, the best fit is also obtained with the presence of an interface, but the difference between the two sigma values is smaller than the previous case. As the accuracy on the measured values of the ellipsometric parameters Tan Ψ and Cos Δ is 1.E-3, such a difference cannot be considered as a physical evidence of the interface presence. However the peculiarities on the k-spectrum are still present. In fig 2, the experimental points have been compared with two simulations, the first one using an interface thickness of 0.7 nm, and an upper layer of 112.8 nm with a negative void fraction of -0.004. The second simulation (0.3 nm, 113.2 nm, -0.004) is closer to the values of the best fit, but does not follow as well the experimental spectrum. It is clearly necessary to measure the parameters with an accuracy of at least 1.E-3 to see these differences. The low value (0.2 nm) of the interface layer found by regression can be explained by high correlation between the parameters of the fit, which can be overcome by the method proposed here.

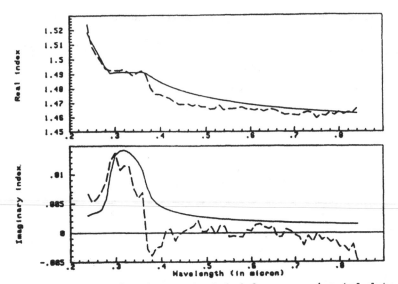

Fig 1 : - - : n,k values calculated from experimental data
——— : n,k values calculated from a simulated spectrum.
SiO2 thickness : 42.5 nm ; void fraction : - 0.02
interface thickness 1.0 nm

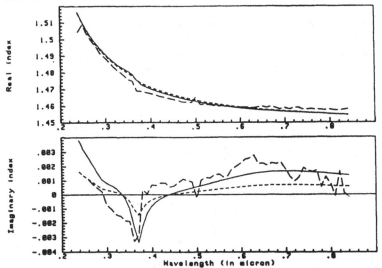

Fig 2 : ——— : n,k values calculated from experimental data
——— : simulation I ; SiO2 thickness : 112.8 nm ;
void fraction : -0.004 ; interface thickness 0.7 nm
——— : simulation II ; SiO2 thickness : 113.2 nm
void fraction : -0.004 ; interface thickness 0.3 nm

In the last case, the differences between the sigma values obtained by fits are even smaller and the inclusion of the interface can be compensated by an increase in the negative void fraction. However, the peculiarities on the index curves are still present (fig 3) and these cannot be produced by simply introducing negative voids, since k will remain zero. The experimental curves in the diagram agree very well with simulated values, using an interface thickness of 1.0 nm and an oxide layer of 204.6 nm with no void fraction. Note the structure in the calculated real index which is exactly followed by the simulation. The experimental imaginary index is within +/- 1.E-3 of the simulated curve over most of the wavelenth range. This, together with fluctuations in k seen in fig 3, indicates the attainable accuracy of the instrument. The spectra were also fitted using a regression program based on a conjugated gradient computation algorithm (9), and the following results obtained :

SIO2 layer : 205.5 nm ; No interface   ; Sigma : 0.0098
SiO2 layer : 205.0 nm ; Interface 0.6 nm ; Sigma : 0.0088

The improvement in sigma is probably due to different weightings and convergence criteria of the analysis programs. It illustrates the difficulty in obtaining interface information from conventional analysis of the spectra.

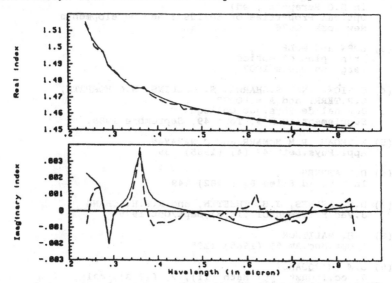

Fig 3 : - - : n,k values calculated from experimental data
           —— : simulation ; SiO2 thickness : 204.6 nm ;
           no void introduced ; interface thickness 1.0 nm)

However, the method presented here based on the apparent n,k values provides increased sensivity to interface properties and can be used for both thick and thin oxides. In order to dismiss the hypothesis of a purely mathematical effect in the method of computing n and k from the ellipsometric parameters, some simulations have been made with calculated spectra, where different levels of noise have been introduced. In all cases, the noise on Tan Ψ and Cos Δ is reproduced on the n,k spectrum in a different way from the peculiarities observed here.

V : CONCLUSION

In parallel with a general approach to the interface problem, using a multilayer regression program, the apparent n,k calculation on the upper layer seems to amplify the disturbance created by an interface. This complementary approach can be quantified by comparison of the experimental peculiarities with simulated curves, containing parameters such as oxide layer thickness, void fraction, and interface thickness. The sensivity of the technique is better than 0.2 nm.

REFERENCES

(1) F.FERRIEU and J.H LECAT
Thin Solids Films 164 (1988) 43

(2) D.E ASPNES
in B.O Seraphin (ed)
Optical Properties of Solids : New Developments
New York, 1976

(3) BORN and WOLF
Principles of optics
Pergamon Press 1987

(4) C.PICKERING, S.SHARMA, S. COLLINS, A.G MORPETH,
G.R.TERRY and A.M.HODGE
Journal de Physique Colloque C4
Supplement au n°9, Tome 49, Septembre 1988.

(5) K.VEDAM ,P.M McMARR and J.NARAYAN
Appl.Phys.Lett 47 (4) (1985) 339

(6) D.E ASPNES
Thin Solid Films 89 (1982) 249 .

(7) D.E. ASPNES, J.B. THEETEN, and R.P.H. CHANG
J.Vac.Sci.Technol.,16(5),Sept/Oct 1979

(8) I.H. MALITSON
J.Opt.Soc.Am 55 (1965) 1205

(9) D.W. MARQUARDT
J.Soc.Indust.Appl.Math (11), 2, (1963), 431

NUMERICAL MODEL OF BOND STRENGTH MEASUREMENTS

Gerald L. Nutt,
Chemistry and Materials Science Department,
Lawrence Livermore National Laboratory,
P.O. Box 808, Livermore, CA. 94550

ABSTRACT

We have reported bond strength measurements of metal/ceramic interfaces using shock waves to separate the bond by spallation[1, 2]. The technique relies on the interpretation of the free surface velocity of the metal film as it is spalled from the substrate.

We answer several questions relating to the details of the interaction of the shock with the interface. Specifically, we examine the role of sound speeds in the measurements. We also calculate the plastic strain in the the bond region and verify the theory relating to the jumpoff velocity of the scab to the bond strength.

INTRODUCTION

The bond strength measurements are a simple extension of a well known method of spall measurement used for homogeneous materials. A plain stress wave is generated in the substrate, propagating in a direction normal to the plane of the interface, as shown schematically in Fig. 1. The compressive stress wave is reflected at the free surface of the metal overlayer as a tensile wave incident on the interface. We model the experiment numerically using the two-dimensional contiuum mechanics computer code DYNA2D[3].

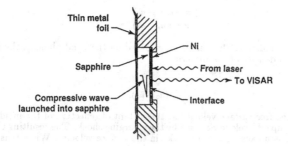

Figure 1. Schematic of exfoliation experiment.

The reflected tensile stress separates the bond, and the newly created free surface at the bond plane communicates the event to the outer free surface. Measurements of the surface velocity history, made with a laser interferometer, are then used to determine the maximum stress at bond separation.

Consider the x, t–diagram shown in Fig. 2. The wave reflected from the surface is a centered rarefaction, propagating along C_- characteristics. The characterisic equations relate the sound speed, flow velocity and axial stress in the region between the free surface and the bond plane:

$$\frac{dz}{dt} = u + c \quad ; \quad d\sigma_z - \rho c du = 0 \tag{1}$$

along C_+, and

$$\frac{dz}{dt} = u - c \quad ; \quad d\sigma_z + \rho c\, du = 0 \qquad (2)$$

along C_-. u and c are the flow velocity and sound speed respectively. σ_z is the axial stress, and ρ, the density.

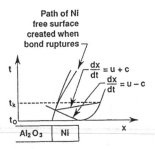

Figure 2. x,t–plot of boundaries and characteristic trajectories in specimen.

One generally assumes the sound speed and density are constant along C_+ and C_-. Integrating along the C_- characteristic from the free surface boundary to the spall event, and then along the C_+ characteristic back to the free surface with constant density and sound speed we get,

$$-\sigma_s = \rho_0 c(u_s - u_0) \qquad (4)$$

$$\sigma_s = \rho_0 c(u_s - u_k), \qquad (5)$$

where the subscript, s, indicates the quantity evaluated on the bond plane at the instant of spall. ρ_0 is the normal density of the material.

Solving for σ_s

$$\sigma_s = \frac{1}{2}\rho_0 c(u_0 - u_k). \qquad (6)$$

Identifying u_0 with the free surface velocity at the instant of reflection of the incident shock wave it is just the "jump off" velocity caused by the emerging shock. The resulting rarefaction will put the Ni in tension causing a "pull back" in the surface velocity. When tension at the sapphire/Ni interface reaches the ultimate strength of the bond, an acceleration wave will move toward the outer free surface. u_k is the signature at the surface of the newly created spall plane. It will appear at the beginning of a velocity jump.

NUMERICAL CALCULATIONS

Experiments were done with a sapphire substrate 1.27 cm in diameter and 0.3175 cm thick. The nickel overlayer and copper flyer are 0.002 cm and 0.0635 cm thick, respectively. Although two-dimensional calculations were made to verify the planarity of the shock wave we will confine our attention to one dimensional calculations representing the conditions along the axis. The bond is represented by a 0.00015 cm thick region of Ni at the sapphire boundary, with spall strength weakened to 0.00177 $Mbar$ so as to correspond with our reported bond strength measurements[2].

The region of Ni representing the rest of the film is given a spall strength of 0.011 $Mbar$ and the Cu 0.005 $Mbar$. The flyer has an initial velocity of 0.025 $cm/\mu sec$. It is separated from the sapphire by a slide line which can open to a void permiting rebound. Zoning in the sapphire is graded to allow a manageable number of nodes while matching the mass of elements across material boundaries.

Hugoniot data is used to define the equation of state of each material. The metals are represented as temperature dependent, elastic-plastic materials with properties as shown in Table 1.

Table 1. Properties defining materials in DYNA2D calculations.

	nickel	sapphire	copper
Density (gm/cm^3)	8.90	3.985	8.93
Shear modulus $(Mbar)$	0.855	1.47	0.477
Bulk sound speed $(cm/\mu s)$	0.465	1.119	0.394
Elastic sound speed $(cm/\mu s)$	0.587	1.32	0.476
Linear Hugoniot coefficient	1.445	1.0	1.489

Fig. 3 shows snap shots of the stress wave as it propagates through the sapphire. The first stress profile in Fig 3 shows two shocks propagating in opposite directions about the point of impact. The next profile, is the stress wave in the sapphire after the shock in the flyer has returned from its trailing surface as a rarefaction. This rarefaction forms the back side of the wave propagating toward increasing z and the interface. The last profile is the wave as it is about to strike the bond.

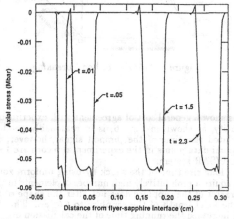

Figure 3. Shock wave propagating through substrate

Fig. 4 is the stress history at the bond. Negative stress values indicate compressive states and positive values are tensile. As the stress swings through the prescribed bond strength, the stress is relaxed to zero indicating spallation.

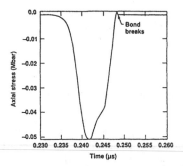

Figure 4. Stress history at bond.

Fig. 5 shows stress profiles in a region about the bond plane before the bond is broken, at the break, and after separation of the metal film from the substrate. Notice that the stress is compressive in the outer part of the film while the bond region is in tension.

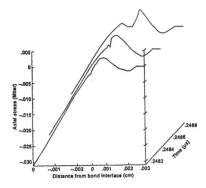

Figure 5. "Movie" of bond breaking.

These calculations show a general sort of agreement with experiment. The bond breaks, and the jumpoff velocity, as shown in Fig. 6, is in reasonably good agreement with experiment. The calculated rise time of the jumpoff signal, however, is much larger than measured. The individual data points in the experimental record are 1 ns apart showing the time resolution of the VISAR system.

We eliminated the long rise time of the shock by using uniform zoning in the sapphire. Normally this would require a prohibitively high number of elements in the calculation. Since the linear elastic character of the sapphire does not alter the shape of the stress pulse with time, we shortened the axial length of the sapphire to 0.06 cm. This trick saved sufficient zones that we neither increased the running time of the calculation nor altered the "physics" of the scabbing process in any way. Mass matching was held to at least 1% throughout.

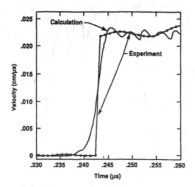

Figure 6. Comparison of calculated free surface velocity with the measurement.

Fig. 7 shows the result of using uniform zoning in the sapphire. Comparison with Fig 5 shows a decrease in rise time from 8 ns to 2 ns. the comparison of free surface velocity history with experiment is also greatly improved. The time resolution of the velocity pullbacks, however, is insufficient to get an accurate measure of the jumpoff–pullback difference.

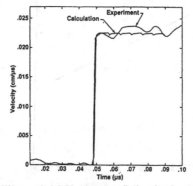

Figure 7. The result of "squaring up" the shock with uniform zoning.

One would expect to be able to calculate the programmed bond strength from the data in Fig. 7 using Eq. (5) if the time resolution were adequate. Fig. 8 shows a calculation designed optimize the experiment for accurate resolution of the pullback. Uniform zoning was deliberately *not* used so that the pullback could be resolved. The flyer velocity has been reduced to improve the signal and a copper pad was placed between the flyer and the sapphire substrate for the purpose of shaping the back side of the wave.

Notice the damping of the ringing in the scab just after spall. This "linear" dissipation is an artifact of finite element calculations arising from the need to eliminate high frequency noise not associated with shocks. It has a strong effect on the pullback velocity as shown by the reduced amplitude of successive oscillations. It is necessary to account for the dissipation if the numerical calculations are to be compared with Eq (5).

Fig. 9 is a plot of the bond strengths calculated from Eq (5) using the pullback velocities of three calculations where the default damping was used, half the default, and finally one

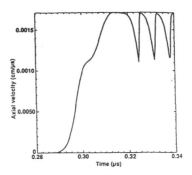

Figure 8. Flyer with velocityof 0.002 *cm/μs* incident
on a copper pad over the sapphire. Nonuniform zoning.

fourth the default value. The intercept, corresponding to zero damping is within 2.3 % of the programmed spall strength.

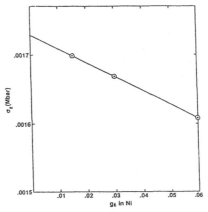

Figure 9. Effect of "linear" damping onbond strength using calculated free surface velocities from model ofFig. 8, in Eq. (5).

Finally, we would like to give estimates of the plastic strain in the bond region as predicted by finite element calculations. Our best calculation shows 8×10^{-5} as the maximum plastic strain in the Ni. All the strain occurs in a compressive stress state arising from the initial shock, not during bond rupture.

CONCLUSION

By numerical simulation we have confirmed the assumptions underlying the previously reported[1, 2] of bond strength measurements. The interpretation of the free surface velocity has been shown to be correct. Insight is provided by the detailed examination of the interaction of the stress wave with the bond. Rise times of the tensile wave at the bond is indeed short, possibly less than the 1 ns time resolution of our measurements. Plastic strain in the sample, although it exists, is small.

ACKNOWLEDGEMENTS

The author wishes to express his thanks for the help and encouragement received from Kenneth Froeschner, Wayne King, and William Lai.

References

1. G. L. Nutt, W. Lai, K. E. Froeschner and W. E. King, Direct Measurement of Interface Bond Strengths, Acta Metall., Acta Metallurgica/Scripta, 1990, to be published.
2. G. L. Nutt, W. Lai, K. E. Froeschner and W. E. King, Measurement of Bond Strength at Metal/Ceramic Interfaces, *Interfaces Between Polymers, Metals, and Ceramics*, edited by B.M. DeKoven, A.J. Gellman, R. Rosenberg (Materials Research Society Proceedings) 153, Pittsburgh, PA 1989)
3. J.O. Hallquist, User's Manual for DYNA2D - An Explicit Two-Dimensional Hydrodynamic Finite Element Code With Interactive Rezoning, Lawrence Livermore Laboratory, UCID-18756, Rev. 2 (1984)

* Work performed under the auspices of the U.S. Department of Energy by the Lawrence Livermore National Laboratory under contract number W-7405-ENG-48.

Author Index

Subject Index

MATERIALS RESEARCH SOCIETY SYMPOSIUM PROCEEDINGS

ISSN 0272 - 9172

Volume 75—Photon, Beam and Plasma Stimulated Chemical Processes at Surfaces, V. M. Donnelly, I. P. Herman, M. Hirose, 1987, ISBN 0-931837-41-3

Volume 76—Science and Technology of Microfabrication, R. E. Howard, E. L. Hu, S. Namba, S. Pang, 1987, ISBN 0-931837-42-1

Volume 77—Interfaces, Superlattices, and Thin Films, J. D. Dow, I. K. Schuller, 1987, ISBN 0-931837-56-1

Volume 78—Advances in Structural Ceramics, P. F. Becher, M. V. Swain, S. Sōmiya, 1987, ISBN 0-931837-43-X

Volume 79—Scattering, Deformation and Fracture in Polymers, G. D. Wignall, B. Crist, T. P. Russell, E. L. Thomas, 1987, ISBN 0-931837-44-8

Volume 80—Science and Technology of Rapidly Quenched Alloys, M. Tenhover, W. L. Johnson, L. E. Tanner, 1987, ISBN 0-931837-45-6

Volume 81—High-Temperature Ordered Intermetallic Alloys, II, N. S. Stoloff, C. C. Koch, C. T. Liu, O. Izumi, 1987, ISBN 0-931837-46-4

Volume 82—Characterization of Defects in Materials, R. W. Siegel, J. R. Weertman, R. Sinclair, 1987, ISBN 0-931837-47-2

Volume 83—Physical and Chemical Properties of Thin Metal Overlayers and Alloy Surfaces, D. M. Zehner, D. W. Goodman, 1987, ISBN 0-931837-48-0

Volume 84—Scientific Basis for Nuclear Waste Management X, J. K. Bates, W. B. Seefeldt, 1987, ISBN 0-931837-49-9

Volume 85—Microstructural Development During the Hydration of Cement, L. Struble, P. Brown, 1987, ISBN 0-931837-50-2

Volume 86—Fly Ash and Coal Conversion By-Products Characterization, Utilization and Disposal III, G. J. McCarthy, F. P. Glasser, D. M. Roy, S. Diamond, 1987, ISBN 0-931837-51-0

Volume 87—Materials Processing in the Reduced Gravity Environment of Space, R. H. Doremus, P. C. Nordine, 1987, ISBN 0-931837-52-9

Volume 88—Optical Fiber Materials and Properties, S. R. Nagel, J. W. Fleming, G. Sigel, D. A. Thompson, 1987, ISBN 0-931837-53-7

Volume 89—Diluted Magnetic (Semimagnetic) Semiconductors, R. L. Aggarwal, J. K. Furdyna, S. von Molnar, 1987, ISBN 0-931837-54-5

Volume 90—Materials for Infrared Detectors and Sources, R. F. C. Farrow, J. F. Schetzina, J. T. Cheung, 1987, ISBN 0-931837-55-3

Volume 91—Heteroepitaxy on Silicon II, J. C. C. Fan, J. M. Phillips, B.-Y. Tsaur, 1987, ISBN 0-931837-58-8

Volume 92—Rapid Thermal Processing of Electronic Materials, S. R. Wilson, R. A. Powell, D. E. Davies, 1987, ISBN 0-931837-59-6

Volume 93—Materials Modification and Growth Using Ion Beams, U. Gibson, A. E. White, P. P. Pronko, 1987, ISBN 0-931837-60-X

Volume 94—Initial Stages of Epitaxial Growth, R. Hull, J. M. Gibson, David A. Smith, 1987, ISBN 0-931837-61-8

Volume 95—Amorphous Silicon Semiconductors—Pure and Hydrogenated, A. Madan, M. Thompson, D. Adler, Y. Hamakawa, 1987, ISBN 0-931837-62-6

Volume 96—Permanent Magnet Materials, S. G. Sankar, J. F. Herbst, N. C. Koon, 1987, ISBN 0-931837-63-4

Volume 97—Novel Refractory Semiconductors, D. Emin, T. Aselage, C. Wood, 1987, ISBN 0-931837-64-2

Volume 98—Plasma Processing and Synthesis of Materials, D. Apelian, J. Szekely, 1987, ISBN 0-931837-65-0

MATERIALS RESEARCH SOCIETY SYMPOSIUM PROCEEDINGS

Recent Materials Research Society Proceedings listed in the front.

MATERIALS RESEARCH SOCIETY SYMPOSIUM PROCEEDINGS

MATERIALS RESEARCH SOCIETY MONOGRAPH

Earlier Materials Research Society Symposium Proceedings listed in the back.

Tungsten and Other Refractory Metals for VLSI Applications, Robert S. Blewer, 1986; ISSN 0886-7860; ISBN 0-931837-32-4

Tungsten and Other Refractory Metals for VLSI Applications II, Eliot K. Broadbent, 1987; ISSN 0886-7860; ISBN 0-931837-66-9

Ternary and Multinary Compounds, Satyen K. Deb, Alex Zunger, 1987; ISBN 0-931837-57-X

Tungsten and Other Refractory Metals for VLSI Applications III, Victor A. Wells, 1988; ISSN 0886-7860; ISBN 0-931837-84-7

Atomic and Molecular Processing of Electronic and Ceramic Materials: Preparation, Characterization and Properties, Ilhan A. Aksay, Gary L. McVay, Thomas G. Stoebe, J.F. Wager, 1988; ISBN 0-931837-85-5

Materials Futures: Strategies and Opportunities, R. Byron Pipes, U.S. Organizing Committee, Rune Lagneborg, Swedish Organizing Committee, 1988: ISBN 1-55899-000-3

Tungsten and Other Refractory Metals for VLSI Applications IV, Robert S. Blewer, Carol M. McConica, 1989; ISSN 0886-7860; ISBN 0-931837-98-7

Tungsten and Other Advanced Metals for VLSI/ULSI Applications V, S. Simon Wong, Seijiro Furukawa, 1990; ISSN 1048-0854; ISBN 1-55899-086-2

High Energy and Heavy Ion Beams in Materials Analysis, Joseph R. Tesmer, Carl J. Maggiore, Michael Nastasi, J. Charles Barbour, James W. Mayer, 1990; ISBN 1-55899-091-7

Physical Metallurgy of Cast Iron IV, Goro Ohira, Takaji Kusakawa, Eisuke Niyama, 1990; ISBN 1-55899-090-9

Printed in the United States
By Bookmasters